# Elements of Electronic Instrumentation and Measurement

**Third Edition**

JOSEPH J. CARR

# Elements of Electronic Instrumentation and Measurement

## Third Edition

Prentice Hall
Englewood Cliffs, New Jersey     Columbus Ohio

**Library of Congress Cataloging-in-Publication Data**

Carr, Joseph J.
    Elements of electronic instrumentation and measurement / Joseph J.
Carr.—3rd ed.
            p.    cm.
    Includes index.
    ISBN 0-13-341686-0
    1. Electronic instruments.   2. Electronic measurements.
I. Title.
TK7878.4.C36   1996
621.3815'48—dc20

95-16744
CIP

Cover Image: © Katherine Arion/The Image Bank
Editor: Charles E. Stewart, Jr.
Production Editor: Stephen C. Robb
Design Coordinator: Julia Zonneveld Van Hook
Text Designer: Anne Flanagan
Cover Designer: Proof Positive/Farrowlyne Associates, Inc.
Production Manager: Patricia A. Tonneman
Marketing Manager: Debbie Yarnell
Illustrations: Diphrent Strokes, Inc.
Copyeditor: Maggie Shaffer

This book was set in Times Roman and Times Bold by Clarinda Company and was printed and bound by Quebecor Printing/Book Press. The cover was printed by Phoenix Color Corp.

© 1996 by Prentice-Hall Inc.
A Simon & Schuster Company
Englewood Cliffs, New Jersey 07632

Printed in the United States of America

10 9 8 7 6 5 4 3 2 1

ISBN: 0-13-341686-0

Prentice-Hall International (UK) Limited, *London*
Prentice-Hall of Australia Pty. Limited, *Sydney*
Prentice-Hall Canada, Inc., *Toronto*
Prentice-Hall Hispanoamericana, S. A., *Mexico*
Prentice-Hall of India Private Limited, *New Delhi*
Prentice-Hall of Japan, Inc., *Tokyo*
Simon & Schuster Asia Pte. Ltd., *Singapore*
Editora Prentice-Hall do Brasil, Ltda., *Rio de Janeiro*

# Preface

One reason why electronic instrumentation is such an interesting field is that it combines elements of technologies ranging from the nineteenth to the twenty-first centuries. Modern computer-based electronic instrumentation is now evident in every reasonably equipped laboratory and workshop and in the catalogs and advertisements of all of the manufacturers. Yet, at the root of many space-age instruments is circuitry, such as the Wheatstone bridge, that is found in nineteenth-century textbooks. Although newer techniques are often used as well, the older techniques are still in widespread use in new as well as old instruments. In this text you will find both types discussed.

A famous nineteenth-century physicist once remarked that one does not really understand a physical phenomenon until one can measure it, while engineers and technologists would add "and apply the resultant numbers to something practical." Indeed, one of my own physics professors stated that physics is the science that measures reality (obviously, he was an experimental, not a theoretical, physicist). Electronic instrumentation is actually about measurement, and no book on electronic instrumentation is complete without the theories and process of measurement.

Several different types of material are used in this book. First, there is basic measurement theory, including some statistics. Second, there is material on basic measurement techniques, such as analog meter movements, ac and dc bridges, digital instruments, and so forth. Third, there are chapters on specific generic "workhorse" instruments such as multimeters, oscilloscopes, and signal generators.

A considerable amount of practice material is oriented toward various fields of measurement: electronic communication, audio, components testing, medical electronics, and servicing. Finally, there is a chapter on the general purpose interface bus (IEEE-488GPIB standard), which is used for automated testing systems in which a computer controls an array of instruments, each of which is connected to the GPIB.

This text is a careful blend of theory and practice. Feedback from users of the previous two editions indicated that the strong practical orientation, with inclusion of the theoretical, is exactly what is needed by students who will soon be out in the practical world doing electronics, rather than just studying the subject.

Joseph J. Carr

v

# Brief Contents

# Contents

**3**    dc and ac Deflection Meter Movements     47

# 1

# Introduction to Electronic Instrumentation and Measurement

## OBJECTIVES

1. Understand **variation** and its effect on measurement.
2. Be able to recognize and correctly use **significant figures.**
3. Understand and be able to use **decibel** notation in systems calculations.
4. Be able to use scientific notation.
5. Be able to explain the differences among the following "averages" of a data set: **mean, median, mode, harmonic average, root mean square (rms),** and **root of the sum of squares (rss).**

1–2

## SELF-EVALUATION

Before studying the material in this chapter, try to answer the questions given below. These questions test your knowledge of the subject. If you cannot answer a particular question, then look for the answer as you read the text.

1. _____ _____ is the square root of the variance of a data set.
2. The number of significant figures in a measurement can be increased by multiplying the result by another measured value. True or false?
3. Express 0.0000000826 A in scientific notation.
4. Name four different measures of the "average" of a data set.

1–3

## INTRODUCTION

The British physicist and mathematician William Thompson, Lord Kelvin (1824–1907), reportedly said that one cannot really claim to know much about a thing until one can measure it. Much of the history of science and engineering in general, and of electronics in particular, has been involved in measuring things. Whether a measurement is made in order to troubleshoot an existing circuit, to characterize and define a new circuit, or to find the value of some nonelectronic physical variable (e.g., pressure or temperature), the

1

common thread is the need for using some electronic device to make a measurement. In this book you will learn about electronic measurement devices. Although the entire universe of possible electronic measurement devices could never be included in a single text, or indeed a shelf full of texts, we will discuss the generic types used in a wide variety of applications.

You will also learn about the theory of measurement. While in the naive sense it is possible to measure a voltage and use the information gained for some particular purpose, without knowing much about measurement theory, there comes a point when it becomes necessary to understand a little of what you are doing in order to gain the maximum utility from your practical measurements.

Before beginning our discussions, however, some readers may wish to undertake the quick review of some arithmetic basics found in the remainder of this chapter. For some people this material is simple; for others it is new; and for still others it is a needed refresher. If you feel no need for the material in the following sections, then please feel free to skip ahead to the next chapter. Although a knowledge of calculus is not strictly needed, it would be helpful to understand the basic concepts, at least in a descriptive sense. For those readers who have not had an introductory course in calculus, a quick overview of the basic principles is provided in Appendix A.

## 1–4        SIGNIFICANT FIGURES

Much of our everyday experience deals with exact numbers of things: 6 stamps, 7.50 dollars, and 7 people. These items can be counted and an exact numerical representation provided; all figures are significant in such cases. But in other situations, you may take measurements that are subject to errors. For example, you might measure the height of a person as 67, 68, or 69 in, depending on how straight the person stands. Or what about the answer you give when asked a person's weight? The scale may register 162 lb, but one of the balance weights may not be perfect, or perhaps the scale dial at rest sticks a little off zero. How many people do you know who own a perfectly accurate watch that never needs to be reset? None! These flaws are implicitly resolved when we apply the concept of **significant figures** to the measurements. This concept demands that we *impute no more precision or accuracy to a measurement or calculation than the natural physical reality of the situation permits.*

The counting numbers (1, 2, 3, 4, 5, 6, 7, 8, and 9) are always significant. Zero (0) is significant only if it is used to indicate exactly zero, or a truly null case. Zero is not significant if it is used merely as a place holder to make the numbers look nicer on the printed page. For example, if "0.60" is properly written, then it means *exactly* 6/10, not "approximately 0.6"; the zero used here in the hundredths place is significant. If the number is written "0.6," then we may assume that it means 6/10 plus or minus some amount of either error or uncertainty.

When we use numbers to indicate a quantity, the concept of significant figures becomes important. For example, "16 gal" has two significant figures but can reasonably be taken to mean that the quantity of liquid is somewhere between 15 and 17 gal.

But if our liquid-measuring device is better, then we might write "16.0 gal" to indicate precisely 16 gal plus or minus a very small error; that is, perhaps the real value is between 15.9 and 16.1 gal. Consider a pressure gage that is guaranteed to an accuracy of ±5%. A reading of "100 torr" has three figures, meaning that the actual pressure is between [100 − 5%] and [100 + 5%], or 95 to 10<u>5</u> torr (two significant figures).

Consider a practical measurement situation. An experiment uses a digital voltmeter to measure an electrical potential difference of exactly 15 V. The instrument reads from 00.00 to 19.99 V, with an accuracy of ±1%. In addition, digital voltmeters typically have a ±1-digit error in the least significant position due to their design; this problem is called **last digit bobble.** For the digital voltmeter in question,

The "last digit bobble" problem means that a reading of 15.00 V could represent any value between 15.00 − 00.01 (14.99) V and 15.00 + 00.01 (15.01) V. In addition, the error of 1% means that the actual voltage could be ±(15 × 0.01) = ±0.15 V. Thus, the actual voltage could be from (15.00 − 0.15) V to (15.00 + 0.15) V, or a range of +14.84 to +15.16 V.

If both errors are minus,

$$
\begin{array}{rl}
\text{Reading:} & 15.00 \text{ V} \\
& -\ 0.01 \text{ V} \\
& \underline{-\ 0.15 \text{ V}} \\
& 14.84 \text{ V} \quad \text{(worst case)}
\end{array}
$$

or, if both errors are positive,

$$
\begin{array}{rl}
\text{Reading:} & 15.00 \text{ V} \\
& +\ 0.15 \text{ V} \\
& \underline{+\ 0.01 \text{ V}} \\
& 15.16 \text{ V} \quad \text{(worst case)}
\end{array}
$$

Significant figure errors are propagated in calculations. A rule to remember is that *the number of significant figures is not improved by combining the numbers with other numbers.* For example, multiplying a significant digit by a nonsignificant digit yields a result that has at least one nonsignificant digit. Often the number of significant figures decreases in calculation. Suppose we measure a voltage $V$ as 15.65 V and a current $I$ in the same electrical circuit as 0.025 A. The power $P$ is the product $VI$. Let's find that product, placing a little hat (ˆ) over each digit that is not significant, and then carry that notation down wherever a nonsignificant digit is a factor with another digit:

$$
\begin{array}{r}
15.6\hat{5} \\
\times\ 0.02\hat{5} \\
\hline
\hat{7}\hat{8}\hat{2}\hat{5} \\
313\hat{0} \\
000\hat{0} \\
000\hat{0} \\
\hline
00.3\hat{9}\hat{1}\hat{2}\hat{5}
\end{array}
$$

As can be seen, only the 3 and one of the leading zeros are significant. Thus, we are claiming more precision than is truly available if we list the power as "0.39 W" when the 9 is not significant. We might be better advised to list this value as "0.4 W."

The reason why scientists and engineers make such a fuss over significant figures is that it is bad form, and potentially dangerous under some circumstances, to claim more precision or accuracy than is truly the case. For this reason, we typically limit figures to the number of decimal places for which a reasonable expectation of physical reality obtains.

Significant figure rules were perhaps a little easier to understand and use in the days when scientists and engineers calculated on slide rules. Those tools were limited to two or three digits, so one was less tempted to write down a very long number. But in this age of $12, 10-digit scientific pocket calculators, and the nearly universal distribution of personal computers, the distinction often gets lost. Consider a simple electrical problem as an example. One expression of Ohm's law states that the current $I$ flowing in a circuit is the quotient of the voltage $V$ and the resistance $R$. Suppose that 10 V is applied to a 3-$\Omega$ resistance. According to my $12 pocket scientific calculator, the current is 10 V/3 $\Omega$ = 3.333333333 A. Does anyone really think that their ordinary, run-of-the-mill, laboratory ammeter can measure to within $10^{-9}$ A (i.e., 3.33 nA)? In most cases we would be exaggerating to claim more than 3.33 or 3.333 A (at most) with very high quality meters with recent calibration stickers on them! Indeed, on most lower-quality instruments, "3" or "3.3" would be a more reasonable statement of the current reading.

Being mindful of significant figures is a key factor in making good electronic measurements and maintaining the integrity and credibility of the measurement system.

**1–5**     ## SCIENTIFIC NOTATION

**Scientific notation** is a simple arithmetic shorthand that allows one to deal with very large or very small numbers using only a few digits between 1 and 10 and using **powers-of-10 exponents.** The form of a number in scientific notation is

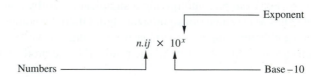

For example, if the age of a college physics professor is 47 years, it could be written

$$\text{Prof's age} = 4.7 \times 10^1 \text{ years} \qquad (1\text{--}1)$$

(Note the units "years" in this equation. The specification of a value is never complete if the units are not included; "47" or "$4.7 \times 10^1$" is *not* the same as "47 *years*" or "$4.7 \times 10^1$ *years*." The only exception occurs when the quantity is nondimensional.)

When the exponent is negative, it is the same as saying $1/10^x$. In other words,

$$10^{-x} = \frac{1}{10^x} \qquad (1\text{--}2)$$

Some of the standard values in powers-of-10 notation, along with their respective prefixes for use with units, are as follows:

$$
\begin{aligned}
1/1{,}000{,}000{,}000 &= 0.000000001 = 10^{-9} \quad \text{(nano)} \\
1/1{,}000{,}000 &= 0.000001 = 10^{-6} \quad \text{(micro)} \\
1/100{,}000 &= 0.00001 = 10^{-5} \\
1/10{,}000 &= 0.0001 = 10^{-4} \\
1/1{,}000 &= 0.001 = 10^{-3} \quad \text{(milli)} \\
1/100 &= 0.01 = 10^{-2} \quad \text{(centi)} \\
1/10 &= 0.10 = 10^{-1} \quad \text{(deci)} \\
1.0 &= 10^{0} \\
10 &= 10^{1} \quad \text{(deka)} \\
100 &= 10^{2} \quad \text{(hecto)} \\
1{,}000 &= 10^{3} \quad \text{(kilo)} \\
10{,}000 &= 10^{4} \\
100{,}000 &= 10^{5} \\
1{,}000{,}000 &= 10^{6} \quad \text{(mega)} \\
1{,}000{,}000{,}000 &= 10^{9} \quad \text{(giga)*}
\end{aligned}
$$

Scientific notation is especially appealing when dealing with numbers for which there are reasonably only a few significant figures. For example, if we measure a human brain wave scalp surface potential[†] as 143.6 μV (microvolts), we may prefer to represent that value as $1.44 \times 10^{-4}$ V.

The prefixes above are used to subdivide units. For example, *milli* means 1/1000 (or 0.001), so a *milli*meter is 0.001 m, and a *milli*ampere is 0.001 A. Similarly, *kilo* means 1000, so a *kilo*meter is 1000 m, and a *kilo*hertz is 1000 Hz.

---

* 1,000,000,000 ($10^9$) is called "1 billion" in the United States, but "1000 million" in the United Kingdom and in most of the rest of the world. The term *milliard* was once applied to $10^9$. To be 1 billion outside the United States, the number would have to be 1,000,000,000,000 ($10^{12}$). Perhaps we should label $10^{12}$ *billiards* so that Carl Sagan could go on TV and talk about "billiards and billiards of worlds."

[†] Electroencephalogram (EEG)

**1–6**          ## UNITS AND PHYSICAL CONSTANTS

In accordance with standard engineering and scientific practice, all units in this text will be in either the CGS (centimeter-gram-second) or the MKS (meter-kilogram-second) systems unless otherwise specified. Because the so-called metric (CGS and MKS) system[‡] depends upon attaching appropriate prefixes to the basic units, the common metric prefixes are listed in Table 1–1. Other tables are as follows: Table 1–2 gives the standard physical units; Table 1–3 shows physical constants of interest, including those used in problems in this and other chapters; and Table 1–4 lists some common conversion factors.

**1–7**          ## WHAT IS "AVERAGE"?

"Isn't average just, well, *average*?" That's a common question, and the answer is not always so obvious as it might seem. There are several different kinds of "average," and all of them are valid in the right situations. The word **average** refers to the *most typical value,* or *most expected value,* in a collection of numerical data. When you collect data, there are a number of ways that the results can vary from one observation to another (even when conditions are supposed to be the same).

First, of course, there is old-fashioned measurement and observational error. Not all rulers are truly the same, and not all applications of the same ruler to the same object turn out the same. Nor is it probable that even the same pair of perfect eyes will correctly read the scale every time a measurement is taken. In short, there will always be some **variability** in the measurements from one trial to another.

---

[‡] Properly known as the International System of Units (or Système international d'unités, abbreviated internationally as "SI").

**TABLE 1–1**
Metric prefixes

| Metric prefix | Multiplying factor | Symbol |
|---|---|---|
| tera | $10^{12}$ | T |
| giga | $10^{9}$ | G |
| mega | $10^{6}$ | M |
| kilo | $10^{3}$ | k |
| hecto | $10^{2}$ | h |
| deka | $10^{1}$ | da |
| deci | $10^{-1}$ | d |
| centi | $10^{-2}$ | c |
| milli | $10^{-3}$ | m |
| micro | $10^{-6}$ | μ |
| nano | $10^{-9}$ | n |
| pico | $10^{-12}$ | p |
| femto | $10^{-15}$ | f |
| atto | $10^{-18}$ | a |

**TABLE 1–2**
Physical units

| Quantity | Unit | Symbol |
|----------|------|--------|
| Capacitance | farad | F |
| Electric charge | coulomb | C |
| Conductance | mhos | ℧ |
| Conductivity | mhos/meter | ℧/m |
| Current | ampere | A |
| Energy | joule (watt-second) | J (W-s) |
| Field | volts/meter | V/m |
| Flux linkage | weber (volt-second) | Wb (V-s) |
| Frequency | hertz | Hz |
| Inductance | henry | H |
| Length | meter | m |
| Mass | gram | g |
| Power | watt | W |
| Resistance | ohm | Ω |
| Time | second | s |
| Velocity | meter/second | m/s |
| Electric potential | volt | V |

**TABLE 1–3**
Physical constants

| Constant | Value | Symbol |
|----------|-------|--------|
| Boltzmann's constant | $1.38 \times 10^{-23}$ J/K | $K$ |
| Electric charge ($e^-$) | $1.6 \times 10^{-19}$ C | $q$ |
| Electron (volt) | $1.6 \times 10^{-19}$ J | eV |
| Electron (mass) | $9.12 \times 10^{-31}$ kg | $m$ |
| Permeability of free space | $4\pi \times 10^{-7}$ H/m | $U_o$ |
| Permittivity of free space | $8.85 \times 10^{-12}$ F/m | $\varepsilon_0$ |
| Planck's constant | $6.626 \times 10^{-34}$ J-s | $h$ |
| Velocity of electromagnetic waves | $3 \times 10^8$ m/s | $c$ |
| Pi | 3.141592654 | $\pi$ |

**TABLE 1–4**
Common conversion factors

| | | |
|---|---|---|
| 1 in | = | 2.54 cm |
| 1 in | = | 25.4 mm |
| 1 ft | = | 0.305 m |
| 1 mi | = | 1.61 km |
| 1 nautical mi | = | 6080 ft |
| 1 statute mi | = | 5280 ft |
| 1 mi | = | $2.54 \times 10^{-5}$ m |
| 1 kg | = | 2.2 lb |
| 1 neper (Np) | = | 8.686 dB |
| 1 gauss | = | 10,000 teslas (T) |

Next, there will be some actual variability in the events being recorded. Natural phenomena do, in fact, vary for one reason or another. One way to handle these variations is to find the most typical value for the lot. Consider the case where a student observed a red berry bush over a period of time. At one point, the observer counted 28 bunches of berries and found that there were from 1 to 8 berries in the different bunches. What does "average" mean in this case?

There are actually several different kinds of averages, but the most commonly encountered are the **arithmetic mean** (usually called simply the **mean**), the **median,** and the **mode.** These are each a little different from the others, and all of them are correct "averages" when used in the right context. Let's look at these terms a little more closely, using the following data values:

**Data Values**

4  6  5  5  3  6  4  3  3  4  5  3  1  6
5  2  5  2  3  4  4  5  7  7  8  4  6  5

The arithmetic mean is the type of average that most people use day to day. The mean is the sum of all values, divided by the number $n$ of different values, or, to put it in proper form,

$$\overline{X} = \frac{X_1 + X_2 + X_3 + \cdots + X_n}{n} \tag{1-3}$$

The sum of all the 28 values is 125, so what is the "average"?

$$\overline{X} = \frac{125}{28} = 4.46 \tag{1-4}$$

The mean is 4.46, although you shouldn't expect to find that "0.46" berry anyplace. This average is the arithmetic mean.

The **median** is another type of average: it is the *middle value in the data set,* that is, the value where exactly *half of the values are above it and half are below it.* In the present case there are 28 values, which is an even number, so the median will be midway between two of them (with 14 above and 14 below). Figure 1–1 shows the data distribution and is a crude kind of bar graph. Count the $X$s in each category from one end to the middle and then from the other end. Note that there are 14 values between 0 and 4, and 14 values from 5 to 9. Thus, the median value will be halfway between 4 and 5, or 4.5. If there were an odd number of data points, then the middle point—the median—would be the actual data point that had an equal number of points above it and below it.

The **mode** is also an average of sorts and is defined as the *most frequently occurring value* in a data set. The mode of the above data is easily seen in the $X$ chart of Fig. 1–1. There were more bunches with 5 berries than any other number, so that's the mode. So, now we have an arithmetic mean of 4.46, a median of 4.5, and a mode of 5 . . . and they're all the average of the same data set depending on how you define *average.*

**FIGURE 1-1**
Data distribution ($X$ chart) for 28
data values.

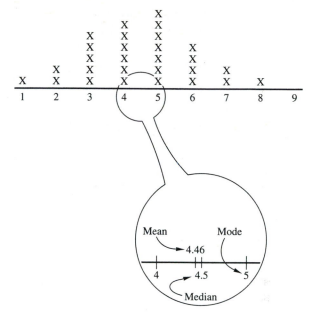

Different averages are used for different situations. If the data are *perfectly symmetrical,* then the mean, median, and mode are the same number. In fact, that's nearly the case in the data above. If the mean, median, and mode are not the same, then the data are not symmetrical around the mean . . . and the difference is a test of that symmetry. In berry-bush data, the distribution is nearly symmetrical, so the mean could be used. But there are other situations where the mean is not terribly useful, especially if one or two data points have very large or very small values compared with the rest of the data.

The mean is best used when the data are symmetrical; the median is often used when the data are highly asymmetrical due to outliers; and the mode is used to answer questions such as "What is the most common cause of death?" or "What is the most popular TV show on Friday night?"

What other averages are there? There's the **geometric mean** and the **harmonic mean.** The geometric mean is often used when the data are not very symmetrical, especially in biological studies. Suppose that you have $48 to spend, and you spend one-half of your available money each day for 5 days. The data would tabulate as follows:

| Day | Amount |
| --- | --- |
| 1 | $48 |
| 2 | 24 |
| 3 | 12 |
| 4 | 6 |
| 5 | 3 |

The arithmetic mean is

$$\frac{48 + 24 + 12 + 6 + 3}{5} = \frac{93}{5} = 18.6$$

If we graph these values as shown in Fig. 1–2(a), the line connecting the tops of the bar graphs is not straight. To find the geometric mean, we need to find the **logarithm** of each value, add these log values, and take the **logarithmic mean.** Then we take the **antilog** of the log-mean. The log-mean is

$$\frac{\log 48 + \log 24 + \log 12 + \log 6 + \log 3}{5}$$

$$= \frac{1.68 + 1.38 + 1.08 + 0.778 + 0.477}{5}$$

$$= \frac{5.395}{5} = 1.079$$

Now take the antilog of the answer:

$$\log^{-1}(1.079) = 11.99$$

The logarithmic chart shown in Fig. 1–2(a) is not a straight line. If we want to straighten out that line, we use **semilog paper** as shown in Fig. 1–2(b).

The other mean—the harmonic mean—is a bit more complicated and is used when data are expressed in *ratios,* such as miles per hour, dollars per dozen, and so forth. The expression for harmonic mean reflects the fact that it is the *reciprocal of the mean of the reciprocals of the data:*

$$\text{H.M.} = \frac{1}{\dfrac{\left(\dfrac{1}{X_1} + \dfrac{1}{X_2} + \dfrac{1}{X_3} + \cdots + \dfrac{1}{X_n}\right)}{n}} - \qquad (1\text{–}5)$$

For example, let's compare the price of eggs in the local store over one past month:

| Week | Price ($/dz) |
| --- | --- |
| 1 | $2.29 |
| 2 | 1.98 |
| 3 | 1.56 |
| 4 | 2.04 |

The arithmetic mean is

$$\frac{\$2.29 + \$1.98 + \$1.56 + \$2.04}{4} = \frac{\$7.87}{4}$$

$$= \$1.9675 \approx \$1.97$$

**FIGURE 1–2**
(a) Linear graph of data, (b) semi-logarithmic graph of same data.

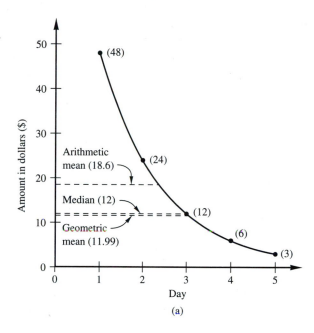

(a)

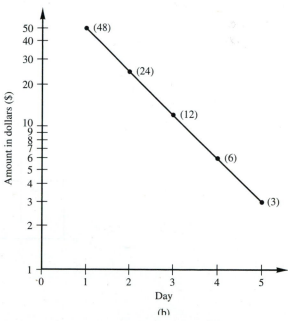

(b)

But the harmonic mean is

$$\text{H.M.} = \cfrac{1}{\cfrac{\left(\cfrac{1}{\$2.29} + \cfrac{1}{\$1.98} + \cfrac{1}{\$1.56} + \cfrac{1}{\$2.04}\right)}{4}}$$

$$= \cfrac{1}{\cfrac{\left(0.437 + 0.505 + 0.641 + 0.490\right)}{4}}$$

$$= \frac{2.073}{4} = \frac{1}{0.518} = \$1.929 \approx \$1.93$$

**1–7–1**        ### Integrated, Root Mean Square (rms), and Root Sum Squares (rss) Averages

Other "averages" are sometimes used in science, engineering, and technology: **integrated average, root mean square (rms),** and **root sum squares (rss).**

The **integrated average** is the *area under the curve of the function* (Fig. 1–3), divided by the segment of the range over which the average is taken:

$$\overline{X} = \frac{1}{T} \int_{t1}^{t2} X \, dt \qquad\qquad \textbf{(1–6)}$$

The integrated average is often found in electronic circuits where either an RC low-pass filter or an RC operational amplifier circuit called a **Miller integrator** (see Appendix A) is used, and where the filter has an RC product much greater than the period of the

**FIGURE 1–3**
Integrated average.

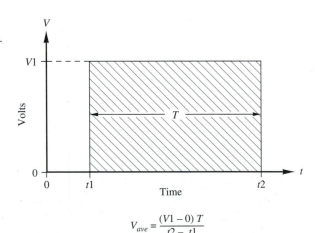

$$V_{ave} = \frac{(V1 - 0)\,T}{t2 - t1}$$

applied input waveform. The output of the circuit is proportional to the time average of the input signal.

The **root mean square (rms)** value is used extensively in electrical circuits and certain other technologies. For example, a sine wave alternating current (ac) wave may be compared with the direct current (dc) voltage level that will produce the same amount of heating in an electrical resistance. The value of the ac wave that is the dc heating equivalent is the root mean square (rms) value. The definition of **rms** is

$$V_{rms} = \sqrt{\frac{1}{T} \int_{t1}^{t2} [V(t)]^2 \, dt} \qquad (1\text{--}7)$$

where  $V_{rms}$ = the rms value

  $T$ = the time interval $t1$ to $t2$

  $V(t)$ = a time-varying voltage function

For the special case of the sine wave, the rms value of voltage is $V_p/\sqrt{2}$, or $0.707V_p$, where $V_p$ is the peak voltage (see Fig. 1–4). For wave shapes other than sinusoidal, however, Equation (1–7) will evaluate differently.

The **root sum squares (rss)** is used in cases where different data are combined to form a single number, even though the data are in no way correlated with one another. For example, noise signals in electronic circuits are errors and come from several different

**FIGURE 1–4**
Peak value of a sine wave voltage.

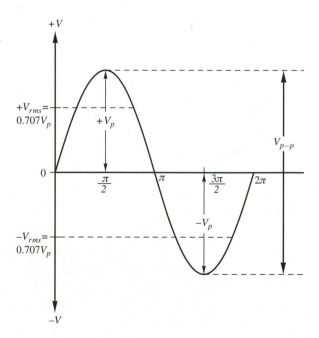

sources. Suppose we have $n$ independent noise voltage sources ($vn_1$, $vn_2$, . . . , $vn_n$). Where these sources are truly independent of each other, they cannot be simply combined in a linear additive manner but rather must be combined using the rss method:

$$V_{rss} = \sqrt{\sum_{i=1}^{n} (vn_i)^2} \tag{1-8}$$

$$V_{rss} = \sqrt{(vn_1)^2 + (vn_2)^2 + (vn_3)^2 + \cdots + (vn_n)^2} \tag{1-9}$$

The rss method is sometimes used to define a single-valued error term from a number of uncorrelated error terms or, alternatively, to find a single standard error of a number of measurements of the same value.

■ **Example 1–1**

An electronic amplifier circuit contains five independent noise sources that produce the following decorrelated noise signal voltage levels: $VN_1 = 25$ nanovolts (nV), $VN_2 = 56$ nV, $VN_3 = -33$ nV, $VN_4 = -10$ nV, and $VN_5 = 62$ nV. What is the rss value of a composite noise signal?

*Solution*

$$\begin{aligned}
V_{rss} &= \sqrt{(VN_1)^2 + (VN_2)^2 + (VN_3)^2 + (VN_4)^2 + (VN_5)^2} \text{ nV} \\
&= \sqrt{(25)^2 + (56)^2 + (-33)^2 + (-10)^2 + (62)^2} \text{ nV} \\
&= \sqrt{625 + 3136 + 1089 + 100 + 3844} \text{ nV} \\
&= \sqrt{8794} = \mathbf{93.8 \ nV}
\end{aligned}$$

Note that the rss value is not the same as the summation of the components.

---

Although it is common in measurement, and in science experiments in general, to quote the "average" value of the data acquired, you must be careful to use the correct average (or most reasonable average) and to correctly interpret what "average" means in the context of the experiment.

## 1–8    LOGARITHMIC REPRESENTATION: DECIBELS

The subject of decibels frequently confuses newcomers to electronics, and even many old-timers seem to have occasional memory lapses regarding the subject. For the benefit of both groups, and because the subject is so vitally important to understanding measurement systems, we will review the decibel.

The decibel measurement originated in the telephone industry and was named after telephone inventor Alexander Graham Bell. The original unit was the "bel." The prefix "deci" means 1/10, so the "decibel" is one-tenth of a bel. The bel is too large for most common applications, so it is rarely if ever used. Thus, we will concentrate only on the more familiar decibel.

The **decibel (dB)** is simply a means of logarithmically expressing the ratio between two signal levels, for example, the "output over input" signal ratio (i.e., "gain") of an amplifier. Because the decibel is a ratio, it is also dimensionless, despite the fact that "dB" looks like a dimension to some people. Consider the voltage amplifier as an example of dimensionless gain; its gain is expressed as the output voltage over the input voltage ($V_o/V_{in}$). It is dimensionless because the units are V/V, which "cancel out."

■ **Example 1–2**

A voltage amplifier outputs 6 V when the input signal has a potential of 0.5 V. Find the voltage gain $(A_v)$.

*Solution*

$$A_v = \frac{V_o}{V_{in}}$$

$$= \frac{6 \text{ V}}{0.5 \text{ V}} = \mathbf{12}$$

Note that the "volts" units appeared in both numerator and denominator, so they "canceled out," leaving only a dimensionless "12" behind.

---

In order to analyze system gains and losses using simple addition and subtraction rather than multiplication and division, a little math trick is used on the ratio. We take the base-10 logarithm of the ratio and multiply it by a scaling factor (either 10 or 20). For voltage systems, such as our voltage amplifier, the expression becomes

$$dB = 20 \log\left(\frac{V1}{V2}\right) \tag{1–10}$$

In the example given earlier we had a voltage amplifier with a gain of 12 because an input of 0.5 V produced a 6-V output. How is this same gain (i.e., the $V_o/V_{in}$ ratio) expressed in decibels?

$$dB = 20 \log(V_o/V_{in})$$
$$= 20 \log(6/0.5)$$
$$= 20 \log(12) = \mathbf{21.6}$$

Although we have changed the ratio by converting it to a logarithm, *the decibel is nonetheless nothing more than a means for expressing a ratio.* Thus, a voltage gain of 12 can also be expressed as a gain of 21.6 dB.

A similar expression can be used for current amplifiers, where the gain ratio is $I_o/I_{in}$:

$$dB = 20 \log\left(\frac{I_o}{I_{in}}\right) \tag{1–11}$$

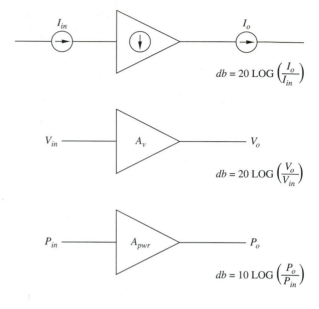

**FIGURE 1–5**
Three basic equations for calculating decibels.

$$db = 20 \, \text{LOG}\left(\frac{I_o}{I_{in}}\right)$$

$$db = 20 \, \text{LOG}\left(\frac{V_o}{V_{in}}\right)$$

$$db = 10 \, \text{LOG}\left(\frac{P_o}{P_{in}}\right)$$

For power measurements we need a modified expression to account for the fact that power is proportional to the square of the voltage or current:

$$\text{dB} = 10 \, \log\left(\frac{P_o}{P_{in}}\right) \tag{1–12}$$

These three basic equations [(1–10), (1–11), and (1–12)] for calculating decibels are summarized in Fig. 1–5.

## 1–8–1  Adding It All Up

So why bother converting seemingly easy-to-handle, dimensionless numbers like voltage or power gains to a logarithmic number like decibels? Fair question. The answer is that it makes calculating signal strengths in a system easier. To see this effect, let's consider the multistage system in Fig. 1–6. Here we have a hypothetical electronic circuit in which there are three amplifier stages and an attenuator pad. The stage gains are as follows:

$$A1 = V1/V_{in} = 0.2/0.010 = 20$$
$$\text{Atten} = V2/V1 = 0.1/0.2 = 0.5$$
$$A2 = V3/V2 = 1.5/0.1 = 15$$
$$A3 = V_o/V3 = 6/1.5 = 4$$

The overall gain is the product of the stage gains in the system:

$$A_v = A1 \times \text{Atten} \times A2 \times A3$$
$$= (20)(0.5)(15)(4) = \mathbf{600}$$

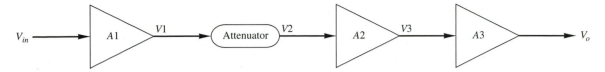

**FIGURE 1–6**
Three-stage amplifier with attenuator pad.

When converted to dB, the gains are expressed as

$$A1 = 26.02$$
$$\text{Atten} = -6.02$$
$$A2 = 23.52$$
$$A3 = 12.04$$

The overall gain of the system (in dB) is the sum of these numbers:

$$A_{v(dB)} = A1 + \text{Atten} + A2 + A3$$
$$= (26.02) + (-6.02) + (23.52) + (12.04)$$
$$= 55.56 \text{ dB}$$

The system gain calculated earlier was 600, and this number should be the same as above:

$$A_{dB} = 20 \log(600)$$
$$= \mathbf{55.56 \text{ dB}}$$

They're the same.

One convenience of the decibel scheme is that gains are expressed as positive numbers and losses as negative numbers. Conceptually it seems easier to understand a loss of "−6.02 dB" than a loss represented as a "gain" of +0.50.

**1–8–2**

### Converting between dB Notation and Gain Notation

We sometimes face situations where gain is expressed in dB, and we want to calculate the gain in terms of the output/input ratio. For example, suppose we have a +20-dB amplifier with a 1-mV (1 mV = 0.001 V) input signal, as shown in Fig. 1–7. What is the expected output voltage? It's 20 dB higher than 0.001 V, right? Yes, that's true, but your meter or oscilloscope is probably calibrated not in decibels but rather in volts. (*Note:* Some instru-

**FIGURE 1–7**
20-dB amplifier with 1-mV input signal.

ments are indeed calibrated in logarithmic units or decibels. For example, audio volt-meters are often calibrated in both volts and dB or VU.) By using a little algebra we can rearrange the expression [dB = 20 log($V_o$/$V_{in}$)] to solve for output voltage, $V_o$. The new expression is

$$V_o = V_{in}(10^{dB/20}) \qquad (1\text{–}13)$$

which is also sometimes written in the following alternate form:

$$V_o = V_{in} \exp(dB/20) \qquad (1\text{–}14)$$

In the example above we want to calculate $V_o$ if the gain in dB and the input signal voltage are known. We can calculate $V_o$ from the above equations, using the values given (20 dB and 1 mV):

$$\begin{aligned}
V_o &= V_{in} \exp(dB/20) \\
&= (0.001) \exp(20/20) \\
&= (0.001) \exp(1) \\
&= (0.001)(100) = \mathbf{0.01 \ V}
\end{aligned}$$

For those who don't want to make the calculation, Table 1–5 shows common voltage and power gains and losses expressed both ways.

Again we see the convenience of decibel scales over gain ratios. If we want to calculate the system gain of a circuit that has a gain of 10,000 and an attenuation of 1/1000 in series, then we can do it either way:

$$A_v = (10,000)(0.001) = 10$$

or,

$$A_v = (+80 \text{ dB}) + (-60 \text{ dB}) = +20 \text{ dB}$$

**TABLE 1–5**
Common gains/losses expressed in decibels

| Ratio (Out/In) | Voltage gain (dB) | Power gain (dB) |
|---|---|---|
| 1/1000 | −60 | −30 |
| 1/100 | −40 | −20 |
| 1/10 | −20 | −10 |
| 1/2 | −6.02 | −3.01 |
| 1 | 0 | 0 |
| 2 | +6.02 | +3.01 |
| 5 | +14 | +7 |
| 10 | +20 | +10 |
| 100 | +40 | +20 |
| 1,000 | +60 | +30 |
| 10,000 | +80 | +40 |
| 100,000 | +100 | +50 |
| 1,000,000 | +120 | +60 |

**1–8–3**          ### Special dB Scales

Various user groups have defined special dB-based scales that meet their own needs. They make a special scale by defining a certain signal level as "0 dB" and referencing all other signal levels to the defined 0-dB point. In the dimensionless dB scale, 0 dB corresponds to a gain of unity (see Table 1–5). But if we define "0 dB" as a particular signal level, then we obtain one of the special scales. Several such scales commonly used in electronics are as follows:

**dBm.**   Used in RF measurements, this scale defines 0 dBm as 1 mW of RF signal dissipated in a 50-$\Omega$ resistive load.

**Volume Units (VU).**   The VU scale, used in audio work, defines 0 VU as 1 mW of 1000-Hz audio signal dissipated in a 600-$\Omega$ resistive load.

**dB** (now obsolete).   This scale (once used in telephone work) defined 0 dB as 6 mW of 1000-Hz audio signal dissipated in a 500$\Omega$ load. (*Note:* One source listed 400 Hz as the reference frequency.)

**dBmv.**   Used in television antenna coaxial cable systems with a 75-$\Omega$ resistive impedance, the dBmV system uses 1000 $\mu$V (1 mV) across a 75-$\Omega$ resistive load as the 0-dBmv reference point.

Consider the case of the RF signal generator. In RF systems using standard 50-$\Omega$ input and output impedances, all power levels are referenced to 0 dBm being 1 mW (i.e., 0.001 W). To write signal levels in dBm, we use the modified power dB expression:

$$\text{dBm} = 10 \ \log\left( \frac{P}{1 \ \text{mW}} \right) \qquad (1\text{–}15)$$

■  **Example 1–3**

What is the signal level 9 mW as expressed in dBm?

*Solution*

$$\text{dBm} = 10 \ \log(P/1 \ \text{mW})$$
$$= 10 \ \log(9/1)$$
$$= \textbf{9.54 dBm}$$

Thus, when we refer to a signal level of 9.54 dBm, we mean an RF power of 9 mW dissipated in a 50-$\Omega$ load.

---

Signal levels less than 1 mW show up as negative dBm. For example, 0.02 mW is also written as −17 dBm.

**1–8–4**     **Converting dBm to Voltage**

Signal generator output controls and level meters are frequently calibrated in microvolts or millivolts (although some are also calibrated in dBm). How do we convert volts to dBm, or dBm to volts?

**Converting Microvolts to dBm.**   Use the expression $P = V^2/R = V^2/50$ to find milliwatts, and then use the dBm expression given in Equation (1–15).

■  **Example 1–4**

Express a signal level of 800 μV (i.e., 0.8 mV) rms in dBm.

*Solution*

$$P = V^2/50$$
$$= (0.8)^2/50$$
$$= 0.64/50 = 0.0128 \text{ mW}$$
$$\text{dBm} = 10 \log(P/1 \text{ mW})$$
$$= 10 \log(0.0128 \text{ mW}/1 \text{ mW})$$
$$= \mathbf{-18.9}$$

**Converting dBm to Microvolts or Millivolts.**   Find the power level represented by the dBm level, and then calculate the voltage using 50 Ω as the load.

■  **Example 1–5**

What voltage exists across a 50-Ω resistive load when −6 dBm is dissipated in the load?

*Solution*

$$P = (1 \text{ mW})(10^{\text{dBm}/10})$$
$$= (1 \text{ mW})(10^{-6 \text{ dBm}/10})$$
$$= (1 \text{ mW})(10^{-0.6})$$
$$= (1 \text{ mW})(0.25) = \mathbf{0.25 \text{ mW}}$$

If $P = V^2/50$, then $V = (50P)^{1/2} = 7.07(P^{1/2})$, so

$$V = (7.07)(P^{1/2})$$
$$= (7.07)(0.25^{1/2}) = \mathbf{3.535 \text{ mV}}$$

(*Note:* Because power is expressed in milliwatts, the resulting answer is in millivolts. To convert to microvolts, multiply the result by 1000.)

**1–9**     **THE BASICS OF MEASUREMENT**

Measurements are "the assignment of numerals to represent [physical] properties" (Herceg, 1972). Measurements are made to fulfill one or more of several different goals:

obtain information about a physical phenomenon, assign a value to some fundamental constant, record trends, control some process, correlate behavior with other parameters in order to obtain insight into their relationships, or figure out how much to pay for a beef cow. A measurement is an act that is designed to "derive quantitative information about" some physical phenomenon "by comparison to a reference" (Herceg, 1972) or standard. The physical quantity being measured is called the **measurand.**

All measurements are subject to a certain amount of *variation* caused by small *errors* in the measurement process, and by actual variation in the measured parameter. In this context the idea of "error" does not mean "mistake," but rather a normal random variation due to inherent limitations of the system. There are many different causes of random variation. Some variations are dependent on the particular type of measurement being made (and are a function of the type of meter being used), while others are inherent in the process being measured.

Consider an ordinary meter stick as an analogy. It is divided into 100 large divisions of 1 cm each, and 1000 small divisions of 1 mm each. Or is it? In truth, the distance from 0 to 100 cm is not exactly 100.0000 cm but rather will be $100 \pm \varepsilon$ cm, where $\varepsilon$ is some small error. In addition, the spaces between divisions (either 1 mm or 1 cm) are not exactly the same size but vary somewhat from mark to mark. As a result, when you measure the size of, say, a printed circuit board, there will be some error.

When measurements are made, random variation causes the data obtained to **disperse.** Consider a practical situation. Suppose we want to measure the mass of an electronic component, say, a capacitor. The design specification weight is supposed to be 1.45 g, but when a sample of 30 identical capacitors are weighed on a very good scale, the results shown in the crude bar graph of Fig. 1–8 and listed below are obtained:

**Sorted Data Points (Low-to-High Sort)**

1.16   1.18   1.2   1.26   1.26   1.3   1.3   1.3   1.33   1.35
1.4   1.4   1.4   1.4   1.4   1.43   1.48   1.5   1.5   1.5
1.5   1.53   1.6   1.6   1.62   1.65   1.7   1.7   1.7   1.81

Even though the capacitors are all identical, there is a dispersion of the data points. Not one capacitor in the sample lot actually meets the 1.45-g specification exactly, but rather all are dispersed around 1.45 g.

If you examine a large number of capacitors and plot their weights as in Fig. 1–8(a), you will find that a pattern emerges. As long as certain conditions are met, the various weights will form the familiar "bell-shaped" curve [Fig. 1–8(b)] that is variously called, depending on country and context, the **normal distribution curve,** the **Gaussian curve,** and the **Laplacean.** Regardless of what it is called, the normal distribution pattern of Fig. 1–8(b) is extremely common in science and technology.

The normal distribution curve plots frequency of occurrence against some other parameter. Note in Fig. 1–8(a) that even with only a few data points some values are beginning to "stack up" higher than other values. When thousands of capacitor weights are measured, it is likely that the resultant bar graph will very nearly resemble the bell-shaped curve. The **mean** is usually designated by the Greek letter mu ($\mu$) when the entire population of values is plotted, or by "$x$-bar" when only a sample of the population is taken.

**FIGURE 1–8**
Weights of 30 capacitors plotted (a)
in bar graph and (b) in line graph
(normal distribution curve).

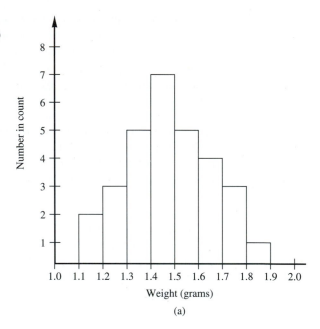

(a)

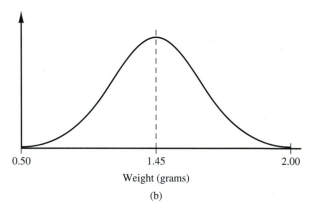

(b)

One of the uses of the normal distribution curve is that it offers us a measure of the dispersion of the data. This property of the data can be summed up as the **variance** and the **standard deviation** of the data. For the entire population, variance is denoted by $\sigma^2$, and standard deviation by $\sigma$. Variance is defined by

$$\sigma^2 = \frac{\sum\limits_{i=1}^{N} (X_i - \overline{X})^2}{N} \tag{1–16}$$

and standard deviation, which is the square root of variance, by

$$\sigma = \sqrt{\frac{\sum_{i=1}^{N}(X_i - \overline{X})^2}{N}} \qquad (1\text{--}17)$$

Equations (1–16) and (1–17) define the variance and the standard deviation for the entire population of data. If a small sample is taken, then replace $\sigma^2$ with $s^2$, $\sigma$ with $s$, and $N$ in the denominators with $N - 1$:

$$s^2 = \frac{\sum_{i=1}^{N}(X_i - \overline{X})^2}{N - 1} \qquad (1\text{--}18)$$

and

$$s = \sqrt{\frac{\sum_{i=1}^{N}(X_i - \overline{X})^2}{N - 1}} \qquad (1\text{--}19)$$

If the process is truly random, then the value that one would report in making the measurement is the mean value, $\mu$. But it is also necessary to specify either the variance or the standard deviation. We know that any particular measurement may or may not be $\mu$. We also know that 68.27% of all values lie between $\pm\sigma$ (see Fig. 1–9), 98.45% between $\pm2\sigma$, and 99.73% between $\pm3\sigma$.

**FIGURE 1–9**
Standard deviation.

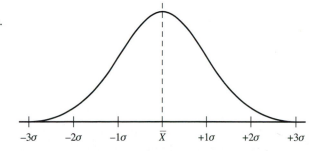

## SUMMARY

1. The reason for recognizing and using only the significant figures of a measurement is so that we impute no more precision or accuracy to the measurement than is justified by the situation.
2. The number of significant figures in a measurement is not improved by mathematically combining them with other numbers.
3. Scientific notation is an arithmetic shorthand that represents very large and very small quantities using powers-of-10 exponents.
4. In the metric system a base unit (e.g., the **meter** as the unit of length) is combined with a prefix that either multiplies or divides the units for larger or smaller quantities (e.g., *kilo*meters and *milli*meters).
5. The **average** of a number set is considered the most typical, or most expected, value. There are three types of averages in common use—mean, median, and mode—as well as other averages that are less commonly used—geometric, harmonic, and logarithmic means, and root mean square (rms) and root sum squares (rss).
6. Signal levels are often represented on a logarithmic scale called the **decibel (dB)** scale. Gains and losses expressed in dB can be directly added or subtracted in making circuit and system calculations. The decibel is an expression of the ratio of two signals. Standardized decibel notations are often used by setting one signal level to a standard value.
7. When measurements are made, the quantity being measured is called the **measurand.**

## RECAPITULATION

Now go back and try to answer the questions at the beginning of this chapter. When you are finished, answer the questions and work the problems below. Place a mark beside each problem or question that you cannot answer, and then go back and reread the appropriate sections of the text.

## QUESTIONS

1. State the number of significant figures used in each of the following:

   **a)** 1     **b)** 22     **c)** 5     **d)** 245.

2. State the number of significant figures in each of the following:

   **a)** 2.4     **b)** 4.32     **c)** 256.322     **d)** 32.

3. Zero is always a significant figure. True or false?

4. Identify the least significant digit and the most significant digit in each of the following:

   **a)** 12.25     **b)** 9.99     **c)** 2.56     **d)** 234.0.

5. A digital instrument display bounces between two readings, 14.56 and 14.57 V. This phenomenon is called _____ _____ _____.

6. A digital voltmeter reads 18.85 V. If the accuracy is 1%, and considering the phenomenon in Question 5, in what range is the actual voltage likely to lie?

7. A 0-to-1999-mV digital voltmeter is used to measure two voltages, one 1230 mV and the other 758 mV. A calculator shows that the quotient of these voltages is 1.622691293. If the voltmeter has an accuracy limit of 1%, what is a more reasonable expression of this ratio?

8. Give the power-of-10 equivalent for the following metric prefixes:

   a) pico      b) milli      c) micro      d) deci
   e) kilo      f) mega      g) giga.

9. The most expected value of a single data set is called the _____.

10. Which of the following represents a 6-dB loss:

    a) 6 dB      b) −6 dB?

11. Errors in measurement are the same as operator mistakes. True or false?

12. When a fixed value is measured, random variation causes the data obtained to _____ around a mean value.

13. The dispersion of a data set is measured by the _____ and the _____ _____.

## PROBLEMS

1. Multiply 2.345 and 2.22.

   a) What is the product?
   b) How many significant figures are there in the answer?

2. Express the following in scientific notation:

   a) 1.00            b) 125            c) 0.0000127
   d) 1,500,000      e) 10,000        f) 56,000.

3. Express the following in regular notation:

   a) $2 \times 10^{-2}$      b) $3.4 \times 10^5$      c) $10^{-4}$.

4. Convert the following to meters:

   a) 25.4 mm      b) 2.54 cm      c) 0.100 km      d) 123 cm.

5. A human electrocardiogram produces a peak signal of 1.6 millivolts (mV) riding on a 1.45-V dc pedestal. Combine these two voltages and express them in

   a) volts      b) millivolts.

6. Find the **arithmetic mean** of the following data: 12.2, 5, 8.5, 14.32, 7.5, 6.5, 6.5, 5.5.

7. Find the **median** of the following data: 5, 6, 4, 3, 2, 9, 7.

8. Find the **mode** of the following data: 5, 6, 4, 3, 4, 2, 4, 7, 2, 9, 4.

9. You have $480 to spend. You spend one-half of your money each day. Tabulate the amount spent and then calculate the logarithmic mean.

10. The following prices are found for a certain brand of canned peas over a period of six weeks: $1.96, $2.05, $1.75, $1.94, $2.25, $2.10. Calculate both the arithmetic mean and the harmonic mean.

**11.** Find the rms value of a sine wave ac signal that has a peak voltage of 45.3 V.

**12.** Find the rss value of the following signal levels: 25.6 µV, 22 µV, 56.5 µV, and 33 µV.

**13.** An amplifier has an output signal of 3.6 V rms when a sine wave input signal of 0.1 V rms is present. State its gain in decibels.

**14.** An amplifier has a gain of 20 dB. This represents an output-to-input ratio of _____.

**15.** An amplifier has three amplification stages and a 1-dB attenuator in cascade. Assuming that all impedances are properly matched, what is the overall gain if the amplification factors are 5 dB, 10 dB, and 6 dB? Express your answer both in decibels and in nondecibel form.

**16.** A double-balanced modulator is rated to accept signals up to +7 dBm. What is this signal level expressed in

**a)** watts

**b)** volts rms? Assume that all impedances are 50 Ω and are matched.

**17.** A signal level of 145 µV is found across a 50-Ω load. What is this signal level expressed in dBm?

**18.** The following table shows data collected by measuring data with a voltmeter:

| | | | |
|------|------|-------|------|
| 36.3 | 33.4 | 30.3 | 34.2 |
| 35.3 | 33.78 | 32.8 | 33.1 |
| 39.0 | 30.4 | 36.4 | 38.4 |
| 32.6 | 39.4 | 29.23 | 35.3 |
| 35.9 | 40.1 | 37.0 | 33.8 |

Calculate the (a) mean voltage, (b) variance, and (c) standard deviation.

# BIBLIOGRAPHY

Campbell, N. R. *Foundations of Science*. New York: Dover, 1957. Cited in Mandel, John, *The Statistical Analysis of Experimental Data*. New York: John Wiley & Sons, 1964. Dover paperback edition, 1984.

Carr, J. J. *The Art of Science*. Solana Beach, CA: HighText Publications, 1992.

Herceg, E. E. *Handbook of Measurement and Control*. Pennsauken, NJ: Schaevitz Engineering, 1972.

# 2

# Some Basic Measurement Theory

**OBJECTIVES**

1. Be able to list the basic categories of measurements.
2. Know the meanings of the terms **accuracy, precision, resolution, reliability,** and **validity** as applied to measurements.
3. Understand and evaluate the role and nature of measurement **error.**
4. Know how to minimize error in measurements.

2–2 **SELF-EVALUATION**

Before studying the material in this chapter, try to answer the questions given below. These questions test your knowledge of the subject. If you cannot answer a particular question, then look for the answer as you read the text.

1. Measurement errors usually result from human error. True or false?
2. Blood pressure measurement by use of an external cuff is an example of a _____ measurement.
3. If a voltage changes when a voltmeter is connected into the circuit, the resultant error is an example of _____ _____ error.
4. A procedure that yields consistent results from one measurement to the next is an example of a(n) _____ definition.

2–3 **INTRODUCTION**

In this chapter you will learn the fundamentals of measurement—electronic and otherwise. The material starts out with a discussion of the various categories of measurement (e.g., direct, indirect, null, etc.). Also discussed are issues such as **precision** and **accuracy** (often erroneously confused), **resolution, validity,** and **reliability** of a measurement. There is also a discussion of **measurement error** and how to avoid some of the most serious and most common errors.

## 2–4    CATEGORIES OF MEASUREMENT

There are three general categories of measurement: **direct, indirect,** and **null.** Electronic instruments are available based on all three categories.

**Direct measurements** are made by holding the measurand up to some calibrated standard and comparing the two. A good example is a meter-stick ruler used to cut a piece of cable to the correct length. You know that the cable must be cut to a length of 24 cm, so you hold a meter stick (the standard or reference) up to the piece of cable (Fig. 2–1), set the "0 cm" point at one end, and make a mark on the cable adjacent to the "24" mark on the meter stick. Then you make your cut at the appropriate point.

**Indirect measurements** are made by measuring something other than the actual measurand. Although frequently considered "second best" from the perspective of measurement accuracy, indirect methods are often used when direct measurements are either difficult or dangerous. For example, you might measure the temperature of a point on the wall of a furnace that is melting metal [Fig. 2–2(a)], knowing that the exterior temperature is related to the interior temperature by a certain factor[Fig. 2–2(b)].

A minicomputer manufacturer once used an indirect temperature measurement to ease the job of service technicians. They created a small hole at the top of a rack-mounted cabinet where the temperature would be less than 39°C when the temperature on the electronic circuit boards was within specification. They used this method for two reasons:

1. The measurement point was easily available (while the boards were not) and thus disassembly was not required.
2. The service technician could use an ordinary household medical fever thermometer (30°C–42°C) as the measurement instrument. No special laboratory thermometers were needed.

Perhaps the most common example of an indirect measurement is human blood pressure, measured by an occluding cuff placed around the arm, a process called **sphygmomanometry** [Fig. 2–3(a)]. Research around 1905 showed that the cuff pressures at two easily heard events[onset and cessation of "Korotkoff sounds"; Fig. 2–3(b)] are correlated to the systolic ($P_s$) and diastolic ($P_d$) arterial blood pressures. Direct blood pressure measurement may be more accurate but is dangerous because it is an invasive surgical procedure.

**Null measurements** are made by comparing a calibrated source to an unknown measurand and then adjusting either one or the other until the difference between them is zero. An **electrical potentiometer** is such an instrument; it is an adjustable, calibrated

**FIGURE 2–1**
Measuring cable with a meter stick.

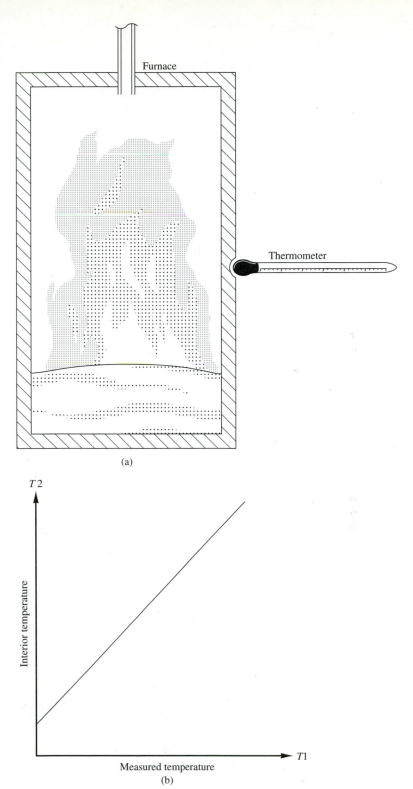

(a)

(b)

**FIGURE 2–3**
Indirect measurement of blood
pressure.

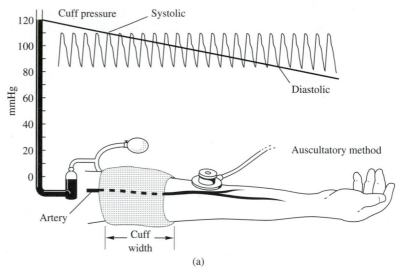

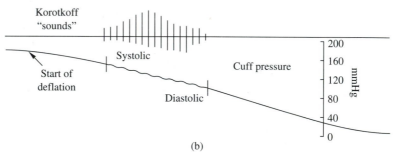

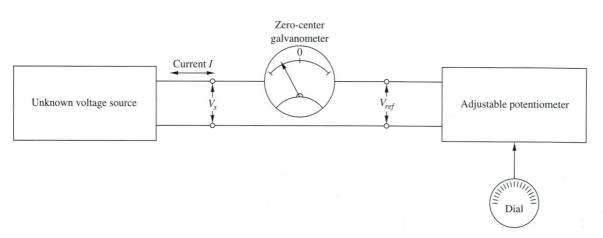

**FIGURE 2–4**
Example of null measurement.

voltage source and a comparison meter (zero-center galvanometer). As shown in Fig. 2–4, the reference voltage from the potentiometer is applied to one side of the zero-center galvanometer (or one input of a difference-measuring voltmeter), and the unknown is applied to the other side of the galvanometer (or remaining input of the differential voltmeter). The output of the potentiometer is adjusted until the meter reads zero difference. The setting of the potentiometer under the nulled condition is the same as the unknown measurand voltage.

**2–5**       ## FACTORS IN MAKING MEASUREMENTS

The "goodness" of measurements involves several important concepts. Some of the more significant of these are **error, validity, reliability, repeatability, accuracy, precision,** and **resolution.**

**2–5–1**      ### Error

In all measurements there is a certain degree of error present. The word **error** in this context refers to normal random variation and in no way means "mistakes" or "blunders." We will discuss error in greater depth shortly.

     If measurements are made repeatedly on the same parameter (which is truly unchanging), or if different instruments or instrument operators are used to make successive measurements, the measurements will tend to cluster around a central value ($X_o$ in Fig. 2–5). In most cases it is assumed that $X_o$ is the true value, but if there is substantial inherent error in the measurement process, then the measured value may deviate from the true value ($X_i$) by a certain amount ($\Delta X$), which is the error term. The assumption that the central value of a series of measurements is the true value is valid only when the error term is small. That is, as $\Delta X \rightarrow 0$, $X_o \rightarrow X_i$.

**2–5–2**      ### Validity

The **validity** of a measurement is a statement of how well an instrument actually measures what it purports to measure. For example, an electronic pressure sensor may actually be measuring the deflection of a thin metallic diaphragm (of known area, of course),

**FIGURE 2–5**
The accuracy of a measurement is indicated by the size of $\Delta X$.

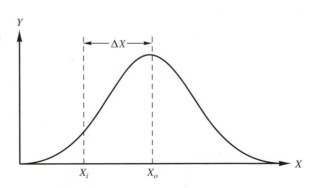

which is in turn measured by the strain applied to a strain gage element cemented to the diaphragm. What determines the validity of a sensor measurement is the extent to which the measurement of the deflection of that diaphragm relates to applied pressure . . . and over what range or under what conditions. In many measurement devices the output readings are meaningful only under certain specified conditions or over a specified range.

**2–5–3**      ### Reliability and Repeatability

The **reliability** of a measurement is a statement of its **consistency** when the values of the measurand are discerned on different trials—when the measurand may take on very different values. In the case of the pressure sensor discussed above, a deformation of a diaphragm may change its characteristics sufficiently to alter future measurements of the same pressure value.

Related to reliability is the idea of **repeatability,** which refers to the ability of an instrument to return the same value when repeatedly exposed to the exact same stimulant. Neither reliability nor repeatability is the same as accuracy, for a measurement may be both "reliable" and "repeatable" while being quite wrong.

**2–5–4**      ### Accuracy and Precision

The **accuracy** of a measurement refers to the freedom from error, or the degree of conformance between the measurand and the standard. **Precision,** on the other hand, refers to the exactness of successive measurements; it is also sometimes considered the degree of refinement of the measurement. Accuracy and precision are often confused with one another, and these words are often erroneously used interchangeably. One way of stating the situation is to note that *a precise measurement has a small standard deviation and variance under repeated trials, while in an accurate measurement the mean value of the normal distribution curve is close to the true value.*

Figure 2–6 shows the concepts of accuracy and precision in various measurement situations. In all of these cases the data form a normal distribution curve when repeatedly performed over a large number of iterations of the measurement. Compare Figs. 2–6(a) and 2–6(b). Both of these situations have relatively low accuracy because there is a wide separation between $X_o$, the measured value of the measurand, and $X_i$, the actual value. The measurement represented in Fig. 2–6(a) has relatively high precision compared with Fig. 2–6(b). The difference is illustrated by the fact that Figure 2–6(a) has a substantially lower variance and standard deviation around the mean value, $X_o$; this curve shows poor accuracy but good precision. The variance of Fig. 2–6(b) is greater than the variance of the previous curve, so it has both low accuracy and poor precision. Somewhat different situations are shown in Figs. 2–6(c) and 2–6(d). In both of these distributions the accuracy is better than in the previous cases because the difference between $X_o$ and $X_i$ is reduced (of course, a perfect, error-free measurement has a difference of zero). In Fig. 2–6(c), the measurement has good accuracy and good precision, while in Fig. 2–6(d), the measurement has good accuracy but poor precision [its variance is greater than in Fig. 2–6(c)].

The standard deviation (which is the square root of variance) of a measurement is a good indication of its precision, which also means the inherent error in the measurement.

**FIGURE 2–6**

Examples of accuracy and precision. (a) Low accuracy but good precision, (b) low accuracy and poor precision, (c) good accuracy and good precision, (d) good accuracy but poor precision.

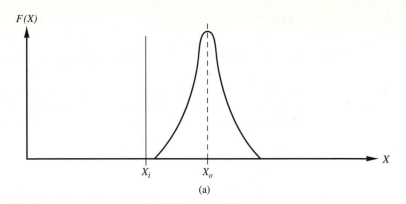

(a)

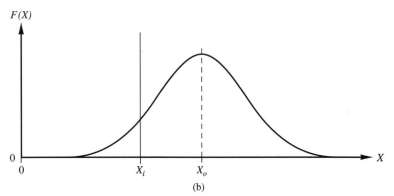

(b)

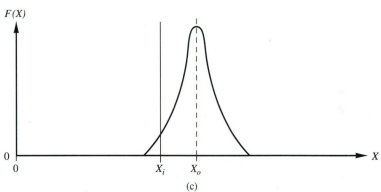

(c)

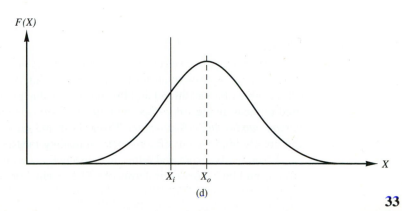

(d)

Several tactics help reduce the effects of error on practical measurements:

1. Make the measurement a large number of times, and then average the results.
2. Make the measurement several times using different instruments, if feasible.
3. When using instruments such as rulers or analog meters, try making the successive measurements on different parts of the scale. For example, on rulers and analog meter dials, the distance between tick marks is not truly constant because of manufacturing error. The same is also true of electrical meter scales. Measure lengths using different points on the scale as the zero reference point (e.g., on a meter stick use 2, 12, 20, and 30 cm as the zero point), and then average the results. By taking the measurements from different sections of the scale, you help make both the individual errors and the biases that accumulate average to a lower overall error.

**2–5–5**  **Resolution**

This term refers to the degree which a measurand can be broken into identifiable adjacent parts. An example can be seen on the standard television test pattern broadcast by some stations in the early morning hours between "broadcast days." Various features on the test pattern will include parallel vertical or horizontal lines of different densities. One patch may be 100 lines per inch, another 200 lines per inch, and so forth, up the scale. The resolution of the video system is the maximum density *at which it is still possible to see adjacent lines with space between them.* For any system there is a limit above which the lines are blurred into a single entity.

In a digital electronic measuring instrument, the resolution is set by the number of bits used in the data word. Digital instruments use the binary (base-2) number system in which the only two permissible digits are 0 and 1. The binary "word" is a binary number representing a quantity. For example, binary 0001 represents decimal 1, while binary 1001 represents decimal 5. An 8-bit data word, the standard for many small computers, can take on values from $00000000_2$ to $11111111_2$ and thus can break the range into $2^8$ (256) distinct values, or $2^8 - 1$ (255) different segments. The resolution of that system depends on the value of the measured parameter change that must occur in order to change the least significant bit in the data word. For example, if 8 bits are used to represent a voltage range of 0 to 10 V, then the resolution is $(10 - 0)/255$, or 0.039 V (i.e., 39 mV) per bit. This resolution is often specified as the "1 – LSB" resolution.

**2–6**  **MEASUREMENT ERRORS**

No measurement is perfect, and measurement apparatus are never ideal, so there will always be some error in all forms of measurement. An **error** is a deviation between the actual value of a measurand and the indicated value produced by the sensor or instrument used to measure the value. Let me reiterate: *error is inherent and is NOT the fault of the person making the measurement.* Error is not the same as mistake! Understanding error can greatly improve our effectiveness in making measurements.

Error can be expressed either in "absolute" terms or by a relative scale. An absolute error would be expressed in terms of "$X \pm x$ cm," or some other such unit, while a relative error expression would be "$X \pm 1\%$ cm." In an electrical circuit a voltage might be stated as "4.5 V $\pm$ 1%." Which expression to use may be a matter of custom, convention, personal choice, or best utility, depending on the situation.

2–7          **CATEGORIES OF ERRORS**

There are four general categories of error: **theoretical error, static error, dynamic error, and instrument insertion error.**

2–7–1          **Theoretical Errors**

All measurements are based on some measurement theory that predicts how a value will behave when a certain measurement procedure is applied. The measurement theory is usually based on some theoretical model of the phenomenon being measured, that is, an intellectual construct that tells us something of how that phenomenon works. Often the theoretical model is valid only over a specified range of the phenomenon. For example, a nonlinear phenomenon that has a quadratic, cubic, or exponential function can be treated as a straight-line (linear) function over small, carefully selected sections of the range. Electronic sensor outputs often fall into this class.

Alternatively, the actual phenomenon may be terribly complex, or even chaotic, under the right conditions, so the model is therefore simplified for many practical measurements. An equation that is used as the basis for a measurement theory may be only a first-order approximation of the actual situation. For example, consider the **mean arterial pressure (MAP)** that is often measured in clinical medicine and medical sciences research situations. The MAP approximation equation used by clinicians is

$$\overline{P} = \text{diastolic} + \frac{\text{systolic} - \text{diastolic}}{3} \tag{2-1}$$

This equation is actually only an approximation (and holds true mostly for well people, not some sick people to whom it is applied) of the equation that expresses the mathematical integral of the blood pressure over a cardiac cycle, that is, the time average of the arterial pressure. The actual expression is written in the notation of calculus, which is beyond the math abilities many of the people who use the clinical version above:

$$\overline{P} = \frac{1}{T} \int_{t1}^{t2} P(t) \, dt \tag{2-2}$$

The approximation [Equation (2–1)] works well but is subject to greater error than Equation (2–2) due to the theoretical simplification of the first equation. It usually does no good for a repair technician to tell an irate intensive care unit (ICU) nurse that the MAP reading on an instrument is correct when it differs from the calculated value. The actual measured MAP is based on the integral calculus equation (an analog electronic Miller integrator circuit is used), while the calculated value is based on the simplification.

Because some patients lack a considerable part of the time-dependent pressure curve (which is sometimes why they are in the hospital), the actual MAP will be lower than the approximation value.

**2–7–2**     **Static Errors**

Static errors include a number of different subclasses that are all related in that they are always present even in unchanging systems (and thus are not dynamic errors). These errors are not functions of the time or frequency variation.

**Reading Static Errors.**   These errors result from misreading the display output of the sensor system. An analog meter uses a pointer to indicate the measured value. If the pointer is read at an angle other than straight on, then a **parallax reading error** occurs. Another reading error is the **interpolation error,** that is, an error made in guessing the correct value between two calibrated marks on the meter scale (Fig. 2–7). Still another

**FIGURE 2–7**
Interpolation error.

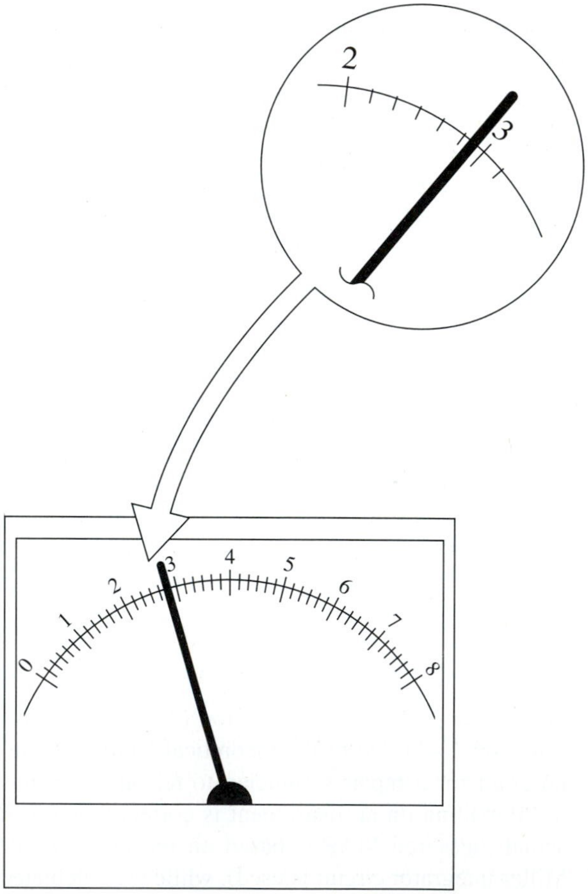

reading error occurs if the pointer on a meter scale is too broad and covers several marks at once.

A related error seen in digital readouts is the **last-digit bobble error.** On digital displays the least significant digit on the display will often flip back and forth between two values. For example, a digital voltmeter might read "12.24" and "12.25" alternately, depending on when you looked at it, despite the fact that absolutely no change occurred in the voltage being measured. This phenomenon occurs when the actual voltage is midway between the two indicated voltages. Error, noise, and uncertainty in the system will make a voltage close to 12.245 V bobble back and forth between the two permissible output states (12.24 and 12.25) on the meter. An example where bobble is of significant concern is the case where some action is taken when a value changes above or below a certain amount—and the digital display bobbles above and below the critical threshold.

**Environmental Static Errors.**   All sensors and instruments operate in an environment, and that environment sometimes affects the output states. Factors such as temperature (perhaps the most common error-producing agent), pressure, electromagnetic fields, and radiation must be considered in some electronic sensor systems.

**Characteristic Static Errors.**   These static errors are still left after reading errors and environmental errors are accounted. When the environment is well within the allowable limits and is unchanging, and when there is no reading error, a residual error will remain that is a function of the measurement instrument or process itself. Errors found under this category include zero offset error, gain error, processing error, linearity error, hysteresis error, repeatability error, resolution error, and so forth.

Also included in characteristic errors are any design or manufacturing deficiencies that lead to error. For example, not all of the "ticks" on a ruler are truly 1.0000 mm apart at all points along the ruler. While it is hoped that the errors are random, so that the overall error is small, there is always the possibility of a distinct bias or error trend in any measurement device.

For digital systems one must add to the resolution error a **quantization error** that emerges from the fact that the output data can take on only certain discrete values. For example, an 8-bit analog-to-digital (AID) converter allows 256 different states, so a 0-to-10-V range is broken into 256 discrete values in 39.06-mV steps. A potential that is between two of these steps is assigned to one or the other according to the rounding protocol used in the measurement process. An example is the weight sensor that outputs 8.540 V, on a 10-V scale, to represent a certain weight. The actual 8-bit digitized value may represent 8.502, 8.541, or 8.580 V because of the ±0.039-V quantization error.

2–7–3          **Dynamic Errors**

Dynamic errors arise when the measurand is changing or in motion during the measurement process. Examples of dynamic errors include the inertia of mechanical indicating devices (such as analog meters) when measuring rapidly changing parameters. A number of limitations in electronic instrumentation fall into this category, especially cases where a frequency, phase, or slew rate limitation is present.

**2–7–4**        **Instrument Insertion Errors**

A fundamental rule of making engineering and scientific measurements is that *the measurement process should not significantly alter the phenomenon being measured.* Otherwise, the measurand is actually the altered situation, not the original situation that is of true interest. Examples of this error are found in many places. For example, pressure sensors tend to add volume to the system being measured, so they slightly reduce the pressure indicated below the actual pressure. Similarly, a flowmeter might add length, a different pipe diameter, or turbulence to a system being measured. A voltmeter with a low impedance of its own could alter resistance ratios in an electrical circuit and produce a false reading (Fig. 2–8). This problem is especially common when cheap analog volt-ohm-milliammeters, which have a low sensitivity and hence a low impedance ($R_m$ in Fig. 2–8), are used to measure a voltage in a circuit. The meter resistance $R_m$ is effectively shunted across the circuit resistance across which the voltage appears.

**FIGURE 2–8**
False reading caused by instrument insertion error.

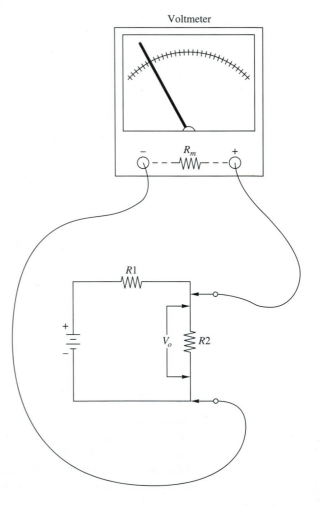

Instrument insertion errors can usually be minimized by good instrument design and good practices. No measurement device has zero effect on the system being measured, but one can reduce the error to a very small value by appropriate selection of methods and devices.

## 2–8    DEALING WITH MEASUREMENT ERRORS

Measurement error can be minimized through several methods, some of which are lumped together under the rubric "procedure," and others under the legend "statistics."

Under "procedure" one can find methods that will reduce, or even minimize, error contributions to the final result. For example, in an electrical circuit, use a voltmeter that has an extremely high input impedance compared with circuit resistances. The idea is to use an instrument (whether a voltmeter, a pressure meter, or whatever) that least disturbs the thing being measured.

A significant source of measurement error in some electronic circuits is ground loop voltage drop and ground plane noise. Figure 2–9 shows how this problem occurs. The circuit seen by the user of the instrument is a dc voltage source driving a resistor voltage divider. An unseen ground current ($I_G$) flows through the resistance of the ground plane ($R_G$) to produce a voltage drop $I_G R_G$. This voltage may add to or subtract from the reading of output voltage $V_o$, depending on its phase and polarity.

A way to reduce total error is to use several different instruments to measure the same parameter. Figure 2–10 shows an example where the current flow in a circuit is being measured by three different ammeters: $M_1$, $M_2$, and $M_3$. Each of these instruments will produce a result that contains an error term decorrelated from the error of the others and not biased (unless, by selecting three identical model meters, we inherit the characteristic error of that type of instrument). We can estimate the correct value of the current flow rate by taking the average of the three:

$$M_o = \frac{M_1 + M_2 + M_3}{3}$$

(2–3)

**FIGURE 2–9**

Ground loop voltage drop and ground plane noise.

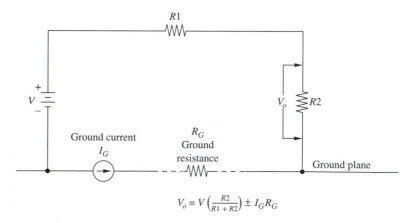

$$V_o = V\left(\frac{R2}{R1 + R2}\right) \pm I_G R_G$$

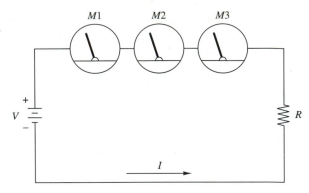

**FIGURE 2–10**
Current measured by three ammeters to reduce measurement error.

You must be careful either to randomize the system in cases where the sensor or instruments used tend to have large error terms biased in one direction, or to calibrate the average error so that it may be subtracted out of the final result.

## 2–9     ERROR CONTRIBUTIONS ANALYSIS

An **error analysis** should be performed in order to identify and quantify all contributing sources of error in a system. A determination is then made regarding the randomness of those errors, and a **worst-case analysis** is made. Under the worst case, you assume that all of the component errors are biased in a single direction and are maximized. You then attempt to determine the consequences (to your purpose for making the measurement) if these errors line up in that manner, even if such an alignment is improbable. The worst-case analysis should be done on both the positive and the negative side of the nominal value. Then you create an **error budget** to allocate an allowable error to each individual component of the measurement system in order to ensure that the overall error is not too high for the intended use of the system.

If errors are independent of each other, are random rather than biased in one direction, and are of the same order of magnitude, then one can find the **root of the sum of the squares (rss)** value of the errors and use it as a composite error term in planning a measurement system. The rss error is

$$\varepsilon_{rss} = \sqrt{\Sigma \, \varepsilon_i^2} \qquad (2\text{–}4)$$

The rss error term is a reasonable estimate or approximation of the combined effects of the individual error components.

A collection of repetitive measurements of a phenomenon can be considered a **sampled population** and treated as such. If we take $N$ measurements ($M1$ through $M_n$) of the same parameter and then average them, we get

$$\overline{M} = \frac{M1 + M2 + M3 + \cdots + M_n}{N} \qquad (2\text{–}5)$$

The average value obtained in Equation (2–5) is the **mean arithmetic average.** This value is usually reported as the correct value for the measurement, but when taken alone it does not address the issue of error. For this purpose we add a correction factor by quoting the **standard error of the mean:**

$$\sigma_{\bar{m}} = \frac{\sigma_m}{\sqrt{N}} \qquad\qquad \text{(2–6)}$$

which is reported in the result as

$$M = \overline{M} \pm \sigma_{\bar{m}} \qquad\qquad \text{(2–7)}$$

Any measurement contains error, and this procedure allows us to estimate that error and thereby understand the limitations of that particular measurement.

2–10        **OPERATIONAL DEFINITIONS IN MEASUREMENT**

Some measurement procedures suggest themselves immediately from the nature of the phenomenon being measured. In other cases, however, there is a degree of ambiguity in the process, and it must be overcome. Sometimes the ambiguity results from the fact that there are many different ways to define the phenomenon, or perhaps no single way is well established. In cases such as these, one might wish to resort to an **operational definition,** that is, a procedure that will produce consistent results from measurement to measurement, or when measurements are taken by different people.

An operational definition, therefore, is a procedure that must be followed; it specifies as many factors as are needed to control the measurement so that changes can be properly attributed only to the unknown variable. The need for operational (as opposed to absolute) definitions arises from the fact that things are only rarely so neat, clean, and crisp as to suggest their own natural definition. By its very nature the operational definition does not ask "true-or-false" questions, but rather it asks, "What happens under given sets of assumptions or conditions?" What an operational definition can do for you, however, is to standardize a measurement in a clear and precise way so that the measurement remains consistent across numerous trials. Operational definitions are used extensively in science and technology. When widely accepted, or promulgated by a recognized authority, they are called **standards.**

An operational definition should embrace what is measurable quantitatively, or at least in nonsubjective terms. For example, in measuring the "saltiness" of saline solution (saltwater), you might taste it and render a subjective judgment such as "weak" or "strong." Alternatively, you can establish an operational definition that calls for you to measure the electrical resistance of the saline under certain specified conditions:

1. Immerse two 1-cm-diameter circular nickel electrodes, spaced 5 cm apart and facing each other, to a depth of 3 cm into a 500-ml beaker of the test solution.
2. Bring the solution to a temperature of 4°C.

3. Measure the electrical resistance $R$ between the electrodes using a Snotz® model 1120 digital ohmmeter.

4. Find the conductance $G$ by taking the reciprocal of resistance ($G = 1/R$).

You can probably come up with a better definition of the conductance of saline solution that works for some peculiar situation. Keep in mind that only rarely does a preferred definition suggest itself naturally.

The use of operational definitions results in both strengths and weaknesses. One weakness is that the definition might not be honed finely enough for the purpose at hand. Sociologists and psychologists often face this problem because of difficulties in dealing with nonlinearities such as human emotions. But such problems also point to a strength. We must recognize that scientific truth is always tentative, so we must deal with uncertainties in experimentation. Sometimes, the band of uncertainty around a point of truth can be reduced by using several operational definitions in different tests of the same phenomenon. By taking different looks from different angles, we may get a more refined idea of what's actually happening.

When an operational definition becomes widely accepted and is used throughout an industry, it may become part of a formal **standard** or test procedure. You may, for example, see a procedure listed as "performed in accordance with NIST XXXX.XXX"* or "ANSI Standard XXX."† These notations mean that whoever made the measurement followed a published standard.

## 2–11    AFTERWORD: WEIGHING A COW

"Ever wondered how to weigh a bull?" John's question took me by surprise. My friend was the agriculture aide to a well-known United States senator from a western state. John doesn't exactly look like the Outlaw Josie Wales, or even the Marlboro Man for that matter. His boyish face belies the fact that he spent the first 22 years of his life riding horses to rope, brand, and curse cattle on his father's western United States ranch. (Yes, there are still real cowboys, and they ride real horses.) "No," I averred, "weighing bulls hasn't weighed very heavily on my mind lately." "No, really, I'm serious," he retorted. "There are some interesting measurement principles found in weighing cattle for market."

John had just returned to Washington from his home state, where he had seen a new digital electronic scale at the cattle market. It was a new computer-based model.

The problem is simple: until after they are slaughtered for beef, cattle are weighed alive, and it is nearly impossible to get them to stand still long enough for a conventional scale to read a steady value. As a result, the scale makers turn to a method that is based on simple statistics—and these methods are just as applicable to other electronic measurements and to the calibration of electronic instruments.

Now let's get back to the issue of weighing a cow on an electronic scale. The main problem is that the cow won't stand still long enough for a static measurement to be

---

* NIST is National Institute for Standards and Technology, formerly called National Bureau of Standards.
† ANSI is American National Standards Institute.

made. As a result, it is necessary to use statistical methods to weigh the animal. Figure 2–11 shows the cattle scale described to me by my electronically literate, political aide friend. The animal stands on two platforms, one for the front legs and another for the rear legs, that are each connected to an electronic pressure sensor. The voltage outputs of each sensor ($V1$ and $V2$) are each approximately equal to half the weight of the animal. Therefore, the sum of $V1$ and $V2$ is proportional to the total weight of the animal.

These voltages ($V1$ and $V2$) are converted to binary numbers by analog-to-digital (A/D) converters; then the binary numbers are input to a desktop computer. Because very fast A/D converters are now available, it is possible to take several thousand samples of

**FIGURE 2–11**
Weighing a cow on an electronic scale.

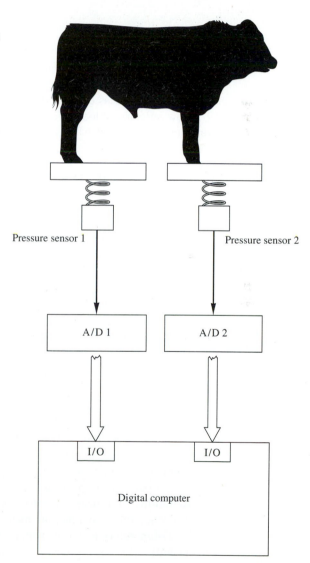

**FIGURE 2–12**
Data from ovens seem to form a
normal distribution curve.

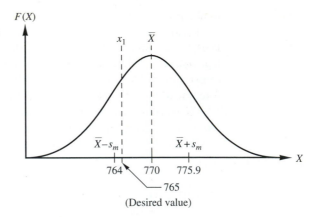

the two half-weights in the 10 to 20 seconds the animal is held on the scale. The average
of each is taken, along with the error factor, and they are then added together. As long as
the error factor is less than the desired resolution, the scale's reading is accepted.

Now let's take a look at another situation that involves measurements with an error
component.

2–12
## DID IT PASS THE TEST?

An adhesives curing oven is supposed to have an equilibrium temperature of 765°F after
10 minutes. A test technician inserted an electronic thermometer and a timer to take the
temperature measurement after exactly 10 minutes; the oven was allowed to cool back to
room temperature between trials. The technician took 20 measurements and found the
mean to be 770°F with a standard deviation of 26.2. Did this oven pass the test? The situation is shown in Fig. 2–12 . . . and the answer seems to be: yes. Why? Because the
error term represented by the standard error of the mean (shown as $\pm s_m$ in Fig. 2–12)
encompasses the desired value of 765 despite the dispersion of the data. We can claim
that the oven failed the test only if the desired value is outside these limits.

## SUMMARY

1. There are three basic categories of measurement: **direct, indirect,** and **null.**
2. Factors involved in measurements include **error, validity, reliability, repeatability,
   accuracy, precision,** and **resolution.**
3. **Error** is inherent in measurements and in no way indicates that mistakes were made.
4. **Validity** of a measurement is a statement of how well the measurement measures
   what it purports to measure.
5. **Reliability** of a measurement is a statement of its consistency.
6. **Accuracy** refers to a measurement's freedom from error.
7. **Precision** refers to the exactness of successive measures of the same unchanging
   quantity.

8. **Resolution** refers to the degree to which a measurand can be broken into identifiable adjacent parts.
9. The basic categories of error are **theoretical error, static error, dynamic error,** and **instrument insertion error.**
10. Measurements that are difficult, complex, or ambiguous are often accomplished with an **operational definition,** that is, a standard procedure for making the measurement. If all users consistently apply the standard method, then the results are comparable.

## RECAPITULATION

Now go back and try to answer the questions at the beginning of this chapter. When you are finished, answer the questions and work the problems below. Place a mark beside each problem or question that you cannot answer, and then go back and reread the appropriate sections of the text.

## QUESTIONS

1. List the three categories of measurements: ———————, ———————, and ———————.
2. A measurement made with a meter stick is an example of a(n) ——————— measurement.
3. Measuring human blood pressure by sphygmomanometry (blood pressure cuff) is an example of a(n) ——————— measurement.
4. A potentiometer and a zero-center galvanometer are used to measure an unknown potential. This is a(n) ——————— measurement.
5. List seven factors in making a good measurement:

   ———————————————, ———————————————,
   ———————————————, ———————————————,
   ———————————————, ———————————————,
   and ———————————————.
6. Random variation in the value reported by successive measurements of the same unchanging quantity is called ———————.
7. The statement of how well a measurement method actually measures the quantity it purports to measure is its ———————.
8. The ——————— of a measurement is a statement of the amount by which a measurement departs from the actual value of the measurand.
9. The ——————— of a measurement is a statement of the dispersion of the values of an unchanging measurand on a large number of successive trials.
10. The standard deviation of a set of measurements made on an unchanging measurand is a good indication of its ———————.
11. On a TV test pattern, a series of lines radiating out from a central point is used to measure the ——————— of the TV set.
12. ——————— is found in all forms of measurement, no matter how accurately or how well the procedure is carried out.

13. List four general categories of measurement error:
    _____, _____,
    _____, and _____.

14. Which of the types of measurement error in Question 13 are normally the fault of the instrument user?

15. The _____ category of error results from the theory of measurement used in a particular case, for example, in measuring mean arterial blood pressure by calculation from the minimum (diastolic) and maximum (systolic) values over one heart cycle.

16. Parallax error is an example of a _____ error in reading an analog meter scale.

17. _____ _____ _____ is an example of a static error in reading a digital voltmeter.

18. List several types of characteristic static errors.

19. For digital systems, overall error must include the _____ error caused by the digital nature of the circuits.

20. A reading of the value of a rapidly changing, near-dc voltage lags because of inertia in the meter movement. This is an example of _____ error.

21. A voltage measurement error is caused by the integration time window of a digital voltmeter. This is a _____ error.

22. If a low-impedance voltmeter is inserted into a high-resistance dc circuit, a _____ _____ error is created.

23. If a set of $n$ measurements are made of an alternating current measurand, and if the errors are independent of each other, an appropriate measurement calculation method is the _____ _____ square.

## PROBLEMS

1. Three voltmeters are connected in parallel across the output of a voltage source. Number 1 reads 12.23 V, number 2 reads 12.1 V, and number 3 reads 12.2 V. What is the probable voltage indicated by this set of readings?

2. Find the mean and the standard deviation of the following set of weight measurements:

    | 1.6 | 1.8 | 1.2 | 1.26 | 1.26 | 1.3 | 1.3 | 1.3 | 1.33 | 1.35 |
    |-----|-----|-----|------|------|-----|-----|-----|------|------|
    | 1.4 | 1.4 | 1.4 | 1.4 | 1.4 | 1.43 | 1.48 | 1.53 | 1.5 | 1.5 |
    | 1.5 | 1.53 | 1.6 | 1.6 | 1.62 | 1.65 | 1.74 | 1.79 | 1.7 | 1.81 |

3. Calculate the standard error of the mean for the sample in Problem 2, and report it in the standard manner.

# 3

# dc and ac Deflection
# Meter Movements

3–1

## OBJECTIVES

1. To learn the principles of operation of the basic dc meter movement.
2. To learn methods for extending the current range of the basic dc meter movement.
3. To learn how to measure voltage by using a dc current meter.
4. To learn the operation of a volt-ohm-milliammeter (VOM).
5. To learn the principles of operation of passive ac current meters.
6. To learn the operation of rectifier-type ac meters.
7. To learn the nature of the values presented by each type of ac meter.
8. To learn the limitations and advantages of each type of meter.

3–2

## SELF-EVALUATION

Before studying the material in this chapter, try to answer the questions given below. These questions test your knowledge of the subject. If you cannot answer a particular question, then look for the answer as you read.

1. Describe in your own words the **d'Arsonval dc meter movement.**
2. What are the principal differences between the **d'Arsonval** and **taut-band** meter movements?
3. How may a dc current meter movement be used to measure an ac voltage?
4. A _____ resistor is used to make a dc current meter into a dc voltmeter.
5. A _____ resistor is used to increase the full-scale range of a dc current meter movement.
6. Describe the **electrodynamometer** and its principal uses.
7. Describe an **iron-vane** meter movement.
8. What type of rectifiers are normally used to make an RF voltmeter?
9. What ac value is read by a rectifier ac voltmeter? How is the scale of such a meter usually calibrated?
10. Describe a conventional ac wattmeter.

## 3–3     THE BASIC ANALOG dc METER

For many decades the basic dc meter movement has been used as the readout device in electronic instruments, even in many instruments that measure ac values. Digital readout devices are common today, but until the mid-1980s they were too expensive for most common instruments. Even today, when digital displays are commonplace and low cost, analog meter movements are still used extensively. For some applications the analog meter is even preferred over the digital (e.g., in finding a peak or null condition, or where a zero-center application is needed).

The two most common dc meter movement configurations are the **d'Arsonval** and the **taut-band** designs. Both are examples of the **permanent magnet moving coil (PMMC)** galvanometer, and both work on the same fundamental principle as the dc motor.

A simplified view of the d'Arsonval meter movement is shown in Figs. 3–1(a) and 3–1(b). A moving coil of wire is mounted in the magnetic field between the poles of a

**FIGURE 3–1**
(a) Construction of a d'Arsonval meter movement. This meter uses a loop of wire rotating in a magnetic field, much like an electrical motor. (b) Side view of the d'Arsonval meter movement.

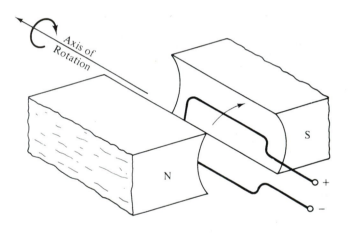

(a)

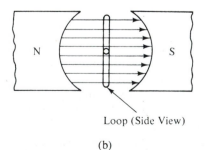

Loop (Side View)

(b)

permanent magnet. A current flowing in a wire generates a magnetic field. The polarity of the field is determined by the direction of the current flow, while the strength of the field is determined by the magnitude of the current.

The coil in a PMMC galvanometer is mounted so that it can rotate in the space between the magnetic poles. A current in the coil creates a magnetic field that either aids or opposes the field of the nearby magnetic poles. A current flow in one direction causes a clockwise rotation, whereas a current flow in the opposite direction causes a counterclockwise rotation. The amount of rotational position change is proportional to the magnitude of the current.

The sensitivity of a galvanometer can be defined in several different ways: **current sensitivity** ($S_I$), **voltage sensitivity** ($S_V$), **megohm sensitivity** ($S_R$), and **ballistic sensitivity** ($S_Q$). The current sensitivity is the ratio of the deflection of the meter movement to the current that produced that deflection, for example, millimeters of deflection per microampere of current (mm/μA). Voltage sensitivity is the deflection per unit of potential, for example, millimeters of deflection per millivolt (mm/mV), when the meter is shunted with a value of resistance called the **critical damping resistance (CDRX).** The megohm sensitivity is the number of megohms in series with the galvanometer, when the galvanometer is CDRX shunted, required to produce one scale division deflection when a potential of 1 V is applied. The megohm sensitivity is usually equal to the current sensitivity. The ballistic sensitivity is a dynamic metric that relates the maximum deflection to the single-pulse charge quantity *(Q)* that produced the deflection. The units are millimeters of deflection per microcoulomb of electric charge (mm/μC).

## 3–4    THE D'ARSONVAL METER MOVEMENT

The basic d'Arsonval meter movement is shown in Fig. 3–2. A side view of the meter, without the permanent magnet, is shown in Fig. 3–2(a), while a frontal view with the magnet in place is shown in Fig. 3–2(b). The coil in Fig. 3–2(a) is wound on an **armature** or **bobbin,** which is mounted on a pair of jewel bearings to reduce friction. The assembly is shown (simplified) frontally in Fig. 3–2(b).

When a current flows in the coil, the armature assembly deflects clockwise[as shown in Fig. 3–2(b)] an amount that is proportional to the strength of the current. The amount of deflection can be marked off in units of current on the dial scale. The coiled **pivot spring** serves to dampen the pointer movement and to return the pointer to the zero position when the current flow in the coil ceases.

The travel of the pointer is limited by high- and low-end mechanical stops just beyond the **zero** and **full-scale limits** printed on the dial scale. The wires to the coil are flexible and are given just enough slack that they will not be stretched anywhere in the pointer's normal range of travel.

## 3–5    THE TAUT-BAND METER MOVEMENT

The taut-band meter movement is essentially the same as the d'Arsonval movement, except for the manner in which the armature coil is mounted (see Fig. 3–3). In the taut-

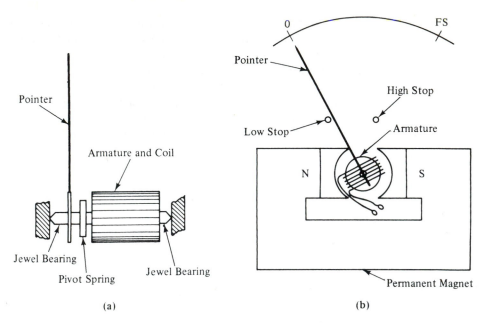

**FIGURE 3–2**

(a) Construction of the coil and pointer assembly in a d'Arsonval meter movement, (b) usual configuration for a d'Arsonval meter movement.

band movement the armature is suspended from fixed supports on a stretched (i.e., "taut") rubber band. The band is twisted as the armature rotates, so no restoring force from a pivot spring is needed.

The taut-band meter movement has two principal advantages over the older d'Arsonval movement: *greater sensitivity* and more *durability*.

**FIGURE 3–3**

Construction of taut-band meter movement.

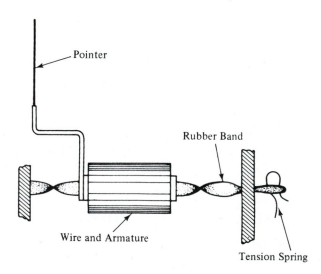

D'Arsonval meter movements are rarely found with full-scale sensitivities of less than 50 μA, whereas taut-band meters are available with full-scale sensitivity of only 2 μA.

D'Arsonval movements are more easily damaged than taut-band movements because of their jewel-bearing construction. Even a short fall to the floor or bench top is sufficient in most cases to break the bearings or dislodge the armature from the bearing cups. The rubber band in a taut-band movement can snap, but this occurs far less frequently than does bearing damage in d'Arsonval movements. While abusing either type of meter is bad business, the taut-band movement is more forgiving and can absorb more punishment than most d'Arsonval movements.

**3–6**      ## TYPES OF ANALOG METERS

The two basic meter movements are available in an almost endless variety of sizes and shapes, but we can classify meter types by the form of scale that is used. Figure 3–4 shows three different configurations.

The meter in Fig. 3–4(a) is the type normally encountered. The zero-current mark is on the left side of the scale, and the pointer deflects to the right when a current flows (assuming the meter is hooked into the circuit correctly!). If the meter is a dc type, then the input terminals of the meter will be marked positive (+) and negative (−) (the negative is sometimes not marked). The terminals must not be reversed, or the pointer may be damaged when it attempts to deflect further to the left.

Figure 3–4(b) shows an example of a zero-center dc microammeter; the zero-current mark on this type of meter is in the exact center of the scale. Current flowing in one direc-

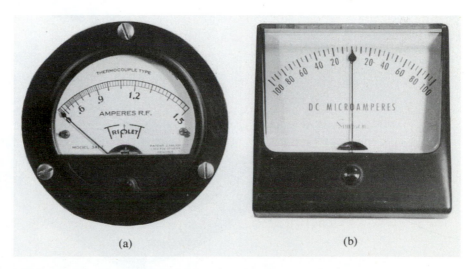

(a)             (b)

**FIGURE 3–4**
(a) Thermocouple RF ammeter, (b) zero-center dc microammeter.

tion will deflect the pointer to the left of center, while current flowing in the opposite direction will deflect the pointer to the right. When the polarity markings on the input terminals are observed, a positive current deflects the pointer to the right, while a negative current deflects the pointer to the left.

The type of meter movement shown in Fig. 3–4(c) is an edgewise panel meter. Its design allows conservation of area on the front panels of equipment in which the meter is used. This saving is at the expense of increased depth, but that is often considered an

**FIGURE 3–4,** *continued*
(c) Two views of an edgewise panel meter movement.

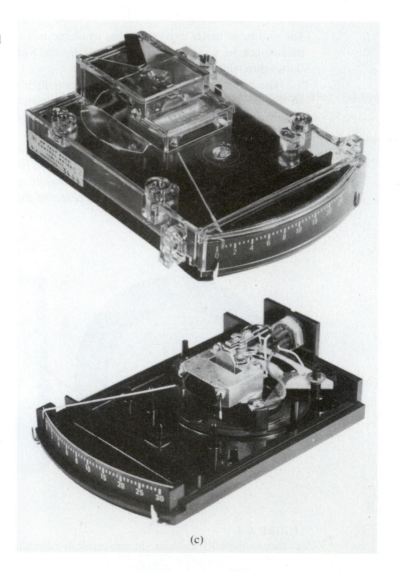

(c)

acceptable engineering trade-off. Edgewise meters are available in both left-zero and center-zero models.

The edgewise panel meter movement has internal photoelectric alarm sensors that indicate when the pointer deflection is below a low limit or above a high limit. Other meter movements with alarm capability have an internal potentiometer element that outputs a resistance proportional to the pointer deflection. In optical alarm sensors, the LOW and HIGH points are set by movable tabs on the meter face. In the resistance type of alarm, external voltage comparators are usually used.

**3–7**         ## USING BASIC dc METERS

There are three basic rules for using dc current meters:

1. Connect the meter in series with the load or circuit in which the current is being measured.
2. Use a meter with a full-scale current rating that is greater than the maximum current expected.
3. Use a meter that has an internal resistance that is low compared with the resistance in which it is being used (i.e., <1:10).

*A current meter is always connected in series with the load!* There are never any exceptions. *Failure to observe this rule may result in permanent, irreversible damage to the instrument.* When a meter is used to measure current in a branch of a larger circuit, then the meter is to be connected in series with that branch.

Figure 3–5 shows the correct connection for a 0-to-1-mA dc current meter. It is necessary to break the circuit (i.e., the wire to resistor $R1$) and then connect the meter in series.

**FIGURE 3–5**
Circuit with 0-to-1-mA milliam-meter.

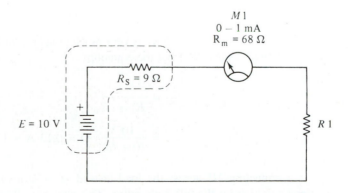

The current flowing in meter $M1$ is given by

$$I_{M1} = \frac{E}{R_S + R_m + R1} \tag{3-1}$$

where   $I_{M1}$ = the meter current in amperes (A)
         $E$ = the open-circuit voltage in volts (V)
         $R_S$ = the power supply source resistance in ohms ($\Omega$)
         $R1$ = the load resistance in ohms ($\Omega$).

In most cases, $R_S$ and $R_m$ are very small with respect to $R1$ and so are considered negligible. If this is true, then Equation (3–1) is reduced to

$$I_{M1} = \frac{E}{R1} \tag{3-2}$$

■   **Example 3–1**

Find the current flowing in meter $M1$ in Fig. 3–5 when $R1 = 15$ k$\Omega$.

*Solution*

Since $R1$ is very much greater than $R_S$ or $R_m$, use Equation (3–2):

$$I_{M1} = \frac{E}{R1}$$

$$= \frac{10 \text{ V}}{1.5 \times 10^{-4}}$$

$$= 6.7 \times 10^{-4} \text{ A} = \textbf{0.67 mA}$$

■   **Example 3–2**

Find the current that would flow in meter $M1$ in Fig. 3–5 if $M1$ were *incorrectly* connected in parallel with $R1$ instead of in series as is proper.

*Solution*

Use Equation (3–1) because $R1$ is very much greater than $R_S$ and $R_m$, so its parallel effect is negligible. Delete $R1$ from the equation:

$$I_{M1} = \frac{E}{R_S + R_m}$$

$$= \frac{10 \text{ V}}{(0 + 68)} \text{ } \Omega = 0.13 \text{ A} = 130 \text{ mA}$$

Connecting $M1$ across the load instead of in series results in a current that is 130 times too high for the full-scale rating. This current will probably destroy the instrument.

Rule number 3 can become troublesome in low-voltage, low-resistance circuits, where the meter's coil resistance $(R_m)$ is a significant portion of the load resistance. Neglecting $R_S$, let's see what would happen if $R1$ were reduced to 150 $\Omega$ and $E$ were only 100 mV (i.e., 0.1 V). By Ohm's law, if $R_m$ were negligible, then we would expect a current of

$$I = \frac{E}{R1}$$

$$= \frac{0.1 \text{ V}}{150 \text{ }\Omega} = 0.00067 \text{ A} = \textbf{0.67 A}$$

But with $R_m$ being a significant portion of $R1$, the actual current is $(0.1 \text{ V})/(68 + 150 \text{ }\Omega)$, or 0.46 mA.

---

No measuring or testing instrument should be used if it significantly alters the circuit performance, and this is the case in the example above. A 31% meter reading error was created by an unwise selection of the meter used to make the measurement. In both cases above, we expected to find 0.67 mA, but in the latter case the reading was only 0.46 mA.

**3–8**     ## OBTAINING HIGHER CURRENT SCALES

The basic dc meter movement has a single current scale (e.g., 0 to 1 mA, 0 to 200 µA, etc.), but it can be used to measure larger currents if a **shunt resistor** in parallel with the meter is used (see Fig. 3–6).

The wire diameter required for the coil in high-current meter movements is excessive and proves impractical, so it has become standard practice to use a low-range meter movement and the shunt resistor.

In some cases the shunt resistor is mounted inside the meter case, whereas in others it is located outside the case (usually mounted directly to the meter's input terminals). An

**FIGURE 3–6**
Principle of the shunt resistor to increase current range.

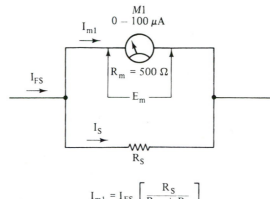

$$I_{m1} = I_{FS} \left[ \frac{R_S}{R_m + R_S} \right]$$

example using the shunt resistor is given in Fig. 3–6. The total full-scale current of the meter with shunt in place is given by

$$I_{FS} = I_{M1} + I_S \tag{3–3}$$

where    $I_{FS}$ = the total current
$I_{M1}$ = the current flowing in the meter
$I_S$ = the current flowing in the shunt

■ **Example 3–3**

The circuit of Fig. 3–6 is used to measure 1 mA full scale. How much current at full scale flows in (a) *M2* and (b) $R_S$?

*Solution*
Solve Equation (3–3) for $I_S$:

$$I_{FS} = I_{M1} + I_S$$

Therefore,

$$I_S = I_{FS} - I_{M1}$$

$$= \left[ 1 \text{ mA} \times \frac{10^3 \text{ μA}}{1 \text{ mA}} \right] - 10^2 \text{ μA}$$

$$= (10^3 - 10^2) \text{ μA} = \mathbf{900 \text{ μA}}$$

There are two methods for calculating the value of resistor *R1* if $R_m$, $I_{M1}$, and $I_{FS}$ are known: **Ohm's law** and the **current divider equation.**

If we know both the full-scale rating of the meter movement [in Example 3–3 it is 100 μA (Fig. 3–6)] and the meter coil resistance $R_m$, then we know by Ohm's law that the voltage drop across *M1* at full scale is $I_{M1} \times R_m$, or in this case

$$E_{M1} = (10^{-4} \text{ A})(5 \times 10^2 \text{ Ω}) = 0.05 \text{ V}$$

Since $R_S$ is connected in parallel with *M1*, the same voltage is also across $R_S$. If we calculate $I_S$ [Eq. (3–3)] and use Ohm's law, we can find the value required of $R_S$. In the case of Example 3–3, in which 900 μA flows in $R_S$,

$$R_S = \frac{E_{M1}}{I_S}$$

$$= \frac{0.05 \text{ V}}{9 \times 10^{-4} \text{ A}} = \mathbf{55.56 \text{ Ω}}$$

The step-by-step procedure for using Ohm's law to determine the value of the shunt resistor is as follows:

1. Determine $I_{FS}$, $I_{M1}$, and $R_m$.
2. Calculate $I_S$, using Equation (3–3).
3. Calculate $E_{M1}$, using Ohm's law.
4. Use Ohm's law again to calculate $R_S$ from $E_{M1}$ and $I_{FS}$, or, in equation form,

$$R_S = \frac{I_{M1}R_{M1}}{I_{FS} - I_{M1}} \qquad (3\text{–}4)$$

The current divider method takes advantage of **Kirchhoff's current law (KCL)**, which tells us that the current in $R_S$ is a fraction of $I_{FS}$:

$$I_S = I_{FS} \times \left( \frac{R_m}{R_m + R_S} \right) \qquad (3\text{–}5)$$

■ **Example 3–4**

A 0-to-50-μA dc meter movement, with a coil resistance $R_m$ of 1250 Ω, is used with a shunt resistor to measure a full-scale current of 500 μA. Calculate the value of $R_S$.

***Solution***

Solve Equation (3–5) for $R_S$:

$$I_S = I_{FS} \times \left[ \frac{R_m}{R_m + R_S} \right]$$

$$(500 \ \mu A - 50 \ \mu A) = (500 \ \mu A) \times \left[ \frac{1250 \ \Omega}{(1250 + R_S \ \Omega)} \right]$$

$$\frac{450 \ \mu A}{(500 \ \mu A)(1250 \ \Omega)} = \frac{1}{(1250 + R_S) \ \Omega}$$

By algebraically rearranging the expression, we obtain

$$R_S = \left[ \frac{(500)(1250)}{450} \ \Omega \right] - 1250 \ \Omega$$

$$= (1389 - 1250) \ \Omega = \mathbf{139 \ \Omega}$$

Although each method for determining the value of the shunt resistor has its own advocates, either method will yield the same answer. You are encouraged to work Example 3–4 using the Ohm's law method.

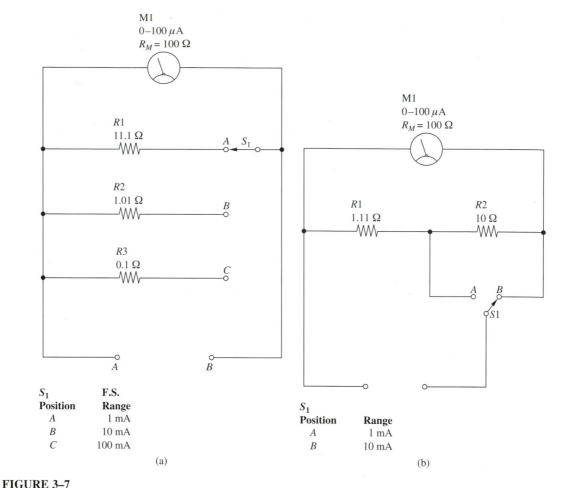

**FIGURE 3–7**
(a) Three-scale, multiple-range ammeter; (b) universal shunt, multiple-range ammeter.

If a switch is provided that permits the selection of any of several different-value shunt resistors, then it is possible to make a **multiple-range ammeter.** Each different shunt resistor is used to set the full-scale meter current to a different value, resulting in different switch-selectable current ranges being covered. Thus, a 0-to-100-μA dc microammeter can be used for full-scale readings of, for example, 100 μA, 500 μA, 1000 μA (1 mA), 10 mA, 50 mA, 100 mA, 500 mA, and 1 A. An example of a simple three-scale, multiple-range milliammeter is shown in Fig. 3–7(a).

A flaw in this design is that during the instant when the switch is being transferred from one position to another, the meter movement is in the circuit *without a shunt* and therefore is subject to damage. A solution to the problem is to use a "make contact before breaking contact" type of switch to select the various shunt resistors.

The circuit of Fig. 3–7(a) is not considered optimum. A superior circuit is the **universal shunt,** also called the **Ayrton shunt,** multiple-range ammeter shown in Fig.

3–7(b). In this circuit there is always a shunt resistor across the meter movement, even during times when the switch contacts are momentarily open, without the need for a make-before-break switch. Nonetheless, in all multiple-range ammeters it is preferable to start the measurement process at the highest current range before connecting the meter into the circuit, and then work down (once power is applied to the circuit) to the point where the deflection of the meter is in the upper half of the scale.

In many multiple-range ammeters there will be a switch-selectable input for current ranges up to 1 A and a separate "hot" input for a 10-A range.

Some multiple-range ammeters are **autoranging.** In these instruments an electronic switch automatically sequences from the highest current range to the lowest current range until a setting is found that results in a reading that is less than full scale. This feature is more common in digital ammeters but is also found in at least a few analog designs.

## 3–9  VOLTAGE MEASUREMENTS FROM dc CURRENT METERS

Voltage can be measured on a dc current meter if a **multiplier resistor** is connected in *series* with the meter movement. This type of circuit is shown in Fig. 3–8(a), with the equivalent circuit shown in Fig. 3–8(b).

Input voltage $E$ will set up current $I1$ in the circuit of Fig. 3–8(a) that has a value of

$$I1 = \frac{E}{R_m + R_{mx}} \qquad (3\text{–}6)$$

where $I1$ = the current in amperes (A)
$E$ = the applied potential in volts (V)
$R_m$ = the meter resistance in ohms ($\Omega$)
$R_{mx}$ = the multiplier resistance in ohms ($\Omega$)

We may calculate the multiplier resistance needed for any given full-scale rating of $M1$ and then solve Equation (3–6) for $R_{mx}$.

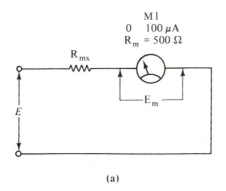

(a)

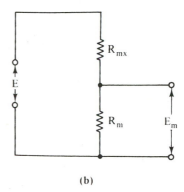

(b)

**FIGURE 3–8**
(a) Use of multiplier resistor to make an ammeter read in volts, (b) equivalent circuit.

■ **Example 3–5(a)**

Calculate the value of the multiplier resistor $R_{mx}$ in Fig. 3–8(a) for a full-scale potential $E$ of 10 V.

*Solution*

Solve Equation (3–6) for $R_{mx}$:

$$I1 = \frac{E}{R_m + R_{mx}}$$

$$10^{-4} \text{ A} = \frac{10 \text{ V}}{(500 + R_{mx}) \ \Omega}$$

$$10^{-5} \text{ A} = \frac{1 \text{ V}}{(500 + R_{mx}) \ \Omega}$$

By rearranging the expression, we obtain

$$R_{mx} + 500 \ \Omega = 10^5 \ \Omega$$
$$R_{mx} = 10^5 \ \Omega - 500 \ \Omega = \mathbf{99{,}500 \ \Omega}$$

We may also use the **voltage divider equation** to calculate the value of $R_{mx}$. We know that the full-scale voltage drop across meter $M1$ will be equal to $I1 \times R1$. Thus,

$$E_m = I1 \times R_m$$
$$= (10^{-4} \text{ A})(5 \times 10^2 \ \Omega) = 0.050 \text{ V}$$

So by the voltage divider equation,

$$E_m = \frac{E R_m}{R_m + R_{mx}} \qquad (3\text{–}7)$$

we find a solution for the problem of Example 3–5(b) which follows.

---

■ **Example 3–5(b)**

Work the problem of Example 3–5(a) using the voltage divider equation method.

*Solution*

$$E_m = \frac{E R_m}{R_m + R_{mx}}$$

$$0.050 \text{ V} = \frac{(10 \text{ V})(500 \ \Omega)}{(500 + R_{mx}) \ \Omega}$$

With a little algebraic manipulation, we find

$$R_{mx} + 500 = \frac{(10)(500)}{0.050}$$

$$R_{mx} = (100,000 - 500)\ \Omega = \mathbf{99,500\ \Omega}$$

The resistance is the same as before, as expected.

---

**3–10**　　　　**VOLTMETER SENSITIVITY (ϕ)**

The **sensitivity** (ϕ) of a voltmeter is specified in terms of **ohms per volt** (Ω/V) and is dependent upon the full-scale current range of the basic dc meter movement used to make the voltmeter. The sensitivity can be found by taking the reciprocal of the meter's full-scale current:

$$\phi = \frac{1}{I_{FS}} \tag{3–8}$$

　　　Table 3–1 gives the sensitivity ratings for dc voltmeters made by using various values of full-scale meter ratings.

**3–11**　　　　**DC VOLTMETER RESISTANCE**

The **resistance** of a dc voltmeter is the product of its full-scale rating and the **sensitivity** rating:

$$R_m = E_{FS} \times \phi\ \Omega \tag{3–9}$$

where　　$R_m$ = the resistance of the voltmeter in ohms (Ω)
　　　　　$E_{FS}$ = the full-scale rating of the meter in volts (V)
　　　　　ϕ = the sensitivity of the meter in ohms per volt (Ω/V)

**TABLE 3–1**
Sensitivity ratings for dc voltmeters.

| Full-Scale Meter Current | Sensitivity (ϕ) |
| --- | --- |
| 1 mA | 1000　Ω V |
| 100 μA | 10 kΩ V |
| 50 μA | 20 kΩ V |
| 20 μA | 50 kΩ V |
| 10 μA | 100 kΩ V |

■  **Example 3–6**

Calculate the input resistance of a dc voltmeter such as in Fig. 3–8(a) on the 0-to-500-V dc scale if the sensitivity is 10,000 Ω/V.

*Solution*

$$R_\mathrm{m} = E_\mathrm{FS} \times \phi$$
$$= (500\ \mathrm{V})(10^4\ \Omega/\mathrm{V}) = 5 \times 10^6\ \Omega = \mathbf{5\ M\Omega}$$

3–12        ## USING VOLTMETERS

Voltage is the *difference in electrical potential between two points* in a circuit. Therefore, to measure voltage, connect the voltmeter *across the load,* that is, in *parallel* with the load or circuit branch that you want to measure. Three basic rules apply:

1. Connect the voltmeter in parallel with the load (see Fig. 3–9).
2. Select a voltmeter that has a full-scale range that is greater than the highest potential expected.
3. Make sure that the voltmeter has an input resistance (Section 3.11) that is *very high* (e.g., >100:1) compared with the circuit resistance.

The voltmeter is a high-impedance device, so any attempt to connect it in series with the circuit will greatly increase circuit resistance, causing improper operation.

Problems of circuit loading will occur if rule number 3 is not obeyed. Indeed, there may be circuits where the >100:1 rule is not enough (electronic multimeters will be needed in those cases; see Chapter 7). Consider what will happen if a circuit is loaded down by the voltmeter resistance [(Fig. 3–10(a)]. From the voltage divider equation we know that *E2* should be

$$E2 = \frac{ER2}{R1 + R2}$$

$$= \frac{(10\ \mathrm{V})(1\ \mathrm{M\Omega})}{1\ \mathrm{M\Omega} + 0.5\ \mathrm{M\Omega}}$$

$$= \frac{10\ \mathrm{V}}{1.5} = 6.67\ \mathrm{V}$$

**FIGURE 3–9**
A voltmeter is always connected in parallel with the load.

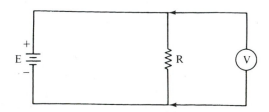

**FIGURE 3–10**
(a) Voltmeter impedance can cause
errors due to circuit loading;
(b) equivalent circuit.

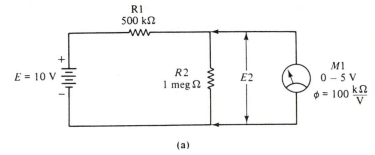

(a)

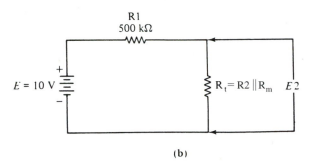

(b)

But at 0 to 5 V full scale, M1 has an impedance of only 500 k$\Omega$. When M1 is connected across R2, then it is equivalent to shunting a 500-k$\Omega$ resistor across resistor R2; the equivalent circuit is shown in Fig. 3–10(b). With $R_m$ in parallel with R2 ($R_m \| R2$), the parallel combination has a resistance of

$$R_{eq} = \frac{(1\ M\Omega)(0.5\ M\Omega)}{1\ M\Omega + 0.5\ M\Omega} = 0.33\ M\Omega$$

Again using the voltage divider equation, but this time with the equivalent resistance $R_{eq}$ substituted for R2, we find that voltage E2, *as indicated on the meter,* is

$$E2 = \frac{ER_{eq}}{R1 + R_{eq}}$$

$$= \frac{(10\ V)(0.33\ M\Omega)}{(1.0 + 0.33)\ M\Omega} = 2.48\ V$$

A reading of 2.48 V, when the actual circuit potential is 6.67 V, is a 63% error, and that error is caused solely by the fact that the measurement was made with a voltmeter that had a resistance that was too low for the circuit being tested.

The general rule is to select a meter with an input resistance that is not less than 10 times greater than the circuit resistance, and if possible to make the meter resistance more than the 100× rule. Even older volt-ohm-milliammeter instruments, of which many are still in service, will offer at least 20 kΩ/V resistance, and many are found with 100 kΩ/V. Beware, however, that many 1000 Ω/V instruments are still in service, and they can be deadly to good measurements. (Indeed, they may upset the circuit enough to cause secondary damage!)

A valid question to ask is, "Why use a passive instrument at all?" In most cases an active instrument such as a digital multimeter is the instrument of choice; but where high electrical or varying magnetic fields are encountered, the electronic voltmeter suffers from input biasing due to electromagnetic interference (EMI; see Chapter 15). People who work on radio transmitters find this problem quite often. The passive meter circuitry of the volt-ohm-milliammeter is far more immune to EMI than even most of the best active meters. Unfortunately, the kind of filtering methods that are used to eliminate EMI also sometimes distort waveforms being measured.

## 3–13    dc ANALOG OHMMETERS

Electrical resistance is measured using an instrument called an **ohmmeter.** According to Ohm's law, the current in a circuit is directly proportional to the applied voltage $(E)$ and inversely proportional to the resistance $(R)$ of the circuit; that is, $I = E/R$ . That relationship can be used to find an unknown resistance $(R_x)$, if a stable voltage is available.

The circuit for a simple **series ohmmeter** is shown in Fig. 3–11(a). There are three resistances in this circuit: $R_m$ is the internal resistance of the voltmeter $(M1)$, $R1$ is the resistance of a zero calibration control potentiometer, and $R_x$ is the unknown resistance. The unknown resistance is calculated from

$$R_x = \frac{E(R_m + R1)}{e} - (R_m + R1) \qquad (3\text{–}10)$$

where   $R_x$ = the unknown resistance in ohms $(\Omega)$
$\qquad R_m$ = the meter's internal resistance in ohms $(\Omega)$
$\qquad R1$ = the resistance of the zero control in ohms $(\Omega)$
$\qquad E$ = the voltmeter reading on $M1$ when terminals $A$ and $B$ are shorted together
$\qquad e$ = the voltmeter reading on $M1$ when the unknown resistance $R_x$ is connected across terminals $A$ and $B$.

Equation (3–10) need not be solved by the ohmmeter user because the scale of the voltmeter would be calibrated not in volts but in units of resistance.

The scale for ohmmeters is reversed in direction from the voltage and current scales. That is, the highest values are on the left side of the scale, while the lowest values are on the right. In the ohmmeter of Fig. 3–11(a), the zero point occurs when maximum current flows (which puts the meter on the right side), that is, when terminals $A$ and $B$ are shorted together. Typical operation of the analog ohmmeter requires that the leads be shorted together with the pointers zeroed with the calibration control. On modern digital voltmeters, by the way, the zeroing is automatic.

A limitation of the series ohmmeter circuit is that it does not work well at low resistances $(R_x < 100 \ \Omega)$. For lower resistance ranges the **parallel ohmmeter,** also called the

**FIGURE 3–11**

Circuits for (a) series ohmmeter,
(b) parallel (short) ohmmeter,
(c) series-parallel ohmmeter.

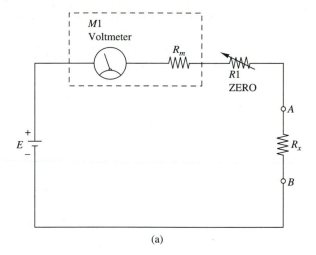

(a)

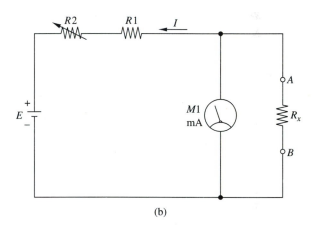

(b)

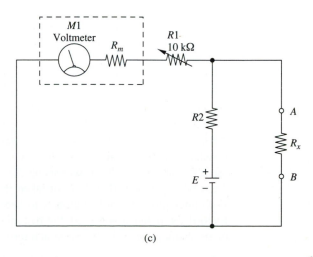

(c)

**shunt ohmmeter,** circuit of Fig. 3–11(b) is used. The meter movement in this case is a milliammeter or microammeter. Resistor $R1$ is used to limit current flow in the meter movement to a safe value. In some instances, $R1$ consists of two resistors in series: a fixed resistor and a variable resistor (e.g., a potentiometer connected as a rheostat). The variable resistor acts as the ZERO control. An assumption needed for this circuit to operate successfully is that it be a near-constant current source. In practical terms this assumption means that $R1 \gg R_m$, so that shunting the unknown resistance across the meter resistance during the resistance measurement affects the current $I$ only a small amount. This requirement usually means that $R1 = 100 \times R_m$, or higher.

The value of the unknown resistance in Fig. 3–11(b) is found from

$$R_x = \frac{I2 R_m}{I1 - I2} \tag{3–11}$$

where   $I1$ = the value of current $I$ when $R_x$ is disconnected (i.e., when terminals $A$ and
        $B$ are open)
      $I2$ = the value of current $I$ when $R_x$ is connected between terminals $A$ and $B$

An example of a **series-parallel ohmmeter** circuit is shown in Fig. 3–11(c). The meter is a voltmeter with a series resistor ($R1$) set to the full-scale voltage. The voltmeter measures the voltage drop across points $A$–$B$, with $R2$ being permanently in the circuit. The value of the unknown resistance is found from

$$R_x = \frac{E R2}{e} - R2 \tag{3–12}$$

where   $E$ = the voltmeter reading with $A$–$B$ shorted together
      $e$ = the voltmeter reading with $R_x$ connected across $A$–$B$

## 3–13–1      OHMMETER ACCURACY VS. BATTERY AGING

The accuracy of the ohmmeters presented above depends on the voltage being sufficiently high to deflect the meter pointer to full scale for zero-calibration purposes. As the battery ages, however, the meter will not be able to come to zero, so any resistance measurement made will be in substantial error. This problem is easily seen when zeroing the meter: if the ZERO control is not capable of causing the meter pointer to deflect to the zero marker, then there is a problem, and the battery must be changed.

Somewhat more subtle, however, is the stage immediately prior to the point where the battery weakness is obvious on the zero scale. As the battery deteriorates, its internal resistance increases, and this resistance can cause an error in the reading produced. It has the effect, for example, of increasing the apparent internal resistance of the meter movement by an amount that was not taken into account when the dial scale was calibrated. For either problem the solution is to keep fresh batteries in the instrument at all times. Even if the meter is not used, the batteries should be refreshed at least every six months or so, or more often if the manufacturer so recommends.

**FIGURE 3–12**
Analog multimeter.

## 3–14    ANALOG MULTIMETERS

A **multimeter** (Fig. 3–12) is an instrument that has several switch-selectable voltage, current, and resistance scales. The main selector switch connects the shunt, multiplier, or range resistors, as required. Most instruments also include a rectifier to allow reading ac values (see Section 3.15).

When using a multimeter, you must follow the same rules as in individual current meters, voltmeters, and ohmmeters, so be very careful. It is all too easy to change the selector switch to a different function without changing the way the meter is connected into the circuit—for example, going from a voltage function to a current function without changing the connection from parallel to series. A typical multimeter might have dc voltage ranges from 0.5 V dc full scale to 1500 V dc full scale, so there is always a possibility of destruction of the instrument if the connection rules are not followed.

It is good laboratory practice to always store a multimeter with the switch in the OFF position; if the meter is a passive volt-ohm-milliammeter (VOM) with no OFF position, then store the it with the selector set to the highest DC VOLTS range position. When using a VOM, begin with the highest range current or voltage (as needed), and work down to a range that gives a readable deflection.

**3–15**       **ANALOG ALTERNATING CURRENT (ac) DEFLECTION-TYPE METERS**

Neither type of basic dc PMMC meter movement will correctly indicate the value of an alternating current (ac), and indeed either might be destroyed by applying ac. There are, however, four basic types of meter movements that will indicate ac values: **thermocouple, hot-wire, electrodynamometer,** and **iron-vane.** We can also make a dc meter movement (d'Arsonval or taut-band) read ac values if a **rectifier** or **integrated circuit rms-to-dc converter** is used between the ac input and the dc meter movement.

**3–16**       **WHICH ac VALUE IS BEING MEASURED?**

The measurement of voltage and current values in dc circuits is straightforward because these values do not vary. In ac circuits, however, the situation is confusing because the ac varies cyclically, there are different values recognized, and all of the values can legitimately be measured. Adding to the confusion is the fact that different types of ac meters inherently display one or the other of the different ac values.

Consider the sine wave in Fig. 3–13. Assuming that there is neither dc offset nor harmonic content (i.e., a "pure" sine wave that is symmetrical on both half-cycles), several different values can be measured (let's consider only the voltage values, for the current values are of the same sort): **peak voltage** ($E_p$), **peak-to-peak voltage** ($E_{p\text{-}p}$), **root mean square voltage** ($E_{rms}$), and **average voltage** ($E_{av}$).

**FIGURE 3–13**
Sine wave relationships.

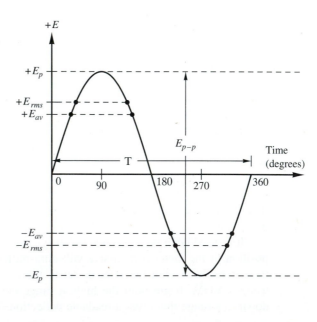

The peak voltage ($E_p$) is the voltage from the zero baseline to the peak of either half of the sine wave; the positive peak is $+E_p$, while the negative peak is $-E_p$. The peak-to-peak voltage is the magnitude of the voltage between the peak voltages:

$$E_{p\text{-}p} = |{+E_p}| + |{-E_p}| \tag{3–13}$$

In the case of a perfect sine wave, with neither a dc offset nor any significant harmonic distortion, the peak-to-peak voltage is equal to twice either the positive or the negative peak voltage.

The root mean square voltage ($E_{\text{rms}}$) is also called the **effective voltage** because it relates to the amount of power that can be delivered to a load. The rms value is equivalent to the dc voltage that will produce the same amount of heating in a resistive load. For any ac waveform, the rms value can be computed from

$$E_{\text{rms}} = \sqrt{\frac{1}{T} \int_0^T E(t)\, dt} \tag{3–14}$$

where   $E_{\text{rms}}$ = the root mean square value of the sine wave voltage
  $E(t)$ = the instantaneous value of $E$ with respect to time
  $T$ = the period of the waveform, that is, the time required for one complete cycle

For the case of the perfect sine wave *(only)*, this equation evaluates to the following well-known relationships:

$$E_{\text{rms}} = \frac{E_p}{\sqrt{2}} = 0.707 E_p \tag{3–15}$$

or

$$E_{\text{rms}} = \frac{\sqrt{2} E_p}{2} \tag{3–16}$$

From these relationships we can also state that the peak voltage $E_p$ is $1.414 E_{\text{rms}}$ and that the peak-to-peak voltage $E_{p\text{-}p}$ is $2.828 E_{\text{rms}}$.

The average value of the sine wave over the entire cycle is zero ($t1 + t2$), but over the half-cycle it is

$$E_{\text{av}} = \frac{2 E_p}{\pi} \tag{3–17}$$

or

$$E_{\text{av}} \approx 0.637 E_p \tag{3–18}$$

The half-wave average value is not much use in most cases, but in the case of some ac voltmeters it can be of great importance in interpreting readings.

For waveforms other than the sine wave, the equation will evaluate differently than the sine wave case. This fact becomes important when dealing with instruments that either are calibrated for the sine wave (the common case) or are inherently read as some other value than what is important for the waveform at hand.

For example, consider the perfectly symmetrical square wave of Fig. 3–14. This waveform possesses both **baseline symmetry** (i.e., the positive and negative excursions are of the same magnitude and are mirror images of each other) and **left-right symmetry** (i.e., the left and right halves of both the positive and the negative excursions are mirror images of each other). The half-wave average is $E_p/2$.

All continuous waveforms can be broken into a mathematical expression of a fundamental frequency $F$ equal to $1/(t1 + t2)$ plus sine and cosine harmonics of $F$. This is called the **Fourier series** of the signal. The Fourier series is not just a mathematical description of the signal. If you employ a device called a **spectrum analyzer,** then you will see an amplitude-vs.-frequency component plot of the Fourier series of the applied waveform. The main difference between signals of differing wave shapes is the particular set of harmonics present, their relative amplitudes, and their relative phase relationships. The undistorted sine wave has only the fundamental component, and no harmonics, while a square wave has a (theoretically) infinite number of even harmonics. The symmetries discussed above tell us something about the way the meter performs when measuring those signals. For the case of the square wave, the positive and negative peaks are equal, as are the durations of those peaks, so over one complete cycle $(t1 + t2)$ the voltage averages out to zero.

**FIGURE 3–14**
Symmetrical square wave with baseline symmetry and left-right symmetry.

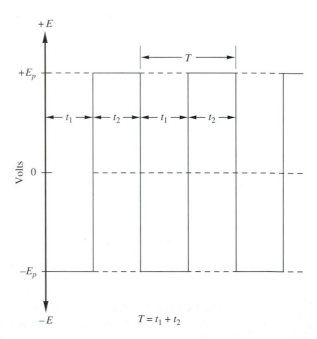

**3–16–1**      **Form Factor (Λ)**

The **form factor** (Λ) of a periodic waveform is the ratio of its rms value to its half-wave average value:

$$\Lambda = \frac{V_{\text{rms}}}{V_{\text{av}}} \qquad (3\text{--}19)$$

For the case of the sine wave, the value of the form factor is

$$\Lambda = \frac{\left[ \dfrac{\sqrt{2}E_p}{2} \right]}{\left[ \dfrac{2E_p}{\pi} \right]} = 1.11 \qquad (3\text{--}20)$$

**3–17**      **THERMOCOUPLE AC CURRENT METERS**

Figure 3–15 shows schematically the **thermocouple ammeter.** This instrument is used primarily to measure radio frequency (RF) currents. It consists of a nonreactive (i.e., resistive) **heating element,** $R$, thermally coupled to a **thermocouple** sensor device (i.e., a sensor that produces an output voltage that is a function of the applied temperature), so that both are in the same thermal environment. Thus, the thermocouple measures the temperature of the heating element, which is proportional to the current flowing in it.

A thermocouple is made by joining two dissimilar metals in a V junction (Fig. 3–15). This assembly effectively forms a temperature-to-voltage transducer; when the thermocouple junction is heated, a voltage appears across terminals $A$ and $B$. This phenomenon, called the **Seebeck effect,** occurs because of the different *work functions* of the two metals (which is why they must be *dissimilar,* i.e., have different work functions) forming the thermocouple. Although the voltage is quite nonlinear over its entire range, it is reasonably linear over the portion of its entire range that is used by the meter. Note, however, that zero-referenced thermocouple meters are not linear in the low end of their scale. It is necessary to select an instrument that will deflect to the upper half of its full-scale range for the current expected in the circuit.

**FIGURE 3–15**
Thermocouple meter circuit.

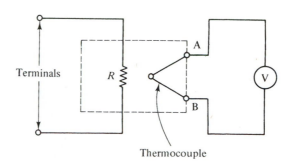

Terminals

$R$

A

B

V

Thermocouple

The alternating current being measured flows through the heater resistance $R$ and causes it to rise in temperature, thereby causing the thermocouple output voltage to rise. A voltmeter connected across terminals $A$ and $B$ measures this millivolt-level potential, but its scale is calibrated in amperes or milliamperes. This operation tells us something of the nature of the thermocouple ammeter measurements: they inherently represent the *rms value of the alternating current being measured.*

Almost any ac meter can be used to measure the rms value of a sine wave because of the well-known relationships given in Section 3.16. But on nonsinusoidal waveforms these relationships are sometimes difficult to predict. Only an inherently rms-reading instrument such as the thermocouple ammeter will read the rms value of such a waveform.

## 3–18       HOT-WIRE AMMETERS

Figure 3–16 shows the **hot-wire** ac meter movement. A resistance wire is stretched between two terminals and is pulled taut by a tension spring. The pointer is mounted to a pivot and is tied to the string and the spring. A current passing through the resistance element causes it to heat and thereby expand. When the wire expands, it deflects the meter pointer, and the amount of deflection is proportional to the rms value of the current flowing in the wire.

## 3–19       THE ELECTRODYNAMOMETER ac METER

The **electrodynamometer** is among the oldest ac-reading instruments still in common use. It was invented by the Siemens brothers in Germany following the development of a similar device by Weber in 1843. By 1910 the configuration of the present instruments had been determined, and these designs are still widely used today (especially in the electrical power industry).

The basic electrodynamometer circuit is shown in Fig. 3–17(a). This type of meter differs from the PMMC types in that it creates its own magnetic field from the current flowing in the windings of coils $L1$ through $L3$. Coils $L1$ and $L2$ are stationary, while coil

**FIGURE 3–16**
Hot-wire ac meter movement.

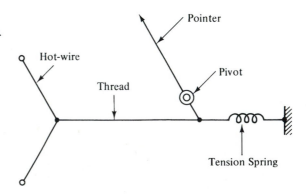

**FIGURE 3–17**
(a) Electrodynamometer circuit,
(b) electrodynamometer used as a
wattmeter.

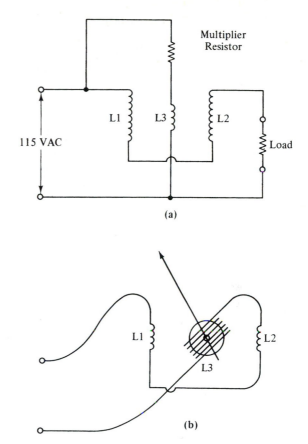

(a)

(b)

$L3$ is free to move and is attached to the meter pointer. The *sense* of the windings is such that the magnetic fields of $L1$ and $L2$ will reinforce each other and will interact with the field of $L3$ to create a rotational force on coil $L3$. The movable coil, $L3$, rotates an amount that is proportional to the strength of the current flowing in the three series-connected coils.

The electrodynamometer is a **square-law device.** That is, the pointer deflects an amount that is proportional to the *square* of the *average* current flowing in its coils [i.e., $(I_{av})^2$]. In most electrodynamometers used to measure current (some are connected to measure power), the dial scale is actually calibrated in terms of rms current, which in sine waves is defined as

$$E_{rms} = \sqrt{i_{av}^2} \tag{3–21}$$

It takes a considerable amount of current to deflect the pointer, so most electrodynamometers have a full-scale rating of 100 mA or more, and there are many in the 0-to-10-A range.

Electrodynamometer voltmeters can be built from a multiplier resistor and a 100-mA meter movement, but these can be used only in low-impedance applications (e.g., power line measurements) because the sensitivity is on the order of only 10 $\Omega$/V, which is far too low for electronic circuit applications.

Extending the current range of an electrodynamometer is usually done with a **current transformer** because suitable shunt resistors would be very low valued, and that renders them difficult to calibrate accurately.

One principal application for electrodynamometers is in the wattmeter used in 60-Hz ac power line systems. An example is shown in Fig. 3–17(b). The stationary coils are designed to carry a heavy current and are connected in series with the load. The movable coil, on the other hand, is wound of a much finer gage of wire, so it will carry a much smaller current (100 mA or less). The small coil ($L3$) is connected across the load and serves as the voltage-measuring element. Coil $L3$ may or may not have a multiplier resistor connected, depending on the design and intended application. When there is no multiplier resistor, the resistance of $L3$ must be sufficient to limit the coil current to a safe value.

The deflection of the pointer connected to $L3$ is proportional to the product of the volts times amperes (V $\times$ A), which is given in watts (units of power in nonreactive circuits).

## 3–20    IRON-VANE METER MOVEMENTS

The **iron-vane** meter movement (Fig. 3–18) consists of a coil of wire to carry the current being measured and two soft iron vanes: one vane is movable, the other stationary. A pointer is attached to the movable vane. When current flows through the coil, it magnetizes the two vanes to the same magnetic polarity, and since like poles repel each other, the movable vane tends to deflect an amount that is proportional to the rms value of the current.

At ac power main frequencies (50 Hz in Europe, 60 Hz in the United States), the iron-vane meter produces accurate readings, but as the frequency of the alternating current increases, **hysteresis** and **eddy current** losses tend to produce nonlinearity problems.

At dc there is another type of error: **residual magnetism** of the vanes by the direct current flowing in the coil. Direct current flows in only one direction, so the vanes will polarize in that direction. Alternating current, on the other hand, reverses direction every half-cycle, so the magnetic field is continually changing, thereby eliminating the buildup of a residual magnetic field in the vanes.

## 3–21    RECTIFIER-BASED ac METERS

Most of the ac-reading meters that you will encounter will be of the rectifier type. **Rectification** is the process of converting a bidirectional ac waveform into a unidirectional waveform that can be read on a dc meter. In a rectifier-type ac meter, the indicating instrument is a dc meter, but there is a rectifier between the dc meter and the ac waveform being measured.

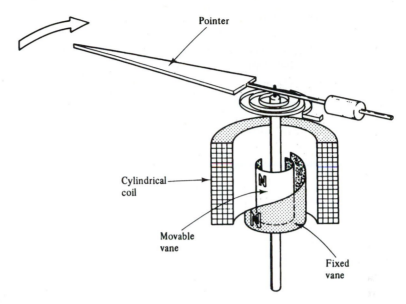

**FIGURE 3–18**
Construction of an iron-vane meter movement.

**3–21–1**     **Rectification**

The mechanism for achieving rectification is the ordinary solid-state diode device [Fig. 3–19(a)] connected in series with the ac path. One attribute of the diode is that it passes current only in one direction. When the anode is positive with respect to the cathode, the device is forward biased, and current will flow. Conversely, when the anode is negative with respect to the cathode, the device is reverse biased, and no current flows. These relationships are shown in Fig. 3–19(b) where the upper waveform is the applied sine wave and the lower waveform is the rectified output waveform. It can be seen from the rectified waveform that only the positive halves of the applied ac waveform were applied to the load; the negative half-waves were cut off. For this reason, the process is called **half-wave rectification,** and the resulting waveform is called **pulsating dc.**

In the ideal half-wave rectified waveform, where we can neglect the voltage drop across $D1$ due to internal resistance and junction potential, the peak voltage $E_p$ is the same as for the sine wave, or $1.414E_{rms}$. The average voltage, however, is approximately equal to $0.45E_{rms}$, or $0.318E_p$. This relationship can be seen in Fig. 3–19(c) where the average voltage is found where the shaded regions ($A$ and $B$) are equal to each other. This occurs at the point equal to one-half the half-wave average voltage ($0.637E_p/2$, or $0.318E_p$).

The half-wave rectified waveform may be unidirectional, but it is far from pure dc and thus will not be usable on many forms of dc meter movement. A measure of the departure of the waveform from pure dc (which would be represented by a horizontal line) is given by the **ripple factor** ($\Gamma$), which for half-wave sine waves is 121%.

**FIGURE 3–19**

Circuit for solid state diode,
(b) applied sine wave (upper) and
rectified output waveform (lower).

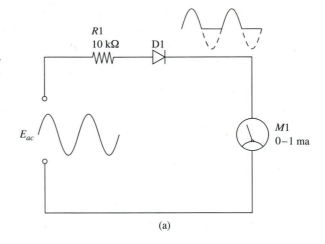

(a)

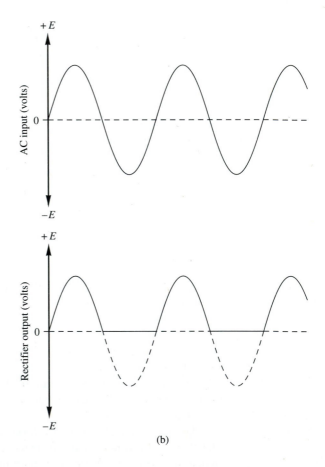

(b)

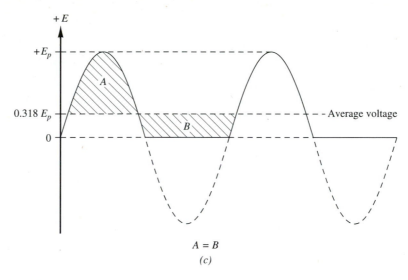

**FIGURE 3–19,** *continued*
(c) Ideal half-wave rectified waveform.

Pulsating dc half-wave rectified waveforms can be smoothed somewhat (i.e., made closer to pure dc) by using a **ripple filter.** This circuit is shown in Fig. 3–20(a). Note that the output meter $M1$ is a voltmeter shunted across a load resistor $R1$ and the ripple filter capacitor $C1$. The simplified explanation of how the ripple filter works is that it stores charge as the pulsating dc waveform is increasing, and then dumps the charge into the circuit as the voltage decreases. The dumped charge fills in some of the space, as shown by the shaded area in Fig. 3–20(b). As a result, the combined waveform of the pulsating dc and the charge dumped from the capacitor[the heavy line in Fig. 3–20(b)] has considerably less ripple than the original waveform, even though it is not pure dc. The value of the capacitor affects the ripple factor, with a high value producing less ripple than a lower value. Although not perfect, as a practical matter, the filtered pulsating dc waveform can be used for some ac meter applications. However, it is also true that the use of a full-wave pulsating dc waveform is even better.

The half-wave rectified waveform ignores one-half of the overall ac cycle, resulting in a value of $\Gamma$ of 121%. *Full-wave rectification,* on the other hand, reduces ripple factor $\Gamma$ to 48%. The lower ripple factor makes full-wave rectified pulsating dc easier to filter and easier to measure in practical circuits, especially in analog meter circuits. Figure 3–21(a) shows the ac waveform along with the full-wave rectified waveform. Note that the negative halves of the wave appear to be "flipped up" to become positive.

Although there are several forms of full-wave rectifiers, the one most commonly found in electronic instrumentation cases is the **bridge rectifier** of Fig. 3–21(b). This circuit uses a group of four diodes arranged in a bridge circuit across the ac generator or source. Let period I be the positive half-cycle of the input ac waveform, and period II be the negative half-cycle [see again Fig. 3–21(a)].

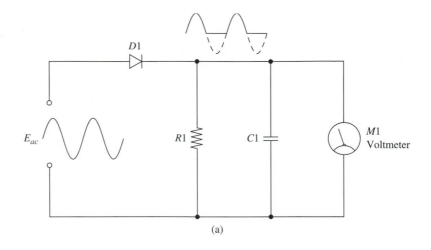

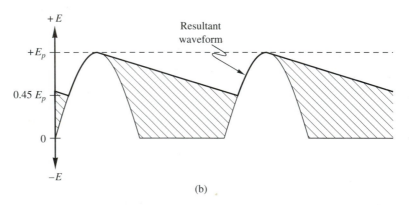

**FIGURE 3–20**
(a) Circuit for ripple filter, (b) "brute force" ripple filter.

In phase I the polarity of the generator is such that diodes $D1$ and $D2$ are forward biased, while $D3$ and $D4$ are reverse biased. Current, flowing from negative to positive, flows out the generator, through $D2$, through load resistor $R$ *(note the flow direction!)*, through $D1$, and then back to the top of the generator.

When the cycle reverses, making the top of the generator negative and the bottom positive, the opposite situation occurs: diodes $D1$ and $D2$ are reverse biased while $D3$ and $D4$ are forward biased. In this case current flows out the top of the generator, through $D3$, through load resistor $R$ (again, note flow direction), through $D4$, and then back to the generator at its bottom.

An important clue to the operation of a full-wave rectifier is that *the current in the load resistor in periods I and II flows in the same direction,* which is what yields the characteristic full-wave rectified waveform in Fig. 3–21(a).

Although the average over the entire ac waveform is zero (assuming positive/negative symmetry), the average of the full-wave rectified wave is nonzero. The average

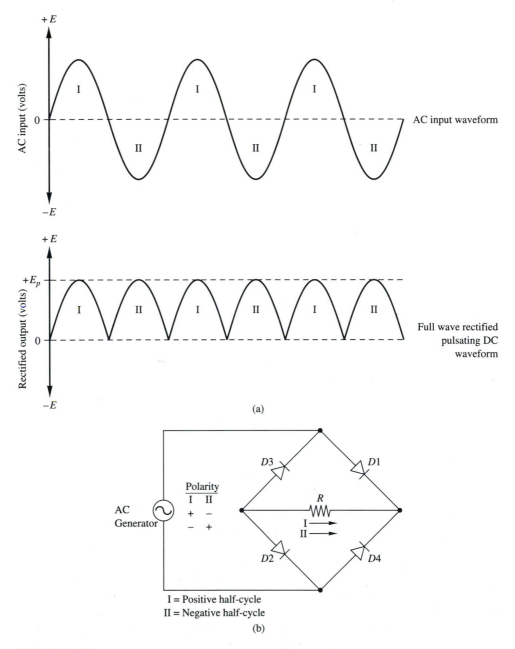

**FIGURE 3–21**

(a) Full-wave rectification, (b) bridge rectifier circuit,

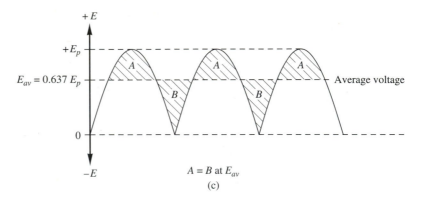

**FIGURE 3–21,** *continued*
(c) Average value is the voltage at which area $A$ = area $B$.

potential (or current) occurs at the point where shaded zones $A$ and $B$ in Fig. 3–21(c) are equal. For a pure sine wave, this situation is such that $E_{av} = 0.637E_p = 0.9E_{rms}$. Because the half-wave symmetry of the ac waveform is not affected by full-wave rectification, the peak voltage of the full-wave rectified waveform remains at $1.414E_{rms}$.

## 3–21–2    Rectifier Circuits for ac Meters

When a rectified ac waveform is applied to the dc (d'Arsonval or taut-band) meter movements, the meter deflection will be proportional to either the average or the peak values of the waveform, depending on the type of rectifier used. We will consider four different circuits: **half-wave average-reading, half-wave peak-reading, full-wave average-reading,** and **full-wave peak-reading.** The principal difference between average-reading and peak-reading instruments is the presence of a ripple filter capacitor across the output of the rectifier. The filter tends to integrate the waveform, producing a reading much closer to the peak value than to the average value.

Figure 3–22(a) shows an average-reading half-wave rectifier instrument. Although several variations on this circuit exist, this circuit is quite widely used because of its simplicity. There is single rectifier diode in series with the signal path to the load. Resistor $R1$ is used to limit the current through meter $M1$, and $R1$ effectively converts it from a current-reading instrument to a voltage-reading instrument.

The circuit of Fig. 3–22(a) could be converted into a peak-reading circuit by connecting a filter capacitor across the series combination of $R1$ and $M1$, as shown in Fig. 3–21(b). Another peak-reading circuit is shown in Fig. 3–22(c); this circuit is a half-wave voltage doubler, and its output is a filtered pulsating dc with an average value that is close to the peak value of the unfiltered pulsating dc waveform.

Full-wave rectifier circuits are shown in Figs. 3–23(a) through 3–23(c). Two different, but related, approaches are shown in these figures: the circuit of Fig. 3–21(a) uses a current meter indicator, while Figures 3–23(b) and 3–23(c) use a voltmeter indicator.

**FIGURE 3–22**
(a) Average reading half-wave
rectifier circuit; (b) peak-reading
circuit with filter capacitor added;
(c) half-wave voltage doubler.

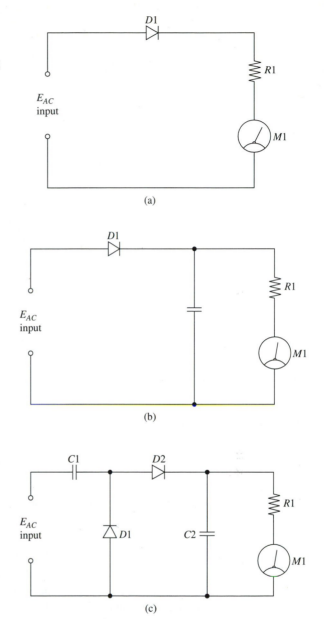

(a)

(b)

(c)

Keep in mind, however, that a d'Arsonval or taut-band voltmeter is nothing but a current meter with a series resistor.

The circuit in Fig. 3–23(a) is an average-reading full-wave rectified instrument because the output of the rectifier is applied directly to the meter without regard for filtering. For peak-reading measurements, the circuit of Fig. 3–23(c) might be preferred because it has a ripple filter capacitor across the load and meter circuit.

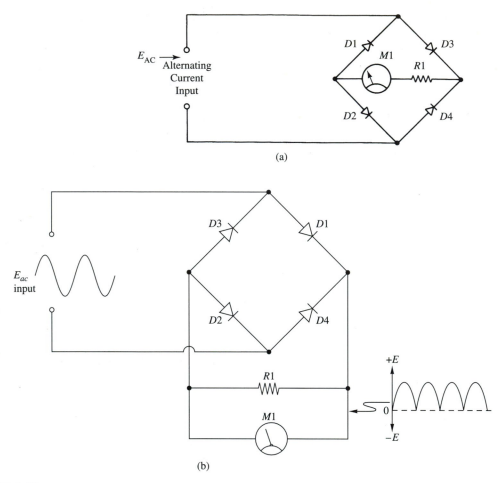

**FIGURE 3–23**
Full-wave rectifier circuits using (a) a current meter indicator, (b) and (c) a voltmeter indicator.

### 3–21–3    Effects of Waveform on Meter Readings

Most ac meters are designed to display the rms value of the current or voltage being measured. Because the instrument measures either peak or average values, depending on the rectifier, a correction factor must be applied, and that correction factor tends to be usable only on sine waves. For an average-reading instrument, the form factor ($\Lambda$) is 2.22 for half-wave rectification and 1.11 for full-wave rectification. The scale will read $\Lambda E_{av}$.

Unfortunately, $\Lambda$ changes with the ac waveform, so such a meter is useful only for the waveform for which it was calibrated (usually a sine wave) unless a correction factor is mentally added to or subtracted from the displayed reading. Figures 3–24(a) and 3–24(b) show two common waveforms: a unipolar **square wave** and a bipolar **triangle wave.** The average and rms values of the unipolar square wave [Fig. 3–24(a)] are equal to each other and to one-half the peak value:

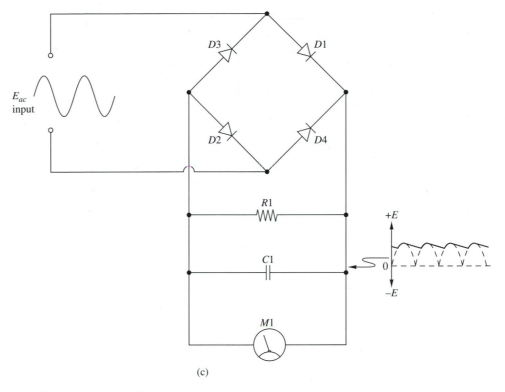

(c)

**FIGURE 3–23,** *continued*

$$E_{av} = E_{rms} = \frac{E_p}{2} \qquad (3\text{–}22)$$

For the triangle waveform [Fig. 3–24(b)], the average and rms values are not equal to each other. The average value $E_{av}$ is 0, while the rms value is

$$E_{rms} = \sqrt{\frac{E_p^2}{3}} \qquad (3\text{–}23)$$

Other factors that affect the indicated value are symmetry of the input waveform across the zero baseline (i.e., is there a dc offset component present?) and harmonic content of the waveform.

Meters of the sort shown in Section 3.21.2 (Figs. 3–22 and 3–23) sometimes exhibit a **turnover effect;** that is, they indicate a different value when the (+) and (−) leads are swapped. This problem is especially severe when the input ac waveform exhibits a lack of zero baseline symmetry.

The effect of waveform on the reading depends somewhat on how the rectifier is operated. Figure 3–25 shows the $E$-vs.-$I$ curve for a standard semiconductor rectifier diode. When the diode is reverse biased ($-E_R$ region), only a small leakage current ($-I_L$)

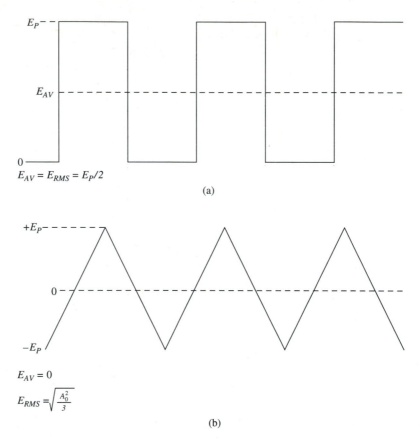

**FIGURE 3–24**
(a) Unipolar square wave, (b) bipolar triangle wave.

flows, which in an ideal diode would be zero. As the applied voltage goes into the forward-biased region, $+E_F$, the forward current begins to rise at a nonlinear rate that is approximately equal to the square of the applied voltage. Rectifiers operated in this low-voltage region are considered **square law devices.** Above some critical potential $E_\gamma$, however, the device enters a more linear **ohmic region** where the forward current is directly proportional to the applied voltage. For silicon diodes, $E_\gamma$ is approximately 0.6 to 0.7 V.

A considerable number of waveform anomalies affect the indicated value of an ac rectifier meter. Table 3–2 shows some of these effects. From the table it appears that, for most applications, the full-wave square law meter is the best and most consistent when nonsinusoidal ac input waveforms are being measured. This situation may require a voltage divider network on higher scales to prevent the rectifier diodes from being operated at a point outside the square law region of the device. It also suggests that peak-reading instruments be avoided unless the peak value is somehow important (transmitter amplitude modulation measurements and RF power measurements in the single-sideband mode are examples).

**FIGURE 3–25**
$E$-vs.-$I$ curve for standard semiconductor rectifier diode.

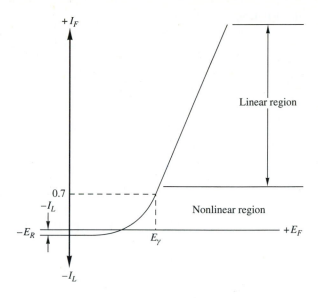

## 3–21–4  True rms-Reading Rectifier Instruments

A meter that reads the true rms value of an ac waveform (instead of merely adjusting the scale according to the form factor $\Lambda$ to indicate $E_{rms}$) will take into account most of the anomalies discussed above. Rectifier instruments are inherently either average-reading or peak-reading devices, depending on the rectifier types. However, the use of a circuit called an **rms-to-dc converter** (Fig. 3–26) will produce a dc output voltage that is equivalent to the rms value of the input waveform. These circuits were once quite expensive and complex, but in modern integrated circuit form they have become quite simple from the external perspective.

**TABLE 3–2**
Effect of waveform anomalies on ac meter readings.

| Effect | Full-Wave Square Law | Half-Wave Square Law | Linear | Peak Reading |
|---|---|---|---|---|
| Turnover effect | No | Yes | No | Yes |
| Phase of harmonics affect reading? | No | Yes | Yes | Yes |
| Effect of 2nd harmonic (%) | 11 | −6 to +27 | 0 to 10 | −25 to +50 |
| Effect of 3rd harmonic (%) | 11 | 12.5 | −10 to +16 | −8 to +50 |

**FIGURE 3–26**
rms-to-dc converter.

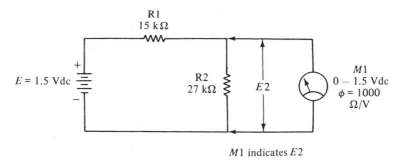

*M*1 indicates *E*2

## 3–22        OTHER ac-READING INSTRUMENTS

Alternating current values can also be measured on the **oscilloscope** (see Chapter 8), on the **audio voltmeter,** and on **electronic multimeters.**

The cathode ray oscilloscope is discussed in detail in Chapter 8, so we will not discuss it here except to state that is allows the user not only to determine the peak and peak-to-peak values of the applied waveform but also to evaluate the *wave shape* and the *period* of the waveform. Many oscilloscopes today are equipped with built-in digital voltmeters (see Chapter 7), so they can provide numerical readouts of the peak, peak-to-peak, rms, and average voltage values, as well as period and frequency information. Some models have separate displays for the numerical data, while others print it right on the CRT with the wave shape.

The audio voltmeter is a rectifier or IC rms module instrument that is preceded by a wideband amplifier. This amplifier greatly increases the sensitivity of the meter, allowing signal levels as low as −100 dBm to be measured. (*Note:* 0 dBm is defined as 1 mW of 1000-Hz signal dissipated in a 600-Ω resistive load.)

Electronic multimeters are instruments that contain dc voltage-, direct current-, ac voltage-, and resistance-measuring capability in a single case, with a common display or meter movement. These instruments today are almost entirely digital in operation and are covered in Chapter 7. Earlier forms of electronic multimeters include devices that are essentially volt-ohm-milliammeters (VOMs) with a field effect transistor (FET) amplifier in the front end and a balanced differential amplifier to drive the meter movement. A switchable frequency–compensated, input voltage divider network was used to reduce higher-level input voltages or currents to the native base range of the instrument.

## SUMMARY

1. The two basic dc meter movement types are the **d'Arsonval** and the **taut-band.** Both are examples of the **permanent magnet moving coil (PMMC)** galvanometer.
2. The range of a current meter can be extended by using a **shunt resistor.**
3. A dc current meter can be used to measure voltage by connecting a **multiplier resistor** in series with the meter.

4. An ohmmeter can be made from a dc current meter if a precision-regulated voltage source is provided.

5. A **multimeter,** called a **volt-ohm-milliammeter** or **VOM,** is an instrument that has switch-selectable voltage, current, and resistance ranges.

6. There are four basic types of ac meter movement: **thermocouple, hot-wire, electrodynamometer,** and **iron-vane.**

7. Direct current (dc) meter movements can be used to measure ac values if a **rectifier** is used in the circuit.

8. Different indications are possible from different types of ac instruments used to measure the same potential difference, because some instruments read **rms** value, some read **average** value, and one reads the **average of the current squared.**

9. Calibration of many ac meters assumes a sinusoidal or near-sinusoidal ac waveform. A **correction factor** is often used when other waveforms are measured.


## RECAPITULATION

Now go back and try to answer the questions at the beginning of the chapter. When you are finished, answer the questions and work the problems below. Place a mark beside each problem or question that you cannot answer, and then go back to the text and reread the appropriate section.


## QUESTIONS

1. An ammeter is connected in _____ with the load.

2. A voltmeter is connected in _____ with the load.

3. Name two types of PMMC galvanometer meter movements:
   _____ and _____.

4. The same basic electrical principles govern the behavior of the PMMC meters and
   _____.

5. The range of an ammeter, milliammeter, or microammeter can be increased if a _____ resistor is used in _____ with the meter movement.

6. A voltmeter can be made from a dc current meter movement if a _____ resistor is connected in _____ with the meter movement.

7. Draw the circuit diagram for a simple dc ohmmeter.

8. A current meter should have an internal resistance that is _____ compared with external circuit resistances.

9. A voltmeter should have an internal resistance that is _____ compared with external circuit resistances.

10. An ohmmeter may be used in circuits with the power turned on. True or false?

11. A thermocouple ac ammeter reads the _____ value of the ac waveform.

12. Thermocouple ammeters are used at frequencies up to _____ MHz.

13. A hot-wire ammeter reads the _____ value of the ac waveform.

14. A _____ is an ac ammeter constructed from a movable

coil in the magnetic fields of two stationary coils. All three coils are connected in
_____ with each other.

15. The ammeter in Question 14 is calibrated in rms values but is actually reading
_____ values.

16. A _____ is used to extend the current range of the electrodynamometer type of ac current meter.

17. Iron-vane meters read the _____ value of the ac waveform.

18. What happens to the iron-vane meter when dc is measured? Can you think of a
method for overcoming this problem?

19. A rectifier meter reads the _____ value of the alternating current or the ac
voltage waveform.

20. A _____ diode is often used as the instrumentation rectifier in audio frequency ac meters.

21. An ac meter reads different values depending on which way the meter probes are
connected. This is called _____ _____.

22. A rectifier-type ac voltmeter is operated such that the voltage applied to the rectifier
diode is less than $V_\gamma$. This is an example of a _____ _____
device.

23. Which type of ac meter rectifier is immune to both turnover effect and the effects of
the phase of harmonics on the reading?

## PROBLEMS

1. You have a 0-to-50-μA dc current meter that has an internal resistance of 4500 Ω.
What value of shunt resistor is needed to make the meter read 0 to 500 μA full scale?

2. A dc current meter is in series with a 12-Ω load that is connected across a 9-V dc battery. How much current flows in the meter if it is assumed that both meter and battery are ideal?

3. A battery has an open-circuit (i.e., no-load) potential of 12 V dc and an internal resistance of 22 Ω. A 0-to-1-mA dc meter with an internal resistance is accidentally (!)
connected across this battery. How much current flows in the meter movement?

4. Find the value of a shunt resistor required to make a 0-to-100-μA dc current meter
read 0 to 1 mA if the meter's internal resistance is 1100 Ω. Use (a) Ohm's law and
(b) current divider methods, and compare the results.

5. Solve Problem 4 for a 0-to-20-μA meter movement.

6. What value of multiplier resistor will make a 0-to-100-μA meter with an internal
resistance of 1300 Ω read 0 to 100 V?

7. Calculate the multiplier resistance needed to make the meter in Problem 6 read 0 to
2 V. Show your work.

8. What is the sensitivity of a voltmeter constructed from a 0-to-30-μA dc meter movement?

9. Calculate the impedance of a dc voltmeter on the 0-to-50-V scale if the sensitivity is
20,000 Ω/V.

10. Refer to Fig. 3–28. (a) What voltage should *ideally* be indicated on meter *M*1? (b) What voltage is *actually* indicated on meter *M*1? (c) What is the percentage of error in this particular measurement?

11. A thermocouple ammeter is used to measure a 5-MHz sine wave signal from a transmitter. It indicates a current flow of 2.5 A in a pure 50-$\Omega$ resistance. What is the peak current of this waveform?

12. An electrodynamometer is used to measure a sine wave current and indicates 1.4 A rms. What is the average value of this waveform?

13. An iron-vane meter indicates 450 mA ac when used to measure a 60-Hz sine wave. What is the rms value of the applied waveform?

14. A full-wave rectified, 60-Hz ac sine wave has a peak current ($I_p$) of 1.35 A. What will an iron-vane meter read if it is used to measure this current?

15. A half-wave rectified, 60-Hz sine wave has a value of 600 mA when measured on an iron-vane meter. What is its peak value?

16. A thermocouple RF ammeter used to measure a 60-Hz square wave current measures 1.2 A. What is the average value of this waveform?

17. Calculate the form factor ($\Lambda$) of a square wave.

18. Calculate the form factor ($\Lambda$) of a triangle wave.

19. Calculate the peak voltage of a sine wave in which $E_{rms} = 74$ V.

20. Find the *average voltage* for (a) a full-wave rectified sine wave, (b) a square wave, and (c) a triangle wave if in each case the peak voltage $E_p$ is 10.0 V.

21. What are the four classes of rectifier circuits for ac instrumentation?

22. What are the form factors ($\Lambda$) of (a) a half-wave rectified sine wave and (b) a full-wave rectified sine wave?

23. Find the rms voltage of a triangle waveform with a peak voltage of 20 V.

# 4

# Bridge Circuits

## 4–1     OBJECTIVES

1. To learn the principles behind the Wheatstone bridge.
2. To learn the different types of ac bridges.
3. To become familiar with the different types of ac null detectors.
4. To learn the applications for the different bridge types.

## 4–2     SELF-EVALUATION

Before studying the material in this chapter, try to answer the questions given below. These questions test your knowledge of the subject matter. If you cannot answer a particular question, then look for the answer as you read the text.

1. Write the null condition equation for a dc Wheatstone bridge.
2. Draw the circuit for a Maxwell bridge.
3. What types of null detectors are suitable for use in an ac bridge?
4. What type of bridge is *best* suited for measurement of (a) high-$Q$ and (b) low-$Q$ components?

## 4–3     dc WHEATSTONE BRIDGES

Figure 4–1(a) shows the classic dc Wheatstone bridge circuit. The circuit is redrawn in Fig. 4–1(b) to more clearly show the fact that we may consider this circuit as two resistor voltage dividers connected in parallel across a voltage source $E$. Current $I5$ through the galvanometer is given by

$$I5 = \frac{E_0}{R5 + R_\mathrm{m}}$$

            **(4–1)**

**FIGURE 4–1**
(a) The Wheatstone bridge,
(b) redrawn to make more clear.

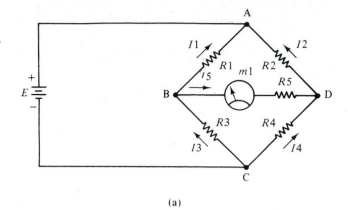

(a)

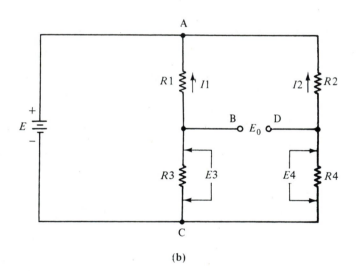

(b)

where    $I5$ = the current flowing in meter $M1$ in *amperes* (A)
$E_0$ = the potential in *volts* (V) between points B and D
$R5$ = the resistance of resistor $R5$ in ohms ($\Omega$)
$R_m$ = the resistance of the meter movement in ohms ($\Omega$)

Voltage $E_0$ is the difference between voltages $E3$ and $E4$:

$$E_0 = E3 - E4 \qquad\qquad\text{(4–2)}$$

But $E3$ and $E4$ are given in terms of $E$ and the resistances:

$$E3 = \frac{ER3}{R1 + R3} \qquad\qquad\text{(4–3)}$$

and

$$E4 = \frac{ER4}{R2 + R4} \qquad (4\text{–}4)$$

By substituting Equations (4–3) and (4–4) into (4–2), we obtain

$$E_0 = E \times \left[ \frac{R3}{R1 + R3} - \frac{R4}{R2 + R4} \right] \qquad (4\text{–}5)$$

Assuming that $R_m$ is negligible, and substituting Equation (4–5) into (4–1), we have

$$I5 = \left[ \frac{E}{R5} \right] \left[ \frac{R3}{R1 + R3} - \frac{R4}{R2 + R4} \right] \qquad (4\text{–}6)$$

■ **Example 3–1**

A dc Wheatstone bridge[Fig. 4–1(a)] uses a 12-V battery for excitation and has the following resistance values in the arms: $R1 = 1.2$ k$\Omega$, $R2 = 1.5$ k$\Omega$, $R3 = 4$ k$\Omega$, $R4 = 3.6$ k$\Omega$, and $R5 = 1$ k$\Omega$. Calculate the meter current.

*Solution*

$$
\begin{aligned}
I5 &= \frac{E}{R5} \left[ \frac{R3}{R1 + R3} - \frac{R4}{R2 + R4} \right] \\
&= \frac{12}{1000} \left[ \frac{4}{(1.2 + 4)} - \frac{3.6}{(1.5 + 3.6)} \right] \\
&= \frac{(12)(0.77 - 0.71)}{1000} = 7.2 \times 10^{-4} \text{ A} = \mathbf{0.72 \text{ mA}}
\end{aligned}
$$

---

**4–4**        **BRIDGES IN THE NULL CONDITION**

The **null condition** in a Wheatstone bridge exists when the output voltage $E_0$ is zero. We may find the sole necessary condition for null by setting Equation (4–6) equal to zero. Thus,

$$\left[ \frac{E}{R5} \right] \left[ \frac{R3}{R1 + R3} - \frac{R4}{R2 + R4} \right] = 0 \qquad (4\text{–}7)$$

For Equation (4–7) to evaluate to zero, either the quantity $E/R5$ or the quantity inside the braces must be zero. The former quantity is never zero in practical situations, so we may conclude that

$$\frac{R3}{R1 + R3} - \frac{R4}{R2 + R4} = 0 \qquad (4\text{–}8)$$

from which we may also conclude that

$$\frac{R3}{R1 + R3} = \frac{R4}{R2 + R4} \qquad (4\text{–}9)$$

Equation (4–9) can be simplified by the following steps. At *null,* $E_0$ is zero, so $I5$ is *also* zero. Therefore,

$$I1 = I3 \qquad (4\text{–}10)$$

$$I2 = I4 \qquad (4\text{–}11)$$

We may then write

$$I1R1 = I2R2 \qquad (4\text{–}12)$$

and

$$I1R3 = I2R4 \qquad (4\text{–}13)$$

The sole necessary condition for bridge null is obtained by dividing Equation (4–13) into Equation (4–12), which yields Equation (4–14):

$$\frac{R1}{R3} = \frac{R2}{R4} \qquad (4\text{–}14)$$

From Equation (4–14) we can compute any resistance value if the other three values are known. Note that it is *not* necessary for the resistor values to be equal—only that the *ratios* $R1/R3$ and $R2/R4$ be equal. Equation (4–14) describes the Wheatstone bridge **null condition** and is essential to understanding many different classes of electronic instrumentation circuits.

■ **Example 4–2**

A Wheatstone bridge has the following arm values: $R1 = 1.6$ k$\Omega$, $R2 = 500$ $\Omega$, $R3$ is unknown, and $R4 = 1800$ $\Omega$. What value of $R3$ brings the bridge into the null condition?

*Solution*

$$R3 = R1R4/R2 \qquad \text{[from Eq. (4–14)]}$$

$$= \frac{(1600)(1800)}{500} = \textbf{5760 } \boldsymbol{\Omega}$$

**FIGURE 4–2**
Measuring resistance with a Wheatstone bridge.

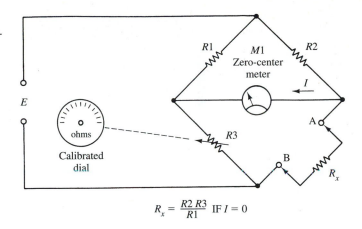

$$R_x = \frac{R2\,R3}{R1} \text{ IF } I = 0$$

## dc BRIDGE APPLICATIONS

Equation (4–14) gives us a means for measuring resistances very accurately—far more accurately, in fact, than is possible on a volt-ohm-milliammeter or even most digital ohmmeters. In Fig. 4–2, the unknown resistance $R_x$ is given by

$$R_x = \frac{R2R3}{R1} \tag{4–15}$$

(when the bridge is in null).

If we make $R1$ and $R2$ fixed and link potentiometer $R3$ to a dial that is calibrated in ohms, then we may measure $R_x$. We could, for example, make $R1$ equal to $R2$, so the ratio $R2/R1$ is unity. In that case $R_x$ will equal $R3$ at the null point. To measure $R_x$, connect it to the bridge terminals A and B, and adjust $R3$ for a *zero* reading on meter $M1$. In some commercial bridges, resistors $R2/R1$ serve as a **range multiplier** and have different values on different resistance ranges.

Another category of Wheatstone bridge applications begins with a balanced bridge and then measures the *amount* of unbalance when one or more of the resistance arms *changes*. This principle is used in many *transducers* in which a stimulus parameter, such as force, temperature, displacement, and so on, changes the value of one or more arms in the bridge. If all arms are equal under the condition of zero stimulus, then $E_0$ will be directly proportional to the value of the stimulus.

## dc NULL INDICATORS

If the bridge is configured such that $E_0$ is always positive or negative, then any dc meter of appropriate range may be used for $M1$. If, on the other hand, the polarity of the bridge output voltage may swing either positive or negative, then meter $M1$ must be a **zero-center galvanometer.**

**FIGURE 4–3**
Zero-center null voltmeter (courtesy of Hewlett-Packard).

Interestingly enough, either a voltmeter or a current meter (milli- or microammeter) may be used for $M1$. In bridges that are passive, a current meter is generally used. Typical ranges fall within the 0–100-$\mu$A to 0–1-mA range. The smaller full-scale current movements are capable of better resolution, *provided* that the minimum change of resistance step used to null the bridge produces current changes that are considerably less than the full-scale current of the meter movement.

Figure 4–3 shows a battery-operated dc null meter that is capable of detecting $E_0$ values down to the *microvolt* range. This instrument is a very sensitive amplified voltmeter that is a special case of the instruments discussed later in this book.

## 4–7     ac BRIDGES

The dc Wheatstone bridge will work with either ac or dc excitation, although in the former case the null detector must be a device that is sensitive to ac. Certain other types of bridge circuits are designed specifically for ac use and have inductive or capacitive *reactances* in one or more of the bridge arms.

## 4–8     ac NULL DETECTORS

The dc galvanometer used with the resistive Wheatstone bridge will not respond to ac, so it cannot be used directly as the null detector in ac bridges. Many rectifier ac meters based on the dc movement are not sensitive enough to indicate the null precisely, so these meters are limited to applications where a relatively shallow (i.e., broad) null can be tolerated.

**FIGURE 4–4**

ac excited Wheatstone bridge using headphones as the null detector.

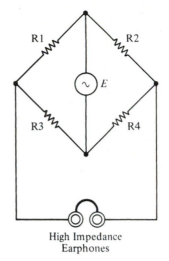

High Impedance
Earphones

Alternating current electronic voltmeters and oscilloscopes, on the other hand, are quite sensitive and thus can indicate the ac null with great precision. These instruments can be used to replace the dc galvanometer-rectifier combination where sensitive measurements are to be made.

In some cases, high-impedance earphones are used as a null detector (see Fig. 4–4) in ac bridges. A surprising degree of sensitivity is possible if the operator has high hearing acuity. In general, highly efficient communications (*not* high-fidelity) earphones with an impedance of 1000 $\Omega$ or higher are used for this purpose.

Problems sometimes occur when one is trying to use instruments with ground-referenced inputs as a null detector, and these problems are especially troublesome if the ac excitation source is also grounded.

There are two basic approaches to solving this problem. One is transformer coupling to the null detector (thereby isolating the detector from the bridge), and the other is the use of a differential amplifier.

Figure 4–5(a) shows the use of a transformer to couple the signal from the $E_0$ nodes of the bridge to an external null indicator such as an oscilloscope or ac voltmeter. In this particular example the null indicator is an ac voltmeter. Transformer coupling is often used in RF bridges to provide isolation. The bridge components are sensitive to body capacitance from the operator or other nearby disturbances. The bridge then may be located at the site of the measurement in relative isolation from the operator, and the output signal may be brought to the null indicator through a coaxial cable transmission line.

A differential amplifier will produce a single-ended output from a differential input signal. The differential amplifier is useful for isolation, as described above, or for increasing the sensitivity of the null indicator by amplification of the bridge output signal.

Many oscilloscopes can be equipped with a differential amplifier plug-in, while even most nonplug-in models can be used in a quasidifferential mode if they are equipped with two channels. To use a two-channel oscilloscope in quasidifferential, use the two inputs as inputs to a differential amplifier, and set the oscilloscope's *mode* switch in *A–B*.

**FIGURE 4–5**
(a) Transformer-coupled ac Wheat-stone bridge, (b) differential amplifier output for an ac Wheat-stone bridge.

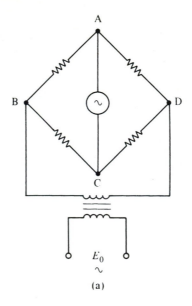

(a)

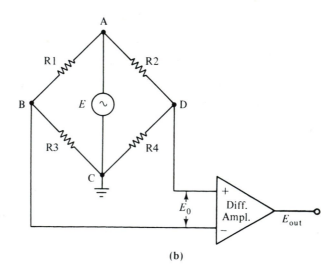

(b)

## 4–9        PHASE DETECTORS

The ac bridges of the following sections require a sensitive phase detector to indicate the output condition. This detector need not be able to indicate absolute phase (i.e., degrees of phase angle $\theta$) but only **relative phase** (i.e., whether or not a signal is **in phase** or **out of phase** with a reference signal).

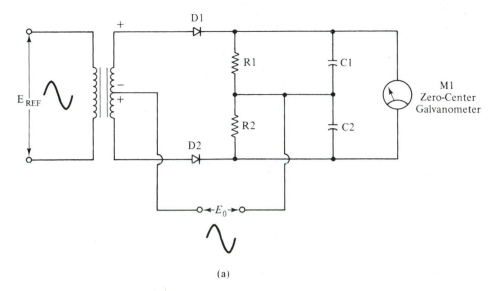

(a)

**FIGURE 4–6**
(a) Phase detector circuit.

Figure 4–6(a) shows a typical phase detector circuit. Signal $E_{ref}$ is a sine wave and is usually the same signal that is used for bridge excitation. This arrangement allows us to measure changes in phase angle $\theta$ due to the reactive component in each bridge arm.

Under *no-signal* conditions diode $D1$ will conduct on one-half of the reference signal cycle, while $D2$ will conduct on the alternate half-cycle. The effect of $C1$, $C2$, and the natural mechanical damping of meter movement $M1$ is to *average* these alternations over the entire cycle, producing a net potential across $M1$ of zero.

When an input signal is applied to the circuit, it algebraically adds to $E_{ref}$, so it *aids* $E_{ref}$ when *in phase* and *opposes* $E_{ref}$ when *out of phase*.

Signal voltage $E_0$ is applied to diodes $D1$ and $D2$ in parallel, while $E_{ref}$ is applied in push-pull. On one-half of its cycle, $E_0$ will forward-bias *both* diodes simultaneously, whereas on the opposite half-cycle it will reverse-bias both diodes.

If $E_{ref}$ and $E_0$ are in phase, then the positive half-cycles *coincide,* so with the polarities shown, the anode of $D1$ sees a large positive potential created by the summation of $+E_{ref}$ and $E_0$, while $D2$ sees a lower potential; $-E_{ref}$ and $E_0$ partially or totally *cancel* [see Fig. 4–6(b)] at $D2$. This situation causes the output of $D1$ to be *more positive* than the output of $D2$, so meter $M1$ swings positive. On the second half-cycle the polarities are reversed, but the output relationships seen by $M1$ are the same.

For an out-of-phase input signal $E_0$, exactly the opposite occurs [Fig. 4–6(c)], so the output of $D2$ is more positive than $D1$, and thus meter $M1$ swings negative.

This phase detector, then, responds to the **relative phase** of the input signal by creating a dc output that is positive for an in-phase condition and negative for an out-of-phase condition. The amplitude of $E_{ref}$ is constant, but the amplitude of $E_0$ changes with

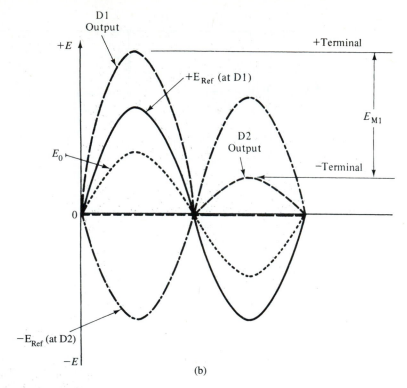

**FIGURE 4–6,** *continued*
(b) Signal relationships in phase detector (in-phase).

bridge balance conditions. Thus, we may conclude that the circuit of Fig. 4–6(a) is sensitive to *both* phase and magnitude.

## 4–10     TYPES OF ac BRIDGES

Although the resistive Wheatstone bridge is technically an "ac bridge" if the excitation source is ac, we usually limit the designation to those bridge types that use a reactive element such as a capacitor or an inductor in one or more bridge arms. Such bridges may be used for *impedance* measurements. In this chapter we will consider the **Maxwell, Hay,** and **Schering** bridge circuits. In Chapter 10 we will take a look at a special class of ac bridges used for radio frequency (RF) work.

The general form of an ac bridge is shown in Fig. 4–7; it is the standard Wheatstone bridge configuration with the resistance elements in the arms replaced by complex impedances of the following general type:

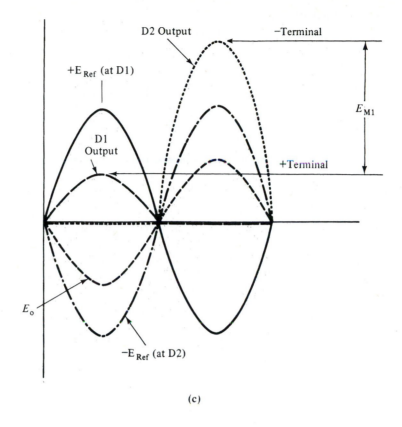

(c)

**FIGURE 4–6,** *continued*
(c) Signal relationships in phase detector (out-of-phase).

$$Z = \sqrt{R^2 + (X_L - X_c)^2} \qquad\qquad \textbf{(4–16)}$$

where   $Z$ = the complex impedance in ohms ($\Omega$)
         $R$ = the resistance in ohms ($\Omega$)
         $X_L$ = the inductive reactance in ohms ($\Omega$)
         $X_c$ = the capacitive reactance in ohms ($\Omega$)

The output signal from this type of bridge will have only a *magnitude* component if the inductive and capacitive reactances cancel, but it will have both *magnitude* and *phase* components if $X_L$ does not exactly balance $X_c$.

## 4–11        MAXWELL'S BRIDGE

Figure 4–8(a) shows the **Maxwell bridge circuit** in which $Z2$ and $Z3$ of the general configuration have been replaced by pure resistances, $Z1$ has been replaced by a parallel combination $R1/C1$, and $Z4$ by the series combination $L1/R4$.

**FIGURE 4–7**
Generalized circuit for complex ac bridges.

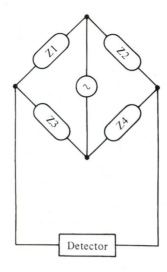

The null condition for a Maxwell bridge circuit is given by the following equations:

$$L1 = R2 \times R3 \times C1 \tag{4–17}$$

$$R4 = \frac{R2 \times R3}{R1} \tag{4–18}$$

The Maxwell bridge is often used to measure unknown values of inductance (e.g., $L1$) because the balance equations are totally *independent* of frequency. The bridge is also not too sensitive to resistive losses in the inductor. Additionally, it is much easier to obtain easily calibrated capacitor standards for $C1$ than it is to obtain inductor standards for $L1$. As a result, the principal use of the Maxwell bridge is the measurement of inductances.

The Maxwell circuit is often used in *Q-meters,* that is, instruments that measure the quality factor $Q$ of inductors. The equation for $Q$, however, is frequency sensitive:

$$Q_L = \omega \times R1 \times C1 \tag{4–19}$$

where    $\omega = 2\pi f$
$R1$ and $C1$ are as previously defined.

4–12      **THE HAY BRIDGE**

The **Hay bridge circuit** [see Fig. 4–8(b)] is physically similar to Maxwell's bridge, except that the $R1/C1$ combination is connected in *series*. The Hay bridge is, however, frequency sensitive. The balance equations are as follows:

## Bridge Circuits

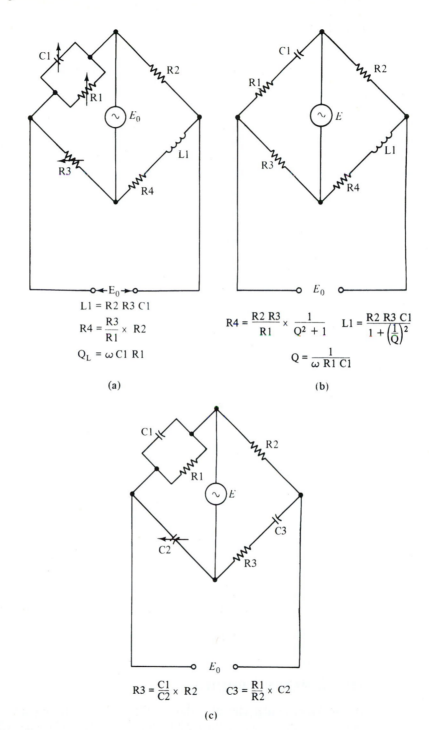

L1 = R2 R3 C1

$$R4 = \frac{R3}{R1} \times R2$$

$$Q_L = \omega \, C1 \, R1$$

(a)

$$R4 = \frac{R2 \, R3}{R1} \times \frac{1}{Q^2 + 1} \qquad L1 = \frac{R2 \, R3 \, C1}{1 + \left(\frac{1}{Q}\right)^2}$$

$$Q = \frac{1}{\omega \, R1 \, C1}$$

(b)

$$R3 = \frac{C1}{C2} \times R2 \qquad C3 = \frac{R1}{R2} \times C2$$

(c)

**FIGURE 4–8**
(a) Maxwell's bridge, (b) the Hay bridge, (c) the Schering bridge.

$$L1 = \frac{R2 \times R3 \times C1}{1 + \left[\dfrac{1}{Q}\right]^2} \qquad\qquad (4\text{--}20)$$

$$R4 = \left[\frac{R2 \times R3}{R1}\right] \times \left[\frac{1}{Q^2 + 1}\right] \qquad\qquad (4\text{--}21)$$

$$Q = \frac{1}{\omega \times R1 \times C1} \qquad\qquad (4\text{--}22)$$

The Hay bridge is preferred for measuring *inductances* with high $Q$ figures, while Maxwell's bridge works best with low-$Q$ inductors.

Note that a frequency-independent version of Equation (4–20) is possible when the $Q$ is very high—that is, greater than about 10—in which case Equation (4–20) can be rewritten in the approximation form:

$$L1 = R2 \times R3 \times C1 \qquad\qquad (4\text{--}23)$$

The error caused by using the approximation of Equation (4–23) is 1% if the $Q$ is 10 and 0.01% if the $Q$ is 100.

*Proof:*   Let

$$R2C3C1 = k = \text{constant}$$

Assume

$$Q = 10$$

a. From Equation (4–20):

$$\begin{aligned}
L1 &= k/[1 + (1/Q)^2] \\
&= k/[1 + (1/10)^2] \\
&= k/(1 + 0.01) = k/1.01 = 0.99k
\end{aligned}$$

b. From Equation (4–22):

$$L1 = k$$

c. $\dfrac{0.99k - k}{0.99k} \times 100 = -1\%$

**4–13**       ## THE SCHERING BRIDGE

The **Schering bridge circuit** is shown in Fig. 4–8(c). This circuit uses a parallel RC network $R1/C1$ for $Z1$, a resistance $R2$ for $Z2$, a capacitive reactance $C2$ for $Z3$, and a series RC network $R3/C3$ for $Z4$. The Schering bridge balance equations are

$$R3 = \frac{R2 \times C1}{C2} \tag{4-24}$$

$$C3 = \frac{C2 \times R1}{R2} \tag{4-25}$$

The Schering bridge is used primarily for the measurement of *capacitance* and the *power factor* of capacitors. In the latter application no actual $R3$ is connected into the circuit, making the series resistance of the capacitor being tested (e.g., $C3$) the only resistance in that arm of the bridge. The capacitor $Q$ is found from

$$Q_{C3} = \frac{1}{\omega \times R1 \times C1} \tag{4-26}$$

## SUMMARY

1. The **Wheatstone bridge** consists of only resistances in the respective branches; it may be used with either ac or dc excitation.
2. The **Maxwell bridge** and the **Hay bridge** are used for inductance measurements, whereas the **Schering bridge** is used for capacitance measurement. The Wheatstone bridge is used mainly for resistance measurements.
3. The **zero-center galvanometer** is used as a null detector in dc Wheatstone bridges.
4. Alternating-current current meters or ac voltmeters are used as null detectors in ac bridges when there are no phase components in the output signal, while an oscilloscope or a phase detector can be used when there are both phase and magnitude components.

## RECAPITULATION

Now go back and try to answer the questions at the beginning of the chapter. When you are finished, answer the questions and work the problems below. Place a mark beside each problem or question that you cannot answer, and then go back and reread appropriate sections of the text.

## QUESTIONS

1. Draw the circuit for a dc Wheatstone bridge and label the components ($R1$, $R2$, $R3$, etc.).
2. Using the component designations given in Question 1, write the *null condition* equations.
3. Draw the circuit for a *Maxwell* bridge, label the components, and write the null condition equations.
4. Draw the circuit for a *Hay* bridge, label the components, and write the null condition equations.

5. Draw the circuit for a *Schering* bridge, label the components, and write the null condition equations.

6. What type of ac null detector is *best* suited for (a) the Wheatstone bridge, (b) the Hay bridge, (c) the Schering bridge, and (d) the Maxwell bridge?

7. What type of bridge is best suited for measuring capacitance?

8. What type of bridge is best suited for measuring the inductance of high-$Q$ coils?

9. What type of bridge is best suited for measuring the inductance of low-$Q$ coils?

10. What type of bridge is best suited for measuring the power factor of a capacitor?

11. A zero-center galvanometer with a range of 1–0–1 mA is used as the null detector in a dc Wheatstone bridge. What type of circuit or device may be used to increase the *sensitivity* of the galvanometer so that 10 μA will deflect the meter full scale?

12. What type of device or circuit may be used with an ac Wheatstone bridge to drive a ground-referenced null indicator?

## PROBLEMS

1. Is the Wheatstone bridge of Fig. 4–9 balanced? What is the value of $E_0$?

2. $R4$ in Fig. 4–9 is changed to 5.2 kΩ. What is the value of $E_0$?

3. A 20-kΩ resistor is connected between points A and B in Fig. 4–9, and $R4$ is changed to 4.7 kΩ. Calculate the current through the 20-kΩ resistor.

4. What value would $E_0$ in Fig. 4–9 become if $R2$ were shunted by a 100-kΩ resistor?

5. An unknown resistance $R_x$ is connected into Fig. 4–9 in place of $R4$, and $R3$ is adjusted to reestablish the null condition. What is the value of $R_x$ if $R3$ has a value of 576 Ω?

6. What is the value of $L1$ in Fig. 4–10?

7. What is the value of $R4$ in Fig. 4–10?

8. Find the $Q$ of the $R4/L1$ network in Fig. 4–10.

**FIGURE 4–9**
Wheatstone bridge for Problems 1–5.

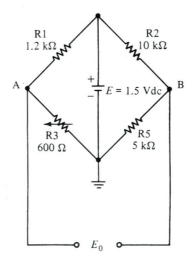

**FIGURE 4–10**
Network for Problems 6–10.

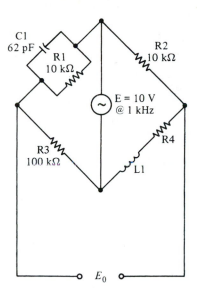

9. A *Hay* bridge is formed by using the *same* components as in Fig. 4–10. An unknown inductor is connected in place of $L1$, and the bridge is adjusted to null. At that point it is found that $C1 = 93$ pF. What is the value of the unknown inductor?

10. What is the $Q$ of the coil in Problem 9?

11. Use the *loop* method of network analysis to derive the expression for $E_0$ in an *unbalanced* dc Wheatstone bridge.

## BIBLIOGRAPHY

Kidwell, Walter. *Electronic Instruments and Measurements.* New York: McGraw-Hill, 1969.

Prensky, Sol. *Electronic Instrumentation,* 2nd ed. Englewood Cliffs, NJ. Prentice-Hall, 1971.

Terman, F. E. and Pettit, J. M. *Electronic Measurements,* 2nd ed. New York: McGraw-Hill. 1952.

Wolf, Stanley. *Guide to Electronic Measurements and Laboratory Practice.* Englewood Cliffs, NJ: Prentice-Hall, 1973.

# 5

# Comparison Measurements

## 5–1    OBJECTIVES

1. To learn the construction and operation of a **precision potentiometer.**
2. To learn how to make **precision measurements** with a potentiometer.
3. To learn the operation of an "automatic potentiometer."

## 5–2    SELF-EVALUATION

Before studying the material in this chapter, try to answer the questions given below. These questions test your knowledge of the subject matter. If you cannot answer a particular question, then look for the answer as you read the text.

1. Describe a **precision potentiometer** in your own words.
2. How do you measure an unknown current using the **equal deflection technique?**
3. What is a **Weston cell?**
4. Describe the operation of a simple **analog-to-digital (A-D) converter** that is based on comparison measurement methods.

## 5–3    COMPARISON MEASUREMENTS

A deflection meter movement calibrated to read a voltage or current is usually not uniformly accurate over its entire range. Most deflection meters produce the best results in the upper half or upper third of their range. High-quality deflection movements are accurate to 1% or 2% of full scale in the upper half of the range, whereas lower-quality instruments are accurate to only 5%.

Comparison measurements, on the other hand, generally use either a null condition or equal deflection of a meter movement to indicate the point where an unknown voltage or current is *equal* to a precise *reference* source. The Wheatstone bridge (Chapter 4) and the **potentiometer** are examples of comparison measurements. In this chapter we will discuss the precision potentiometer and related circuits. Note that in this application the

**109**

word *potentiometer* is used to refer not to the low-cost electronic component often used as a radio volume control but rather to a precision instrument used to make precise, accurate, high-resolution voltage measurements.

Two techniques are commonly employed in potentiometer measurements: **equal deflection** and **null.** In the equal deflection method the unknown and calibration sources are alternately connected to the meter, and the potentiometer is adjusted so that both produce the *same* deflection on a meter movement. This procedure is more accurate than ordinary deflection methods. Although the accuracy of the meter may be in doubt, it is quite resettable; that is, the same potential applied to the meter on different occasions will produce very nearly the same *deflection;* even if the *dial calibration* is in error, the deflection will be very nearly the same.

## 5–4     POTENTIOMETERS

Most electronics students are familiar with the type of potentiometer used to make circuit adjustments, for example, either the radio volume control mentioned above or the trimmer potentiometer in an operational amplifier circuit. But these electronic components are only a crude relative of the type of potentiometer used to make comparison measurements. Most of these instrumentation potentiometers consist of a resistance wire or **element** stretched between two terminals [Fig. 5–1(a)] and a **wiper** that slides along the resistance element and is connected to a third terminal. If a potential, that is, voltage, is applied across the resistance element, then the potential measured between the wiper and either end terminal is a function of the wiper *position.* If the position can be precisely determined, then the voltage between the wiper and end terminals is also precisely known, hence the name *potentiometer.* Although potentiometers are an archaic form of circuit, they are (like the Wheatstone bridge) still used in high-accuracy measurements. Even when purely digital methods are employed, the basic principles of the potentiometer are at the heart of the instrument.

**FIGURE 5–1**
(a) Potentiometer circuit, (b) precision slide-wire potentiometer.

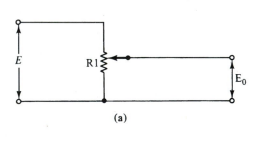

(a)

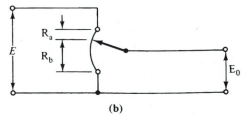

(b)

Precision potentiometers usually employ a very precise resistance wire [Fig. 5–1(b)] as the resistance element. If the wire is formed to have a uniform cross-sectional area and composition, then a very accurate (0.01% to 0.05%) potentiometer can be made. Even if temperature changes cause resistance changes in the wire, the change is essentially uniform, so it does not affect the output calibration. The output voltage in a nonloaded potentiometer is given as the ratio

$$E_0 = \frac{ER_b}{R_a + R_b} \tag{5–1}$$

where   $E_0$ = the output potential in *volts* (V)
     $E$ = the reference potential in *volts* (V)
     $R_a$ = the resistance in ohms ($\Omega$) between terminals *a* and *c* [Fig. 5–1(b)]
     $R_b$ = the resistance in ohms ($\Omega$) between terminals *b* and *c*

■ **Example 5–1**

The overall resistance of the slide wire in Fig. 5–1(b) is 150 $\Omega$. If the wiper is set to the ⅔ mark, then $R_a = 50\ \Omega$ and $R_b = 100\ \Omega$. Find $E_0$ if $E = 1.36$ V dc.

*Solution*

$$E_0 = ER_b/(R_a + R_b) \tag{5–1}$$

$$= \frac{(1.36\ \text{V})(100\ \Omega)}{(50 + 100)\ \Omega}$$

$$= (1.36\ \text{V})(100)/(150) = \mathbf{0.907\ V}$$

If reference potential $E$ is precisely known, and if the relative position of the wiper can be determined, then very precise output voltage ($E_0$) can be created.

Consider what happens when resistance $R_{a\text{-}b}$ changes from 150 $\Omega$ to, say, 156 $\Omega$ due to a change in temperature. By using the changed resistance in Example 5–1, we produce the *same* output voltage:

$$R_{a\text{-}b} \to 156\ \Omega, \qquad \text{so} \quad R_a \to 52\ \Omega \quad \text{and} \quad R_b \to 104\ \Omega$$

$$E_0 = \frac{(1.36\ \text{V})(104\ \Omega)}{(52 + 104)\ \Omega}$$

$$= (1.36\ \text{V})(104)/(156) = \mathbf{0.907\ V}$$

If the resistance change is caused by an environmental factor such as a change in ambient temperature, the potentiometer calibration does not change. Clearly, the precision potentiometer maintains its accuracy even under changes in temperature, something that *cannot* be said of certain other instruments. In other words, the potentiometer is robust against changes in temperature ($\Delta T$).

## 5–5     POTENTIOMETER CIRCUIT

Figure 5–2 shows the circuit for a manual potentiometer used to measure an unknown potential $E_x$.

Resistor $R1$ is the potentiometer and consists of a slide-wire element [as in Fig. 5–1(b)] and a *variable* working voltage $E$. The wiper of $R1$ is mechanically coupled to a dial that measures its position precisely and gives this relative position on a micrometer scale of 0000 to 9999. The indication is 0000 when $R_b$ is zero, and 9999 when $R_a$ is zero (i.e., when $R_b$ is full scale). The output potential $E_0$, then, can be expressed as a decimal fraction of $E$.

Switch $S1$ is a normally open pushbutton switch and is used to turn on the meter circuit. In normal operation $S1$ is just *tapped* by the operator until the meter indicates that the circuit is close to null.

Switch $S2$ is the **sensitivity control** and is normally open until the meter is near the null. At that time, the operator closes $S2$, increasing the sensitivity, and then completes the adjustment.

Switch $S3$ selects either the unknown potential $E_x$ or the precision reference source $E_r$ to balance against $E_0$.

There are three voltage sources in this circuit: $E, E_x,$ and $E_r$. Voltage $E$ is the working voltage. This supply is a variable voltage source. It need not be accurate, but it must be *stable*. In some models the working voltage is a battery connected in series with a rheostat (two-terminal variable resistor), but in many recent models $E$ is a variable, electronically regulated power supply.

Source $E_r$ is a precision reference source. In most older, and many recent, models $E_r$ would be taken from a **Weston cell,** also called a **standard cell** (Fig. 5–3).

The Weston cell is available in two styles: **saturated** and **unsaturated.** These terms refer to the fact that the solution of cadmium sulphate used in the cell is saturated in one

**FIGURE 5–2**
Instrumentation potentiometer.

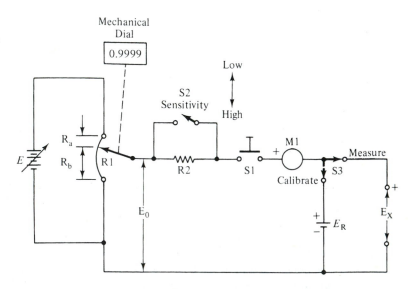

**FIGURE 5–3**
Weston (standard) cell.

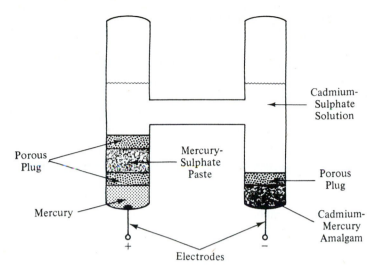

Cadmium-Sulphate Solution

Porous Plug

Mercury-Sulphate Paste

Porous Plug

Mercury

Cadmium-Mercury Amalgam

+    −

Electrodes

type, but not in the other (see Fig. 5–3). The saturated cell is accurate to within 1 μV of its test value at a temperature of 20°C, but the unsaturated type is considered more portable.

The theoretical output voltage for a Weston cell at 20°C is 1.01830 V dc, and the potential *changes* at a rate of approximately 40 μV/°C.

The unsaturated cell is built in a closed container, so it can be made portable. It exhibits nearly the same temperature properties as the saturated version but is known to have a **negative drift** in the terminal voltage of approximately 3 μV/month. Also, at 20°C the output voltage when the cell is new may be any value between 1.018 and 1.020 V. The actual value for any specific cell is given on a label attached to the cell.

The Weston cell, although very accurate, is a delicate device, and there are several restrictions on its use. The most important restrictions are as follows:

1. Never allow more than $10^{-4}$ A to pass through the cell in either charge or discharge modes. This limit restricts the minimum load resistance across the cell to 10.2 kΩ.
2. Always maintain the cell temperature between 4°C and 40°C.

**5–6**

## POTENTIOMETER OPERATION

The instructions for operating the manual potentiometer are given below. Refer to Fig. 5–2 while reading.

### *Calibration*

1. Set $S3$ to *calibrate*, $S2$ to *low* (e.g., $R3$ in-circuit), and potentiometer $R1$ to *half-scale* (5000) or some other point specified by the manufacturer.

2. *Briefly* tap $S1$ and note whether the pointer on meter $M1$ deflects positive or negative. If
   a. $E_0 > E_r$, then $M1$ swings positive.
   b. $E_0 < E_r$, then $M1$ swings negative.
3. Adjust voltage source $E$ in *small steps* according to Step 2(a) or 2(b) until $E_0 = E_r$. If $M1$ deflects positive, then $E$ must be decreased, but if $M1$ deflects in the negative direction, $E$ must be increased.
4. Repeat Step 3, using very small incremental changes in $E$ until $M1$ reads nearly zero; then close $S2$ (i.e., set $S2$ to *high*) and continue repeating Step 3 until no further improvement is possible.
5. The output of potentiometer $E_0$ is now equal to $E_r$. Set $S3$ to *measure* and $S2$ to *low*.

*Use*

1. Perform the calibration procedure above.
2. Using the same tapping method, with the unknown voltage $E_x$ connected, adjust potentiometer $R1$ in small steps until $M1$ reads nearly zero.
3. Close $S2$, and repeat Step 2 until no further improvement is possible. Voltage $E_x$ is calculated from

$$E_x = \frac{E_r(R1 \ \text{dial reading})}{9999} \tag{5–2}$$

■ **Example 5–2**

Find the unknown potential $E_x$ if the reference voltage is 1.0186 V and the micrometer dial reading is 6991.

*Solution*

$$\begin{aligned}
E_x &= E_r(6991)/9999 \\
&= (1.0186 \ \text{V})(6991)/(9999) \\
&= 7.1217 \times 10^{-1} \ \text{V} = \mathbf{0.71217 \ V}
\end{aligned}$$

5–7        **EQUAL DEFLECTION METHODS**

An equal deflection method for measuring voltage is shown in Fig. 5–4(a). The potentiometer is the circuit of Fig. 5–2, and the calibration procedure must be followed before a measurement can be made.

Voltmeter $M1$ must have a very high impedance and must have a full-scale range of approximately, but greater than, $E_0$, and preferably several ranges with $E_0$ being the highest.

Switch $S1$ connects the voltmeter to the unknown voltage *first*. The *range* of the voltmeter is adjusted so that the meter pointer deflects to some specific mark on the scale. Adjust $R$ so that the pointer lands on an easily repeatable point on the upper half of the meter scale.

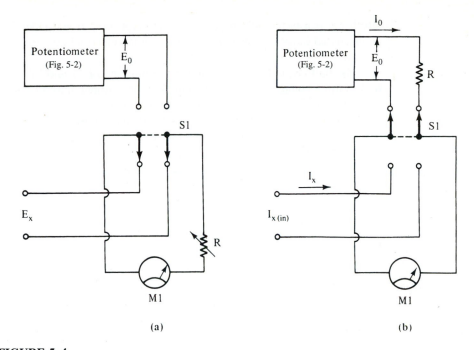

**FIGURE 5–4**

(a) Equal deflection measurement of voltage, (b) equal deflection measurement of current.

Next, the switch (e.g., $S1$) is set to connect the potentiometer to the voltmeter. Adjust the output $E_0$ of the potentiometer ($R1$ in Fig. 5–2) until the pointer of the meter deflects to the *same* mark as $E_x$. Calculate the unknown voltage from the dial setting of the potentiometer.

A very similar technique is used with the circuit in Fig. 5–4(b) to measure currents. Adjust $E_0$ so that $I_x$ and $I_0$ produce the same deflection on meter $M1$. In that case,

$$I_x = I_0$$

and

$$I_0 = \frac{E_0}{R}$$

[$R$ is a precision resistor, and $E_0$ is calculated from Equation (5–2).]

5–8     **AUTOMATED (DIGITAL) COMPARISON CIRCUITS**

There are two types of automated comparison potentiometers: (a) **servomechanisms** or **mechanical** and (b) **electronic.** The servomechanism type is electromechanical and usually has a pen to record the potential's waveform on graph paper. These machines will be discussed in detail in Chapter 10, so only the electronic type will be discussed in this section.

The null condition is indicated by a **comparator circuit** [Fig. 5–5(a)]. In its most basic form a comparator is a very high gain operational amplifier with no negative feedback. The open-loop gain of most operational amplifiers is at least 100,000, so theoretically a differential potential $(E_r - E_x)$ of 10 µV will saturate the amplifier output. In actual practice, however, the potential required to saturate the comparator will be between 10 µV and 10 mV, depending upon the model selected.

**FIGURE 5–5**
(a) Voltage comparator, (b) simplified automatic comparison circuit.

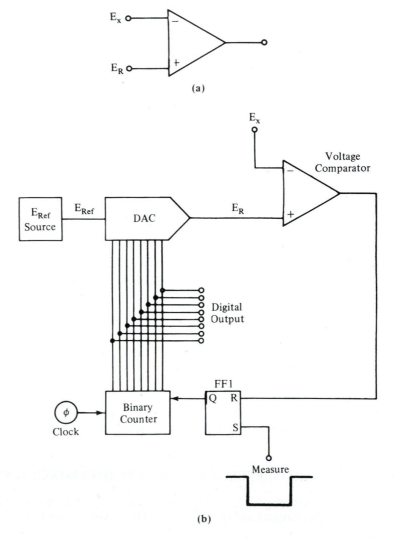

The unknown potential $E_x$ is applied to one input terminal of the comparator, while a step-variable reference voltage $E_r$ is applied to the other input.

Potential $E_r$ is created by a precision dc voltage source and a **digital-to-analog converter (DAC)**, as shown in Fig. 5–5(b). The DAC is a special circuit that produces an output voltage that is the *product* of reference voltage $E_{ref}$ and a fractional binary number applied to the respective digital inputs. The transfer equation for a DAC with an *n*-bit digital input word is

$$E_r = E_{ref} \times \frac{(\text{binary word})}{2^n} \qquad\qquad (5\text{–}3)$$

### ■ Example 5–3

An 8-bit DAC is connected to a 1.000-V reference source. Calculate the full-scale value of $E_r$. (*Hint:* $E_r$ is at full scale when the digital input word is 11111111.)

### *Solution*
$(11111111_2 = 255_{10})$

$$
\begin{aligned}
E_r &= E_{ref} \times (255/256) \qquad\qquad (5\text{–}3)\\
&= (1.00) \times (255/256)\\
&= \mathbf{0.996\ V}
\end{aligned}
$$

Each count increments the DAC output by 3.9 mV, so the closest resolution of an 8-bit DAC connected to a 1.00-V reference source is just under 4 mV. If $Q$ represents the value of each DAC output voltage increment, then there is an uncertainty of $\pm Q/2$, or in this case approximately 2 mV.

---

The operation of the electronic potentiometer is as follows:

1. The unknown voltage $E_x$ is applied to the comparator.
2. The *measure* command causes the control logic section to generate a *reset* pulse to clear the binary counter to 00000000, followed by a *measure start* pulse [Fig. 5–5(c)].
3. The *measure start* pulse sets the $Q$ output of R-S flip-flop FF1 HIGH, which turns on the counter.
4. As the counter increments once for each clock pulse, the DAC output $E_r$ begins to rise.
5. The output of the comparator remains HIGH as long as $E_x$ is greater than $E_r$, but it drops to LOW when $E_x$ equals $E_r$. When this occurs, FF1 is reset, so its $Q$ output goes LOW, thereby inhibiting the counter.
6. The value of $E_x$ is determined by substituting $E_x$ for $E_r$ in Equation (5–3).

The electronic potentiometer is used in digital instruments, A/D converters, and computer-controlled instrumentation processes and control systems. It is also used to auto-zero some analog circuits.

**FIGURE 5–5,** *continued*
(c) Timing scheme for (b).

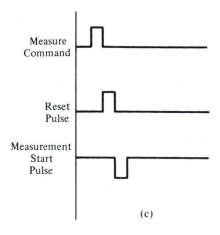

Measure
Command

Reset
Pulse

Measurement
Start
Pulse

(c)

## SUMMARY

1. A **precision potentiometer** is used to create a precision voltage source.
2. The potentiometer may be used in either the **null** mode or **equal deflection** mode.
3. A **digital-to-analog converter (DAC)** can be used to make an automated potentiometer in comparison measurements.

## RECAPITULATION

Now go back and try to answer the questions at the beginning of the chapter. When you are finished, answer the questions and work the problems below. Place a mark beside each problem or question that you cannot answer, and then go back to the text and reread the appropriate sections.

## QUESTIONS

1. Describe the construction of a slide-wire potentiometer.
2. Discuss why a slide-wire potentiometer produces a stable output voltage even when the operating temperature changes.
3. Describe in your own words how to measure an unknown potential $E_x$ by using a Weston cell and a slide-wire potentiometer.
4. Describe a Weston cell; use a diagram if you need it.
5. Describe the *long-term* accuracy of an *unsaturated* Weston cell.
6. What temperature environment is required to maintain the accuracy of the output voltage calibration of a fresh Weston cell?
7. Describe the operation of an automated electronic potentiometer. What circuit elements are required?
8. Describe the principal differences between the *null* and the *equal deflection* measurement techniques.

## PROBLEMS

**1.** A potentiometer is set at the ¾ point such that $R_b = 0.75R$ [Fig. 5–1(b)]. Find $E_0$ if $E = 12.6$ V dc.

**2.** A precision potentiometer is mechanically linked to a micrometer dial that reads from 00000 to 99999. Find $E_0$ if $E = 9$ V dc and the dial is set to 63247.

**3.** A 10-bit DAC is connected to a 2.560-V dc reference. Find the output voltage if the digital word at the input is $1011111111_2$ (i.e., $767_{10}$).

# 6

# The Basics of Digital Instruments

6–1 **OBJECTIVES**

1. To learn the operation of digital counting circuits.
2. To learn the stages and circuits needed to produce totalizer, frequency, and period counters.
3. To learn what factors in counter design affect accuracy.
4. To learn how to deal with signal-related problems that can cause count errors.

6–2 **SELF-EVALUATION**

Before studying the material in this chapter, try to answer the questions given below. These questions test your knowledge of the subject matter. If you cannot answer a particular question, then look for the answer as you read the text.

1. How can a 4-bit (i.e., modulus 16) binary counter be made into a decimal (i.e., modulus 10) counter?
2. What is the function of a **latch**?
3. How can amplitude modulation of a signal affect counting?
4. Some signals will not count because they do not cross _____ of the hysteresis window.
5. Name three factors that affect the accuracy of a counter time base.

6–3 **WHAT IS "DIGITAL"?**

Most electronic instruments of recent design and manufacture are said to be "digital." But what *is* a digital instrument? What can a digital instrument do that an analog instrument cannot? Does being "digital" confer any advantage to the user?

Some people consider an instrument to be digital if the readout devices are digital displays. Such a readout will present the output in the form of lighted digits, such as those found on a calculator or a digital wristwatch. But many such instruments use ordinary

**121**

analog electronic circuits, and they then use an ordinary voltmeter or current meter as the readout device. In medical electronics, for example, a transducer is used to convert blood pressure to an analogous dc voltage that is proportional to the pressure. A dc voltmeter, with the dial scale calibrated in units of pressure (i.e., mmHg or torr), is connected to the output of the pressure amplifier to serve as the readout.

Some manufacturers replace the analog voltmeter with a digital voltmeter (Chapter 7), but the remainder of the instrument remains unchanged. Is such an instrument "digital"? Not really, because only the readout device is a digital circuit. Such an instrument is usually not made any more accurate because of the digital readout, but it may give an indication that is less ambiguous and therefore easier to interpret. In that respect, the digital output meter may prove advantageous, if not more accurate.

Such digital instruments can, however, prove misleading if too many digits are used in the display. For example, if an instrument produces a dc output between 0 and 2 V, with a resolution of 10 mV, then a digital voltmeter capable of measuring voltages of a lower level may tend to give the user more confidence in the end measurement than is justified. One popular DVM is capable of measuring voltages from 0 to 1.9999 V, in 100-$\mu$V steps. Using this meter in an application where other factors would ordinarily indicate the use of a 0-to-1.99-V instrument results in the user's gaining a false sense of precision—a fact that is, incidentally, not lost on some less scrupulous manufacturers. Not that they would lie, mind you, but some are guilty of creative specification writing.

A truly digital instrument uses binary (i.e., base 2) logic elements such as gates, inverters, flip-flops, and so on, to perform the measurement function. These circuits operate with only two voltage levels (see Fig. 6–1), and these represent the binary digits 0 and 1. The medical monitor discussed above would be digital if the transducer passed the signal directly to an **analog-to-digital (A/D) converter** with but minimal intermediate amplification or processing. All further signal-processing functions would be performed in binary logic circuit or software, and the instrument would truly be "digital."

Digital electronics has caused revolutionary changes in electronic instrumentation and measurement devices. Whether a function is implemented in digital logic circuits (i.e., hardwired) or in software, the digital revolution has had a profound and positive effect on instrumentation. There are some cautions, however.

First, do not assume that simply because an instrument is digital, it's automatically better than an analog implementation. The digital solution usually has the potential for superior performance over analog for most instrumentation problems. But issues of accuracy, precision, and functionalism are affected by the instrument as a system. A badly designed digital instrument will perform less well than a properly designed analog instrument.

**FIGURE 6–1**
Digital instruments recognize only two voltage levels: (a) positive logic, (b) negative logic.

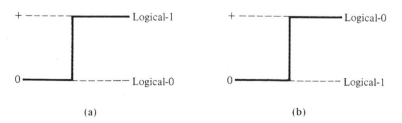

Second, there may be a cost-benefit reason for using an analog implementation over a digital implementation of a circuit function. For example, a bandpass filter circuit can be implemented with a few low-cost capacitors, inductors, and resistors. If there are only a few other functions to perform, then it may well be less cost effective to implement a microprocessor solution instead of an analog solution.

Another example of cost effectiveness is the electrocardiograph (ECG) amplifier used in medical electronics. A proper ECG system will be computerized, of that there is no doubt. The ECG signal consists of a 1-mV analog signal riding on a 1500-mV dc offset pedestal, of which only the 1-mV portion is of clinical interest. There are two implementations of the "front end" of the system. One uses an ac-coupled instrumentation amplifier, while the other uses a 24-bit A/D converter. The ac-coupled amplifier uses the capacitors in the amplifier input to strip the 1-mV signal off the 1500-mV pedestal. A 10-bit A/D converter is then used to produce the binary input to the computer. In the A/D converter implementation, the signal is directly converted to binary by a 24-bit A/D converter, but only the 10 least significant bits are input to the computer. Until 24-bit A/D converters cost less than a 10-bit converter and instrumentation amplifier, many designers will choose the analog implementation.

Electro-optical sensors have a similar problem in that they produce an output current of microamperes riding on a dc pedestal of milliamperes.

Third, there are situations where the digital implementation is either not possible or not practical. For example, a 50-$\Omega$ radio frequency (RF) filter can be made of inductors ($L$) and capacitors ($C$). Suppose we need a 30-MHz low-pass filter with a very high dynamic range. The limiting factors in the LC filter are the thermal noise at the low end and the current at which the inductor self-heats enough to change value or saturate its core. A typical small-wire, air-core inductor will operate at currents up to 500 mA (0.50 A), and the thermal noise level for a 50-$\Omega$ system is on the order of $7 \times 10^{-7}$ A for a 30-MHz bandwidth. By 20 log ($0.50/7 \times 10^{-7}$), the LC coil has a dynamic range of 117 dB. To get that performance digitally, you would need a converter capable of making a 20-bit conversion at a rate fast enough to accommodate a 30-MHz signal.

On the benefits side, however, analog solutions tend to grow out of proportion as the complexity of a function, or a collection of functions, increases. While software digital solutions are also very complex, the hardware implications are minimal. You could add RAM or ROM or use a hard drive to store a larger, more complex program that would be impossibly difficult to implement in analog hardware.

Another distinct advantage of digital implementation, especially where the complexity is in embedded software, is that the instrument is more easily modified. Although software is often difficult and expensive to modify or produce anew, once the software has been finalized, it can easily be retrofitted into existing hardware by reading in a new disk or tape or by changing a ROM chip.

Digital implementation also eases the chore of integrating several instruments together in a system. Data can be transferred back and forth between instruments in a coherent manner much more easily than can analog signals, especially over long distances. For larger instrumentation systems, the use of the general purpose interface bus (IEEE-488 GPIB) makes the chore of integrating instruments easy.

**6–4**      ## BINARY COUNTERS

This text assumes that you are familiar with basic digital logic circuits and the various digital integrated circuits that are available. Be sure that you understand binary arithmetic, inverters, the various types of gates, and the different types of flip-flops used in digital electronics.

The **binary counter** forms the heart of many "discrete logic" digital implementations, so we will use it here to illustrate the basics of digital instrumentation. A binary counter consists of a chain of clocked flip-flops [Fig. 6–2(a)]; each stage is designed to count by 2. A **decoder** is used to indicate the state of the count by examining the unique $Q$ and not-$Q$ states at the output of each stage.

A 4-bit binary counter will count from $0000_2$ to $1111_2$ (i.e., $0_{10}$ to $15_{10}$), while a **binary coded decimal (BCD)** counter counts only from $0000_2$ to $1001_2$ ($0_{10}$ to $9_{10}$). Table 6–1 shows both binary and BCD coding, the latter of which is a subset of 4-bit binary coding.

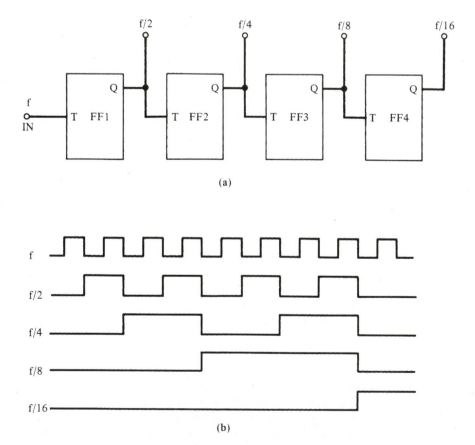

(a)

(b)

**FIGURE 6–2**
(a) Cascading flip-flops create a counter, (b) logic states at outputs in (a).

**TABLE 6–1**
Binary and BCD codes

| Decimal | Binary | BCD | | |
|---------|--------|-----|------|-----|
| 0 | 0000 | | 0000 | |
| 1 | 0001 | | 0001 | |
| 2 | 0010 | | 0010 | |
| 3 | 0011 | | 0011 | |
| 4 | 0100 | | 0100 | |
| 5 | 0101 | | 0101 | |
| 6 | 0110 | | 0110 | |
| 7 | 0111 | | 0111 | |
| 8 | 1000 | | 1000 | |
| 9 | 1001 | | 1001 | |
| 10 | 1010 | 0001 | 0000 | |
| 11 | 1011 | 0001 | 0001 | |
| 12 | 1100 | 0001 | 0010 | |
| 13 | 1101 | 0001 | 0011 | Two 4-bit |
| 14 | 1110 | 0001 | 0100 | counters |
| 15 | 1111 | 0001 | 0101 | required |

**6–5**

## DECIMAL COUNTING UNITS (DCUs)

The heart of any electronic counter is a circuit called the **decimal counting unit,** or DCU. All DCUs, whether discrete or integrated, consist of a **decade counter,** and a **decoder/display-driver,** and most also include a **data latch.**

An example of a DCU is shown in block form in Fig. 6–3. This circuit has been widely used and is constructed of TTL integrated circuit logic elements. The decade counter uses a chain of flip-flops not unlike those in Fig. 6–2(a), but it also contains some internal gates that sense the tenth state (i.e., "9") and reset the counter to 0000 if another input pulse is received.

Very few designers now use individual flip-flops to form decade counter circuits, because there are several complete IC decade counters available in both TTL and CMOS lines. In fact, there are several MSI and LSI integrated circuits available that contain *all* circuitry needed to form a complete electronic counter of several decades.

The four output lines from the decade counter in Fig. 6–3 are coded in the 8421 BCD format, and they change state with every pulse. There are also two other lines: **input** and **reset.** The input line accepts the pulses being counted, while the reset line will reset the counter to 0000 when brought HIGH. For normal counting the line is kept LOW, and to clear the counter it is brought momentarily HIGH.

The decoder is used to convert the 4-bit BCD code to the code needed to correctly drive the readout device. The counter outputs can be used to drive the decoder directly, but this results in a difficult-to-read "rolling" display as the number changes. The rolling effect is created because the display will change with the count. For example, if the count is to be "8," then the display will read "0-1-2-3-4-5-6-7-8" in sequence as the count accumulates to the "8" state.

The rolling effect can be eliminated by using a quad latch circuit. A typical latch is a bank of four cascade type-D flip-flops connected so that their **clock** terminals are tied

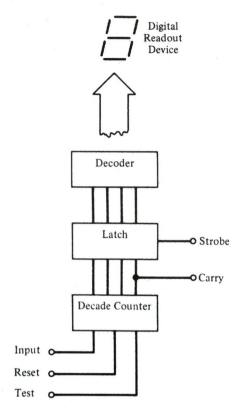

**FIGURE 6–3**
Decade or decimal counter.

together to become a **strobe** line. If the strobe line is high, then data appearing on the input lines are transferred to the output lines. But if the strobe line is low, then the output lines retain the last data present at the input when the strobe was high.

In normal practice the strobe line is held low while the count is being accumulated, and at the end of the count period the strobe line is brought high momentarily, just long enough to transfer the count data from the input lines to the output lines. These data will then be held on the output lines until another strobe pulse updates the output to reflect the result of the next count.

## 6–6        DISPLAY DEVICES

Early electronic counters used a column of ten incandescent lamps for each digit. The decoder was a circuit that examined the BCD data lines and determined which decimal digit was being represented. The decoder would then ground one of ten lines to light up the lamp that corresponded to that decimal digit. In this respect, the decoder is said to behave like a single-pole, ten-position switch.

The ten-lamp column display was used because it was very low in cost compared with the other alternatives then available. One of those alternatives was the Burroughs

Nixie® tube display, shown in Fig. 6–4. The Nixie® tube is still occasionally seen on older or surplus electronic instruments, but it has been considered obsolete for new designs for many years. It is presented here for historical interest only.

The Nixie® tube is a neon gas discharge device similar to a glow lamp such as the NE-2 or NE-51. The anode is a positive electrode common to all of the digits. But there are ten separate cathodes—one for each digit. The cathodes take the form of thin wire elements that are formed into the shape of digits 0 through 9.

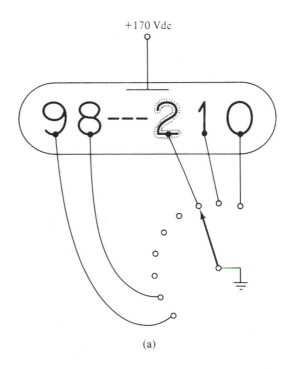

(a)

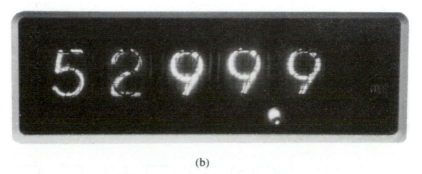

(b)

**FIGURE 6–4**
(a) Nixie® tube operation, (b) Nixie® tubes.

When one of the digits is grounded, the neon gas particles close to its surface are ionized and give off light in the shape of the digit represented. All ten cathodes are arranged to that they face the same viewing plane.

The Nixie® tube requires the same type of BCD-to-1-of-10 decoder as did the lamp column, but it must be able to withstand the high voltages (e.g., 170 V dc) required to ionize the tube's gas.

The most common digital display in modern equipment is the seven-segment readout shown in Fig. 6–5(a). This type of display consists of seven illuminated bars arranged in a figure-eight pattern. It is possible to form all ten decimal digits by lighting appropriate bars [Fig. 6–5(b)]. A BCD-to-seven-segment decoder will ground those terminals *a* through *g* required to form the decimal digit represented in BCD form at the decoder input.

Probably the most popular types as of this writing are the light-emitting diode (LED) and liquid crystal diode (LCD) displays. In the LED type of display the segments are formed of (usually) red LEDs. Other forms of seven-segment readouts are used, however, including *fluorescent* (greenish-blue) and gas discharge (yellow-orange). The liquid crystal display is slow to form and is a dark gray or light gray in color. But it has certain advantages, including the ability to create any alphabetic, numeric, or graphic device on the display along with the seven-segment digital readout. Also (and this is a big advantage to designers of low-current-drain portable equipment), the liquid crystal display draws current only when the crystal is changing state. Once the number is formed, the device draws a leakage current of only a few microamperes. As a result, hearing aid batteries can be used to power CMOS calculators for 1500 hours.

One last type of display is the $5 \times 7$ dot matrix, shown in Fig. 6–6. This type of display is commonly used on athletic scoreboards and bank time-temperature billboards. It consists of 35 light sources (usually LEDs) arranged in five vertical columns, each containing seven light sources. This $5 \times 7$ matrix of light sources can be illuminated to form all numerical and alphabetic characters, and they look more natural than do seven-segment displays. The dot matrix method is also used in cathode ray tube (CRT) computer TV displays.

**FIGURE 6–5**
Seven-segment decimal readout.

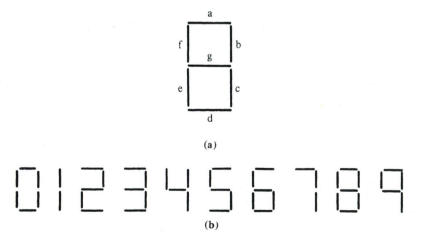

(a)

(b)

**FIGURE 6–6**
5 × 7 matrix.

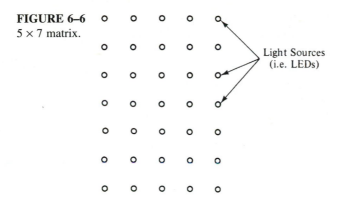

Light Sources
(i.e. LEDs)

In the latter application the points of light making up the matrix are unblanked points on a raster-scanned, but blanked, CRT screen. Each horizontal row of points of light represents one scan of the TV screen by the electron gun inside the CRT.

**6–7**          **DECIMAL COUNTING ASSEMBLIES (DCAs)**

The individual DCU can count only from 0 to 9, but if two DCUs are used in cascade, the assembly can count from 0 to 99, and three DCUs can count from 0 to 999 (etc.).

Figure 6–7 shows a basic **decimal counting assembly (DCA)** consisting of three DCU elements. Three DCUs allow counting to 999, and one more DCU must be added to the assembly in cascade for each additional order of magnitude (i.e., for 0 to 9999 use four DCUs; for 0 to 99,999 use five DCUs, etc.).

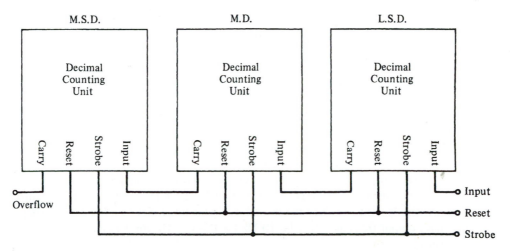

**FIGURE 6–7**
Decimal counting assembly.

The *strobe* and *reset* lines from all DCUs are tied to *common* strobe and reset lines that serve the entire DCA. All three DCUs in the assembly will be affected simultaneously by pulses applied to these lines.

The *carry* output of a DCU is the "D" data output line in any given DCU (i.e., the line weighted "8"), and it is used to drive the input of the next DCU in the cascade chain. Recall from your digital electronics studies that the J-K flip-flops used in decade counters change state only on a negative-going transition of the clock terminal. The BCD line that is weighted "8" is HIGH for both "8" and "9" states, but it drops LOW when the decade counter overflows on the "10" count. It therefore toggles the input flip-flop of the next DCU *only* following the tenth count. This same tenth count also returns the counter to the "0" state.

The original signal to be counted, then, is applied to the input of the DCU that is in the least significant digit position in the DCA (LSD in Fig. 6–7). The *carry* output of the LSD DCU is applied to the input of the next-digit DCU, and finally the output of that DCU is applied to the input of the most significant stage (i.e., MSD). The output of the MSD stage can then serve as a **counter overflow indicator.**

A counter constructed of only a basic DCA is called a **totalizer;** it accumulates input pulses as long as it is turned on and not reset. The DCA will simply totalize counts either until it is turned off or until a reset pulse is applied. If the DCA overflows, the count starts back at zero but continues to increment as pulses are received at the input. Of course, when using a totalizer counter, you must be sure that the number of **events** (i.e., input pulses) expected is less than the maximum number that the counter will accept without overflowing.

**6–8**     **FREQUENCY COUNTERS**

Figure 6–8(a) shows the basic block diagram for a **frequency** counter. The sections include the **DCA, main gate, trigger, input amplifier, main gate flip-flop, time base,** and **display clock.**

The DCA is a totalizer counter as shown in Fig. 6–8(a). The overflow stage is a flip-flop that is SET when the MSD carry output goes high. The overflow flip-flop turns on a lamp to make the operator aware of the overflow condition so that the data can be disregarded.

A frequency counter measures **events per unit of time (EPUT)** (i.e., cycles per second), so the DCA must be turned on only for a given period of time (e.g., 0.1, 1, or 10 s). The main gate, main gate flip-flop, and time base sections are used to allow input pulses into the DCA for the designated period of time.

The time base section consists of a crystal oscillator that produces pulses at a precise rate such as 100 kHz, 1 mHz, 4 mHz, 10 mHz, and so on. A chain of decade dividers is used to reduce the crystal oscillator frequency to a lower frequency. The time base output frequency will be 10 Hz for 0.1-s, 1 Hz for 1-s, and 0.1 Hz for 10-s measuring periods.

The timing diagram for one complete interval of an EPUT counter is shown in Fig. 6–8(b). Pulses $t1$, $t2$, and $t3$ are output from the time base section. When pulse $t1$ goes

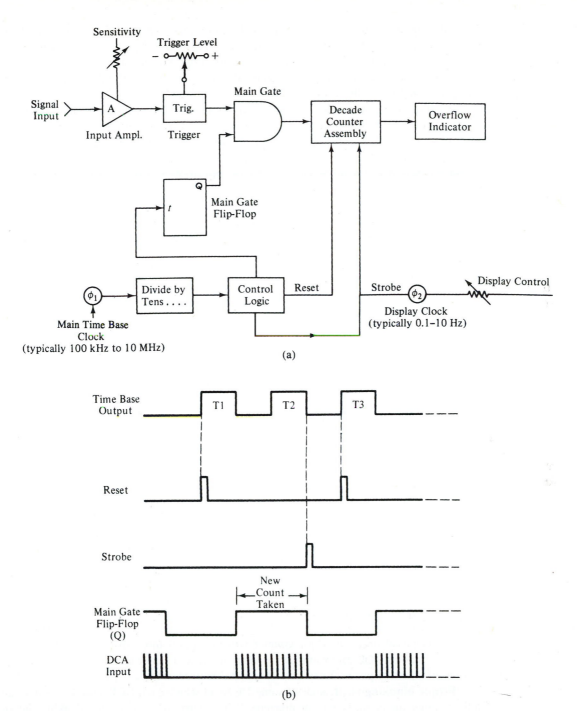

**FIGURE 6–8**

(a) Block diagram of a frequency counter, (b) timing diagram for a counter.

high, the control logic section generates a short pulse to reset the DCA to zero. When $t1$ goes low again, the $Q$ output of the J-K main gate flip-flop will go high. The main (AND) gate has one input tied to the $Q$ output of the flip-flop, and the other input is tied to the signal being counted. As a result, the main gate passes input pulses to the DCA *only* when the $Q$ terminal of the flip-flop is high.

The flip-flop remains set until the negative transition of $t2$ occurs. At that time the $Q$ output of the flip-flop drops low, turning off the flow of pulses into the DCA and causing the control logic section to generate a strobe pulse. This pulse tells the DCU latches to transfer data from the counter to the decoders. The display, then, shows only completed count cycles and will hold the previous count until the end of the next interval. The trigger and input amplifier circuits will be discussed in Section 6.11.

The frequency counter of Fig. 6–8(a) counts **frequency** (i.e., events per unit of time) because the DCA is enabled only for a specific unit of time. The frequency of the input signal is the number of counts accumulated on the DCA divided by the time base period in seconds:

$$f_{(Hz)} = \frac{\text{counts on DCA}}{\text{time (s)}} \qquad (6\text{--}1)$$

■ **Example 6–1**

What is the frequency of a signal in hertz if a five-digit DCA reads 08211 and the time base is set to 1 ms?

*Solution*

$$f_{(Hz)} = \text{counts/time}$$

$$= \frac{8211 \text{ counts}}{0.001 \text{ s}} = \mathbf{8,211,000 \text{ Hz}} \qquad (6\text{--}1)$$

If the time base is 1 s, the DCA reading will be the frequency in hertz, and if the time base is a decade multiple or submultiple of 1 Hz, then the frequency in hertz can be determined by adjusting the decimal point.

## 6–9    COUNTER DISPLAYS

The display used on a frequency counter will be one of those discussed earlier in the chapter, but additional features might be added to make it easier for the operator to use or to conserve battery power in the case of a portable instrument.

**Ripple blanking** is a feature of some DCAs in which each DCU issues a signal that tells whether the count is zero or nonzero. If it is zero, then this ripple blanking output tells the ripple blanking input of the next least significant DCU. In this way the counter is able to blank out (i.e., turn off) nonsignificant zeros to the left of the most significant number. In Example 6–1, a five-stage DCA will read 08211. If ripple blanking were used, the display would read 8211; the leading zero would be suppressed. This makes the display less confusing and saves battery power in portable models.

Another tactic that saves both power and wiring effort is **display multiplexing.** This is a technique common in low-cost calculators and in much portable equipment. All of the seven-segment lines (*a* through *g*) from all of the displays are tied together to form a seven-line **bus.** A timing control circuit will output the seven-segment code for only one digit at a time, along with a pulse that turns on only that digit. Each digit receives its code sequentially, so only one display is turned on at a time. But if the switching occurs rapidly enough, the operator's eye persistence causes the display to *appear* constant.

**6–10**       ## PERIOD COUNTERS

**Period** is defined as the time elapsing between identical features on successive cycles of a waveform. Period can be calculated from the frequency and is the reciprocal of frequency:

$$P_{(s)} = \frac{1}{f_{(Hz)}} \tag{6–2}$$

A period counter can be made by reversing the roles of the input stage and time base; the input amplifier is connected to the main gate flip-flop, and the time base is connected to the main gate [see Fig. 6–9(a)].

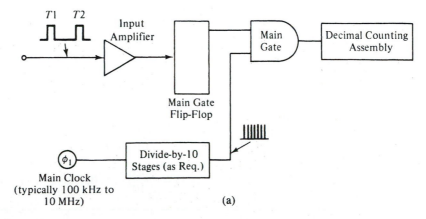

**FIGURE 6–9**

(a) Block diagram of a period counter,

**FIGURE 6–9,** *continued*
(b) timing diagram for period
counter.

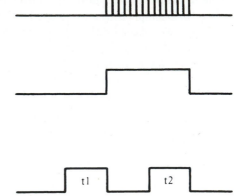

(b)

The timing diagram for a period counter is shown in Fig. 6–9(b). The flip-flop is set (i.e., $Q$ goes high) on the negative-going transition of $t1$, and this allows the gate to pass pulses from the time base to the DCA.

The time base frequency determines the time interval represented by each pulse. For example, a time base frequency of 1000 Hz yields a period counter resolution of 1/1000 s, or 1 ms. Similarly, one 10-kHz time base yields 0.1-ms resolution. The period is given by the count accumulated by the DCA and the time base frequency:

$$P_{(s)} = \frac{\text{DCA count}}{f_{(Hz)}} \tag{6–3}$$

■ **Example 6–2**

Find the period in seconds of a signal if the DCA count is 8026 and the time base frequency is 10 kHz.

*Solution*

$$P_{(s)} = \text{DCA count}/f_{(Hz)} \tag{6–4}$$

$$= \frac{8026 \text{ events}}{(10^4 \text{ events/s})}$$

$$= \frac{8026 \text{ s}}{10^4} = \textbf{0.8026 s}$$

The **resolution** is the smallest time interval that can be measured on the counter and is defined as the reciprocal of the time base frequency, $(1/f)$.

## 6–11     TRIGGER CIRCUITS

The input signal will very rarely be the nice, clean, square waves required for proper operation of the digital logic circuit elements used to make a counter. The signals may also be too low in amplitude to operate the digital logic circuits, or they may be too noisy. Remember, a TTL flip-flop needs to see fast rise and fall times (i.e., good square waves) and amplitudes greater than 1.4 V, or it may not operate properly.

The input signal, then, is passed through two processing stages: an **amplifier** and a **trigger.** The amplifier is a wide-band voltage amplifier with enough gain to build up the minimum allowable signal (usually 25 to 100 mV) to a level great enough to drive the trigger stage (i.e., 500–1000 mV).

The trigger stage is a Schmitt trigger circuit with a built-in hysteresis. This type of circuit is used to clean up irregularly shaped signals by making them into square waves. Figure 6–10 shows the normal operation of a trigger circuit. The output snaps high when the input signal crosses the lower hysteresis limit, and it remains high until the signal crosses the lower limit in a negative-going direction. The hysteresis window is the quantity $E_u - E_L$. Note that the trigger output possesses the shape and amplitude required by the digital circuits that it drives.

**FIGURE 6–10**
Normal operation of the trigger circuit.

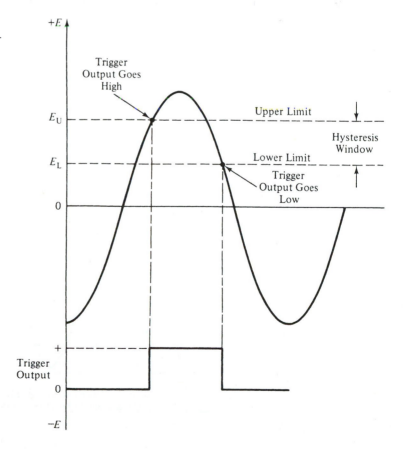

It is a fundamental rule that input signals must cross *both* hysteresis limits, or no count will be entered by the DCA. Figure 6–11(a) shows the required situation; the input sine wave crosses *both* limits. But in Fig. 6–11(b) the sine wave crosses only one of the window limits, so no count is registered on the DCA. Figure 6–12 shows some of the many types of signals that will *not* count when the hysteresis window is positioned symmetrically about zero volts.

Some counters have a **trigger level** control that allows the user to adjust the position of the window over a wide range. Other models use a three-position switch labeled "+," "preset," and "−." The switch allows the window to be placed in any of three locations (see Fig. 6–13). A continuously variable trigger level control allows positioning of the window *anywhere* within the range. Note that neither the continuously variable control nor the three-position-switch type of control varies the width of the window (i.e., $E_u - E_L$), but they do vary the position of the window.

Some counters are equipped with a **trigger amplitude** control, which *does* allow the operator to vary the width of the hysteresis window.

## 6–12    COUNTER ERRORS

Several factors tend to reduce the accuracy of an electronic counter, and these can be grouped as **inherent errors** and **signal-related errors.** The inherent errors are a function

**FIGURE 6–11**
(a) Signal crosses both limits, so count occurs. (b) Only one limit is crossed by signal, so no count will occur.

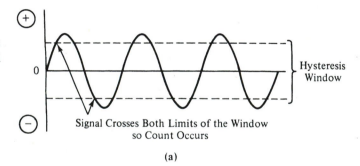

(a)

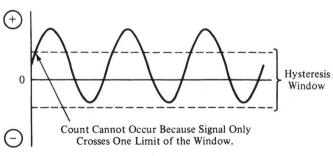

(b)

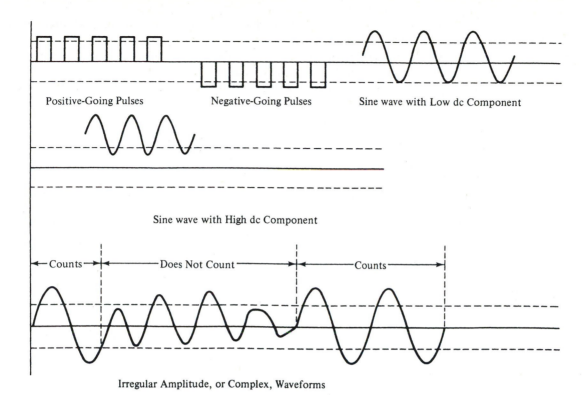

Positive-Going Pulses     Negative-Going Pulses     Sine wave with Low dc Component

Sine wave with High dc Component

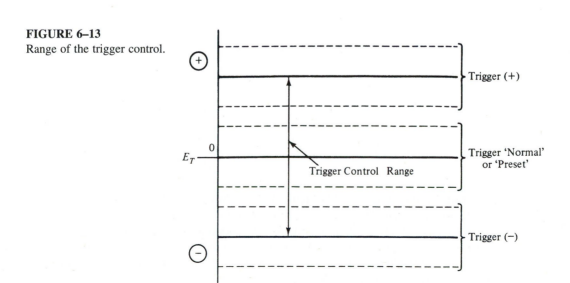

←Counts→ | ←────Does Not Count────→ | ←────Counts────→

Irregular Amplitude, or Complex, Waveforms

**FIGURE 6–12**
Various signals that will not count.

**FIGURE 6–13**
Range of the trigger control.

137

of the quality, age, and history of the individual counter. Little can be done about these errors unless their source is a serious need for recalibration of the time base. Signal-related errors, on the other hand, are often correctable by proper manipulation of sensitivity, trigger level, and trigger amplitude controls.

**6–13     INHERENT ERRORS**

There are two sources of inherent error in all frequency and period counters: time base error and a ±1 count ambiguity.

The time base error is expressed in terms of a percentage, or in parts per million. The error from time base inaccuracies is directly reflected in all measurements of frequency or period. For example, suppose a 1-mHz time base is off by 30 Hz; that is, it is actually 1,000,030 Hz instead of 1,000,000 Hz. This is an error of 30 parts per million (30 ppm), which in percent is

$$\frac{1,000,030 - 1,000,000}{1,000,000} \times 100 = 0.003\%$$

The measurement error due to time base inaccuracy is constant regardless of the frequency being measured. That is, there will be a 0.003% error at 1 kHz and the same 0.003% error at the maximum frequency that the device will measure. For example, a 27-mHz signal would be measured with an error of

$$27 \text{ mHz} \times \frac{30 \text{ Hz}}{\text{mHz}} = 810 \text{ Hz}$$

This means that a counter reading 27,000,000 indicates that the actual frequency is 27 mHz ± 810 Hz. In other words, the actual frequency lies between

$$27,000,000 \text{ Hz} - 810 \text{ Hz} = 26,999,190 \text{ Hz}$$

and

$$27,000,000 \text{ Hz} + 810 \text{ Hz} = 27,000,810 \text{ Hz}$$

If the time base frequency is 30 Hz high, then the counter reading will be low, and if the time base frequency is low, then the counter reading will be high.

Total time base inaccuracy is the sum of several individual errors: **initial error, short-term stability, long-term stability, temperature change,** and **line voltage change.**

The initial error is the calibration error at the time the time base is initially adjusted at the factory or at recalibration in a metrology laboratory. Different methods are used to measure the time base frequency. In many cases the time base oscillator frequency is compared with standard frequency broadcasts of the *National Bureau of Standards* radio stations WWV, WWVB, or WWVH. Alternatively, it might be compared with the output of a cesium or rubidium beam atomic clock in a metrology/calibration laboratory. For high-accuracy time bases, the latter is preferred, although 60 kHz WWVB comparator receivers are very good standards.

The short-term stability is the time base oscillator frequency drift *per day.* Long-term stability is the frequency drift *per month* and is often designated the **aging rate.**

The temperature and line voltage stability specifications refer to the frequency change over the 0°C–50°C temperature range and to the ±10% line voltage change, respectively.

There are four different classes of counter time base: **ac line, room temperature crystal oscillator, temperature-compensated crystal oscillator (TCXO),** and **oven-controlled crystal oscillator.**

The use of the 60-Hz ac line as a counter time base is limited to the very cheapest models and to a few low-grade older units. Even low-cost units today have a crystal oscillator for the time base, and even though the crystal is operated at room temperature, it provides better accuracy than the 60-Hz power mains. Note that power companies will typically quote very high accuracy figures for their power plant's operating frequency, but these are frequency *averages* over a very long time. The short-term accuracy, which is what concerns counter users, is terrible.

The TCXO is an encapsulated oscillator that is specifically compensated against temperature changes; it provides at least an order of magnitude better stability than room temperature oscillators. It is less expensive now than in the past, so even moderately priced counters now offer TCXO stability.

The oven-controlled crystal oscillator places the crystal, and in some cases the rest of the oscillator circuitry, inside an oven or a thermal chamber. Thermostat ovens are considered an order of magnitude better than TCXO designs, while the **proportional control** type of oven is from one to two orders of magnitude better than TCXO.

Table 6–2 lists typical stability specifications for several models of counters by several different manufacturers. Note that the short-term stability is given only for the oven type of time base. The TCXO and crystal oscillator must often be operated for a full 24 hours before the stability reaches the specified level. At operating times less than 24 hours the stability is poorer. Some models use a separate regulated power supply for the TCXO that is not turned off by the main power switch. Rechargeable batteries are used in portable models for the same purpose, so the TCXO is not turned off while the counter is being transported between job sites.

**TABLE 6–2**
Typical Stability Specifications

| | XTAL | TCXO | Oven |
|---|---|---|---|
| Long-Term aging (per mo.) | $5 \times 10^{-7}$ | $2 \times 10^{-7}$ | $5 \times 10^{-10*}$ |
| Short-term aging (per day) | — | — | $10^{-10}$ |
| ±10% line voltage | $10^{-7}$ | $10^{-8}$ | $10^{-9}$ |
| Temp. 0°C–50°C (ambient) | $10^{-6}$ | $10^{-7}$ | $10^{-9}$ |

* After 24-hour warmup.

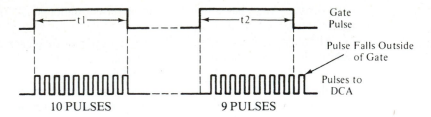

**FIGURE 6–14**
Mechanism for causing one-count ambiguity in counters.

The ±1 count ambiguity is caused by the lack of synchronization between the input signal and time base, as illustrated in Fig. 6–14. During period $t1$, ten pulses are gated into the DCA, while during $t2$ only nine pulses reach the DCA. On some subsequent count, it may be that nine pulses are gated into the DCA. One fundamental rule for all digital counter instruments is that there is an error of ±1 count of the least significant digit. In other words, a counter that reads, say, 10,000 Hz is measuring a frequency that lies between 9999 Hz and 10,001 Hz (i.e., 10 kHz ± 1 Hz).

The ±1 count ambiguity produces an error that is inversely proportional to the frequency being measured and the gate time:

$$\text{Error } (\%) = \frac{\pm 100}{fT} \qquad (6\text{–}5)$$

where $f$ = the frequency being measured in hertz (Hz)
$T$ = the time the gate is open in seconds (s)

■ **Example 6–3**

Find the percentage of error due to ±1 count ambiguity at (a) 2 mHz and (b) 27 mHz for a gate time of 1 s.

*Solution*
(a)                    Error $(\%) = \pm 100/fT$                                    **(6–5)**

$$= \pm 100/(2 \times 10^6 \text{ Hz})(1 \text{ s})$$

$$= \pm \mathbf{0.00005\%}$$

(b)                    Error $(\%) = \pm 100/fT$                                    **(6–5)**

$$= \pm 100/(2.7 \times 10^7 \text{ Hz})(1 \text{ s})$$

$$= \pm \mathbf{0.000004\%}$$

---

The error is ±1 count regardless of the frequency being measured, so the *percentage* of error decreases for higher frequencies. Compare (a) and (b) in Example 6–3.

**6–14**    ## SIGNAL-RELATED ERRORS

Poor signal quality can introduce errors that add to or subtract from the true count. Most of these errors result from hysteresis crossing errors or from noise on the signal.

Trigger errors occur because the input signal crosses the hysteresis window limits *too many* or *too few* times. We saw in Figs. 6–11 and 6–12 that a signal will fail to increment the DCA if it does not cross *both* limits of the hysteresis window, causing too low a count.

Figure 6–15(a) shows how severe **ringing** on a signal can create extra, spurious counts of the DCA if the trigger level control is adjusted so that the ringing portions of the signal cross the limits, creating two *additional* "input" pulses, a two-count error.

The cure is to adjust the trigger level controls so that the ringing portions of the waveform fall outside the window limits [Fig. 6–15(b)].

The same problem exists on sine wave input waveforms (Fig. 6–16) that have a large amount of **harmonic distortion.** The cure is the same, however: readjust the trigger level control so that it is operating over a lower portion of the waveform.

Similarly, impulse noise riding on the signal can have an amplitude sufficient to cross both limits of the hysteresis window. An example of this phenomenon is shown in Fig. 6–17(a) (p. 143), in which a pulse in a symmetrical wave train is carrying impulse noise artifacts. In the case shown, the two noise bursts cross the window limits and thereby force the trigger output to create three pulses instead of just one.

Once again the correction requires readjustment of the trigger level control to a point further down the waveform. In the case of a nonsquare wave, the noise may appear on the leading or trailing edges and still cause the problem. Figure 6–17(b) shows the proper and improper positions for the hysteresis window on such a waveform.

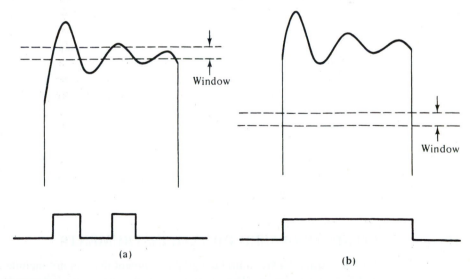

(a)                                          (b)

**FIGURE 6–15**
(a) Ringing causing false counts, (b) proper adjustment of trigger to prevent problem in (a).

**FIGURE 6–16**
2nd-harmonic distortion causing
false counts.

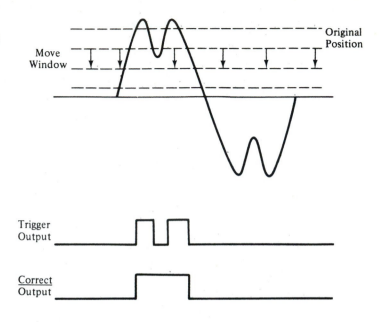

Note that filtering of the noise is usually not feasible because of the bandwidth requirements of the input amplifier.

Figure 6–18 shows a type of error, caused by noise, that is particularly troublesome on period measurements. In this example noise rides on a signal that has a shallow slope, thus creating a band of uncertainty around the signal. The trigger circuit should produce a high output when the signal crosses the upper limit, and should drop low again when the signal crosses the lower limit. But noise impulses adding to, or subtracting from, the signal amplitude could provide premature, or delayed, trigger transitions. The correct duration of the trigger output pulse in Fig. 6–18 is $t4 - t1$, but under worst-case conditions the actual duration may be as much as $t5 - t_0$, and that amount represents a considerable error.

The solution for this problem is to cause the signal to slew through the hysteresis band as *rapidly* as possible. Two methods can be used to implement this solution. One is to narrow the window by adjusting the trigger amplitude control, whereas the other is to increase the waveform's slope by preamplification.

On some types of signal waveforms it is sufficient to adjust the trigger level control so that the counter triggers on the *steepest* portion of the waveform. On sine waves, for example, this point occurs at zero crossings, but on other waveforms it may occur elsewhere on the signal.

**6–15**  **COMPUTER-BASED DIGITAL INSTRUMENTS**

It would be rare today to find a digital instrument that did not include a computer embedded somewhere inside. The range of possible computers is large, but they share some attributes.

**FIGURE 6–17**

(a) Oscillations on the waveform causing false counts, (b) correct and incorrect settings of trigger for signals with noise or oscillations.

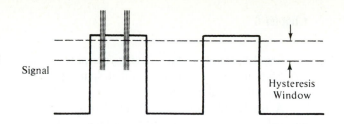

Signal

Hysteresis Window

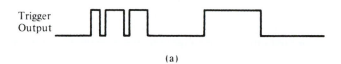

Trigger Output

(a)

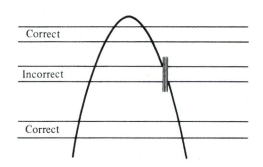

Correct

Incorrect

Correct

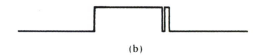

(b)

**FIGURE 6–18**

Counter errors due to hysteresis window and noisy signal.

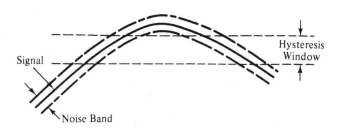

Signal

Hysteresis Window

Noise Band

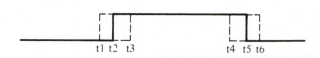

t1  t2  t3          t4  t5  t6

Small "controller chip" devices are programmable in a limited way, but they serve the function of computer quite well, despite a limited repertoire of commands and instructions that they will execute.

A single-chip computer might also be used. These LSI devices typically have an internal central processing unit, internal random-access memory (RAM), and internal read-only memory (ROM). They will also have at least one input/output (I/O) port, and some even come with features such as event counters and analog-to-digital (A/D) converters.

Still other instruments might rely on a single-board computer, that is, a plug-in printed circuit board that contains a computer chip, memory, and I/O as needed for the application.

A number of instruments, when opened and inspected, reveal a standard, off-the-shelf computer backplane (e.g., VMEBus or Futurebus+) motherboard with standard plug-in printed circuit cards (similar to a desktop personal computer) with functions such as CPU, memory (RAM and ROM), I/O, and any analog subsystem that might be needed. There may also be mass storage devices such as tape drives, hard disk drives, or floppy disk drives within the instrument.

Figure 6–19 shows the block diagram of a simple computer-based instrument. Although simple, it conceptually can represent many, if not most, computer-based instruments on the market. Although the block diagram is simple, the complexity of the instrument is found in the embedded software. And it is the software that makes the instrument so flexible and so powerful.

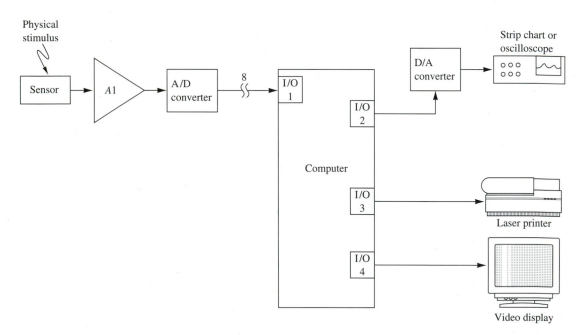

**FIGURE 6–19**
Block diagram of simple computer-based instrument.

Briefly described, this computer processes only a single input signal. This signal ("stimulus") affects a sensor of some kind, which produces an analog output voltage. (If the sensor uses digital output, then the next two stages are deleted.) The sensor signal may be further amplified in an analog subsystem amplifier ($A1$), may be filtered, or may receive some other signal processing prior to input to the computer. The output of the analog subsystem is an A/D converter. It accepts an analog input signal and then converts it to a binary word that represents the same quantity. For example, an 8-bit A/D converter will recognize 256 different states. If 00000000 is used to represent 0 V, then there are 255 states remaining to represent a voltage range. If the input range is 10 V, then each one-step change of state will represent 0.039 V, for a maximum input value of 9.96 V (represented by binary word 11111111).

The processing is done in the internal software. The output devices, also connected to various types of I/O ports, depend on the application. There might be a simple alphanumeric display such as discussed earlier, or there might be a computer video terminal. A laser printer can display not only alphanumeric data but also waveform and graphical data as well. The computer may have to provide analog waveform data to analog displays such as an oscilloscope or paper chart recorder. To accommodate these instruments, a digital-to-analog converter (DAC) is needed.

Computer-based instruments are not the be-all and end-all, for there are still plenty of applications for analog instruments and basic digital instruments. But design is now a lot easier, and implementation a lot cheaper, so the power and flexibility of computer-based instrumentation are finding a home in progressively lower-priced and simpler products. It is perhaps the sheer flexibility of computer-based approaches to instrument design that is of primary importance. A relatively low cost hardware infrastructure will support a huge array of different instruments. If an open systems standards approach is taken, then flexibility and universality are even more within reach.

## 6–16  EXAMPLES OF COMMERCIAL EQUIPMENT

The range of commercial digital instrument and measurement equipment is nearly endless, as a visit to any dealer or trade show will reveal. We will take a look at only a tiny segment of the overall population of possible devices in order to give you some idea of what's available.

Figure 6–20 shows a pair of plug-in digital modules. Both of these instruments are intended to plug into a rack, frame, or chassis that provides the dc power needed to operate the circuits. Some housings are portable and are intended for maintenance personnel in the field. Other housings are fixed and are intended for test stations, troubleshooting labs, or other locations where the equipment need not be portable.

The unit on the left in Fig. 6–20 is a universal digital counter/timer that works very much like the circuits described in previous sections. It is a two-channel device ("A" and "B") that will measure period, frequency, the ratio $A/B$, and the difference $A - B$. It will also totalize count and will measure events $A$ during $B$, that is, the number of input counts that occur in channel A during a period set by B. Time base and period can be set independently.

**FIGURE 6–20**
Plug-in digital modules (courtesy of
Tektronix, Inc.).

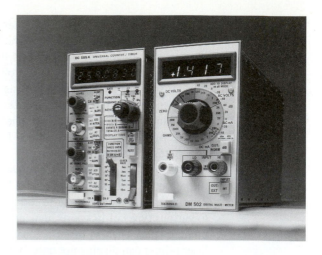

The unit on the right in Fig. 6–20 is an ac-dc multimeter that will measure the usual volts, ohms, and milliamperes, but it provides a digital rather than an analog display. This instrument is intended as a service-grade device for electrical and electronic troubleshooting uses.

Two other multimeters are shown in Figs. 6–21 and 6–22. The device shown in Fig. 6–21 is a hand-held portable intended for field service applications, while the device in Fig. 6–22 is a laboratory bench/rack model that is intended for applications where greater precision and greater accuracy are needed.

Figure 6–23 shows a digital dc power supply intended for troubleshooting and design prototyping work, as well as for automated test equipment applications. This instrument provides digital readout of the voltage settings and the current drawn. It is also digitally programmable. A number of modern instruments will accept a digital input signal that instructs the device on what to do, for example, (for a dc power supply) what voltage and polarity output to provide.

**FIGURE 6–21**
Hand-held portable multimeter for
field service applications (courtesy
of Tektronix, Inc.).

**FIGURE 6–22**
Laboratory bench/rack model providing greater precision and accuracy (courtesy of Tektronix, Inc.).

**FIGURE 6–23**
Digital dc power supply (courtesy of Tektronix, Inc.).

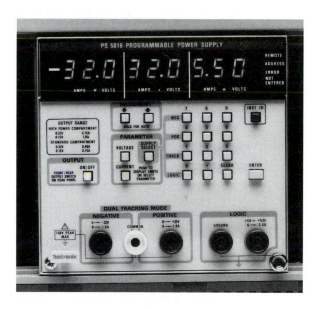

## SUMMARY

1. Digital instruments use binary logic elements to perform circuit functions.
2. Frequency, period, and totalizing counters can be built from flip-flops, decoders, and digital display devices.
3. Frequency is counted by opening a gate to allow input pulses into the DCA for a specified period of time.
4. Period is counted by opening the gate to admit time base pulses into the DCA between two successive cycles of the input signal.

## RECAPITULATION

Now go back and try to answer the questions at the beginning of the chapter. When you are finished, answer the questions and work the problems given below. Place a mark beside each problem or question that you cannot answer, and then go back to the text and reread the appropriate sections.

## QUESTIONS

1. Does replacement of a taut-band voltmeter with a digital voltmeter as the readout device in an electronic instrument generally make it a *digital* instrument?
2. A 0-to-1-V signal normally has a ±1% (i.e., a 10-m V) error. How many digits of a 0-to-999-m V digital voltmeter represent valid data?
3. Draw a block diagram for a simple **decimal counting unit** with **latch.**
4. In a latch circuit such as the 7475, data to the display are updated when the *strobe* line goes ——————.
5. Name three forms of counter decimal displays.
6. Which type of digital display device is used most frequently?
7. Discuss the relative advantages and disadvantages of the Nixie® tube and the seven-segment readout.
8. Draw a 5 × 7 matrix display; indicate how it might display the decimal digit "8."
9. Draw a block diagram for a four-stage decimal counting assembly (DCA) using DCUs. Show any necessary interconnections, and indicate the MSD and LSD.
10. Which line in a BCD output of a counter is used as the **carry?** What is its decimal weight?
11. Another name for an EPUT counter is —————— counter.
12. What type of stages make up the time base section?
13 What changes are necessary to make a frequency counter measure period?
14. Part of the input signal conditioning is a ——————, which has upper and lower hysteresis window limits.
15. A **trigger level** control varies ——————.
16. A **trigger amplitude** control varies ——————.
17. (a) Define in your own words the following terms: short-term stability, long-term stability, ±10% power supply stability, and 0°C–50°C stability. (b) In which stage of a counter do these terms apply?
18. Rank the following in terms of long-term stability, beginning with the least stable: oven-controlled, TCXO, room temperature crystal oscillator, and 60-Hz mains.
19. Which is usually more stable—a proportional controlled oven or a thermostat-controlled oven?
20. Describe how noise on a signal can create an erroneous frequency count.
21. Describe how amplitude modulation of a signal could result in an erroneous frequency count.
22. Describe how noise on a signal can create trigger error when the period of a slow slew rate signal is measured.
23. What steps may be taken to alleviate the problem discussed in Question 22?

## PROBLEMS

1. A DCA reads 12496 when the time base is set to 0.1 s. What is the frequency of the applied signal?

2. A six-decade DCA is used to measure a 1256-Hz signal when the time base is set to 10 s. What reading would you expect on the display?

3. What reading would you expect to see on the DCA display in Problem 2 if the frequency is 1.567 mHz and the time base is 10 ms?

4. What is the period of a 2.56-mHz signal?

5. What is the period of the signal if the DCA count is 12567 and the time base frequency is 10 kHz?

6. Find the period if the DCA count is 8067 and the time base period is 0.1 ms.

7. What DCA reading would you expect if the period of the input signal is 881 ms and the time base frequency is 1 kHz.

8. Find the percentage of error due to ±1 count ambiguity at (a) 100 kHz, (b) 1 mHz, (c) 2 mHz, (d) 20 mHz, and (e) 512 mHz.

9. A 0.005% error represents _____ ppm.

10. A 10 ppm error represents _____ percent.

11. Find the error in percent *and* ppm if a transmitter that is precisely tuned to a frequency of 20.005 mHz produces a reading of 20,004,682 Hz on a counter.

# 7

# Electronic Multimeters

## OBJECTIVES

1. To learn the principles behind the operation of electronic multimeters.
2. To learn the different types of electronic multimeters.
3. To learn how electronic multimeters differ from the VOM.
4. To learn to use and interpret readings from the electronic multimeter.
5. To learn the precautions necessary when you are using electronic multimeters.

7–2

## SELF-EVALUATION

Before studying the material in this chapter, try to answer the questions given below. These questions test your knowledge of the subject. If you cannot answer a particular question, then look for the answer as you read the text.

1. How does an FET VM differ from (a) the VOM and (b) the VTVM?
2. Describe in your own words the operation of an FET VM. Make a drawing if it will aid your discussion.
3. Describe the operation of a **dual-slope integrator** digital multimeter. Use a drawing if necessary.
4. How does a **true rms computer** measure the rms value of an ac waveform?
5. What is meant by a "3½-digit" multimeter?

7–3

## VOMs vs. EMMs

The volt-ohm-milliammeter, or VOM, is a rugged and reasonably accurate analog instrument, but it suffers from certain disadvantages. The main problem is that it lacks both **sensitivity** and **input impedance.** The analog VOM is often the instrument of choice, especially when making measurements in the presence of strong RF fields, but for most applications it is second best. A major problem is its **sensitivity.**

The sensitivity of a VOM, given in the units ohms/volt ($\Omega$/V), has been described earlier as the reciprocal of the meter movement's full-scale current. A 50-$\mu$A movement used with a multiplier resistor to make a voltmeter, for example, has a sensitivity of $1/5 \times 10^{-5}$, or 20,000 $\Omega$/V. On low voltage ranges, then, the impedance of such an instrument tends to be low. In the case of the 20,000 $\Omega$/V meter with a 0-to-0.5-V range, the input impedance is only (0.5 V)(20,000 $\Omega$/V), or 10 k$\Omega$.

The electronic voltmeter (EVM), on the other hand, can have an input impedance as low as 10 M$\Omega$ or as high as 100 M$\Omega$, and the input impedance is *constant* over all ranges instead of being different on each range as in the VOM. The loading of the circuit under test tends to be less with the EVM than with the VOM.

On the negative side, however, many EVMs cannot be used in the presence of strong electric or electromagnetic fields such as those produced by a radio transmitter, television flyback transformers, and so forth. The field will tend to bias the transistors or integrated circuits used in the EVM to a point where they will not operate properly. One popular EVM model will read a constant one-third scale whenever the instrument is used inside an AM broadcasting station transmitter site! The VOM, on the other hand, is practically immune to such effects except on the ac scales.

## 7–4    THE BASIC ELECTRONIC MULTIMETER

An electronic multimeter (Fig. 7–1) might use any of several technologies, including transistors, integrated circuits, or digital circuitry. Some meters use an input amplifier to

**FIGURE 7–1**
Analog electronic voltmeter (courtesy of Simpson Electric).

increase the amplitude of weak signals, while others use an **attenuator** to reduce the amplitude of signals that are above the meter's basic range.

The electronic multimeter might be ac mains powered, or it might be battery powered to allow portability. While the power supplies may differ in type, the one commonality among all electronic multimeters is that they use active devices, so they require a power supply.

The Simpson model 314 analog EVM shown in Fig. 7–1 measures full-scale ac voltages down to 10 mV and dc voltages down to 50 mV. The current ranges are from 10 μA full scale to 1 A full scale. The ohmmeter ranges are from $R \times 1$ to $R \times 1$ MΩ. A function switch provides selection for ac volts, dc volts, alternating and direct current, and two ohms ranges, high and low (more on this subject later).

Electronic voltmeters that use an analog meter movement almost invariably use an electronic balance circuit such as that shown in Fig. 7–2 to drive the meter.

The different types of electronic voltmeters (EVMs) go by a variety of names that tend to reflect the technology used in the balance circuit. The original EVMs (pre-1965) used vacuum tubes, so they were called **vacuum tube voltmeters,** while solid-state mod-

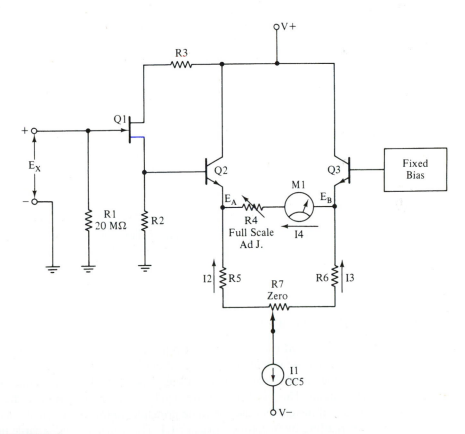

**FIGURE 7–2**
Typical solid-state electronic voltmeter circuit.

els with JFET input stages were known as **transistor voltmeters (TVMs)** or **FET voltmeters (FET VMs).** The basic circuit operates on the same principles, however, regardless of the technology employed.

The circuit in Fig. 7–2 uses a **differential amplifier** consisting of transistors $Q2$ and $Q3$ to form a balanced bridge circuit. Field effect transistor $Q1$ serves as a source follower and is used to provide impedance transformation between the input and the base of $Q2$.

Current $I1$ is generated by a **constant current source,** which may be a high-value resistor to the $V$-source, or one of the electronic CCS sources that are known. Since $I1$ is constant, we know that the following relationship holds true:

$$I1 = I2 + I3 = k \tag{7–1}$$

The bias on $Q2$ and $Q3$ is such that $I2 = I3$ when $E_x$ is zero. Under that condition, $E_a = E_b$, so the current through dc meter movement $M1$ is zero; $I4 = 0$.

The bias on $Q3$ is fixed by a stable reference supply, but the bias on $Q2$ is a function of the voltage drop across resistor $R2$, which in turn is controlled by $E_x$. When an unknown voltage $E_x$ is applied, the bias on $Q2$ increases, and that causes voltage $E_a$ also to increase. Now, $E_a$ is greater than $E_b$, so current $I4$ is no longer zero. The magnitude of current $I4$ flowing in $M1$, and hence the deflection of the $M1$ pointer, is proportional to $E_x$.

The value of $E_x$ that causes maximum deflection of $M1$ is the **basic range** of the instrument and is usually the lowest range on the selector switch in nonamplified models. Higher ranges can be accommodated through the use of an **input attentuator** such as that shown in Fig. 7–3, while lower ranges can be accommodated by a preamplifier.

The attenuator in Fig. 7–3 is a resistance voltage divider. The full-scale voltage from the probe appears across the divider, so the voltage at each tap is a progressively lower fraction of the full input voltage. If the basic range of the balance circuit is 50 mV, then a 1000-V input potential will be reduced to 50 mV when the selector switch is connected to the bottom tap. Precision resistors are used for $R1$ through $R10$, and most of these are of the **wirewound** variety. Such resistors have significant *inductance,* so if the meter is to be used on ac, some sort of compensation is required. The compensation takes the form of capacitors $C1$ through $C10$. Several of these capacitors are made variable, and these capacitors are adjusted, with a square wave applied to the input, for best "squareness" as viewed on an oscilloscope.

**7–5**          ## ac MULTIMETERS

Most basic EVM circuits such as that in Fig. 7–2 are for use with dc potentials only. To accommodate ac potential some sort of converter is required.

In the simplest case, shown in Fig. 7–4(a), a full-wave rectifier is used to drive the meter circuit. This type of converter tends to suffer from most of the same faults as passive ac voltmeters that use the same principle. Some circuits make the circuit essentially **peak reading** by including capacitor $C1$. This capacitor will charge up to the peak value of the ac waveform due to the action of the rectifier.

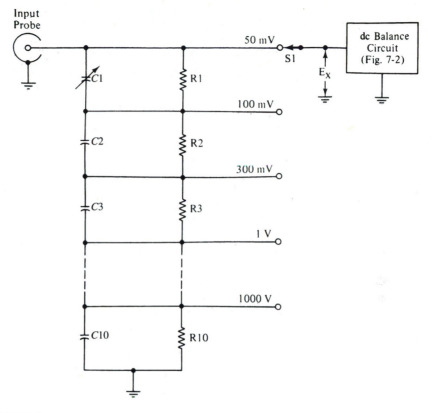

**FIGURE 7–3**
Input attenuator circuit.

A somewhat superior rectifier/averager circuit is shown in Fig. 7–4(b). In this case the rectification is provided by an operational amplifier circuit called an **ideal rectifier** or a **precision rectifier.** A normal pn junction diode is nonlinear at low forward voltages. Silicon diodes do not become linear until the applied forward voltage exceeds approximately 0.6 to 0.7 V, while germanium diodes require 0.2 to 0.3 V of forward bias. In the circuit of Fig. 7–4(b) these nonlinearities are "servoed out" by the operational amplifier and its negative feedback loop.

The pulsating dc at point A in the circuit is smoothed out (i.e., *averaged*) by the low-pass filter $R7–R9/C3–C5$. The relatively pure dc output from the filter is applied to an analog balance circuit such as in Fig. 7–2 or to a dc digital voltmeter.

Although most ac voltmeters are calibrated in terms of **volts rms,** most will not actually read the rms value unless the input signal has a pure sine wave shape. The reasons for this were discussed in Chapter 3.

Some electronic voltmeters, however, use an **rms-to-dc converter** [block diagrammed in Fig. 7–4(c)] that computes the rms value of most common ac waveforms.

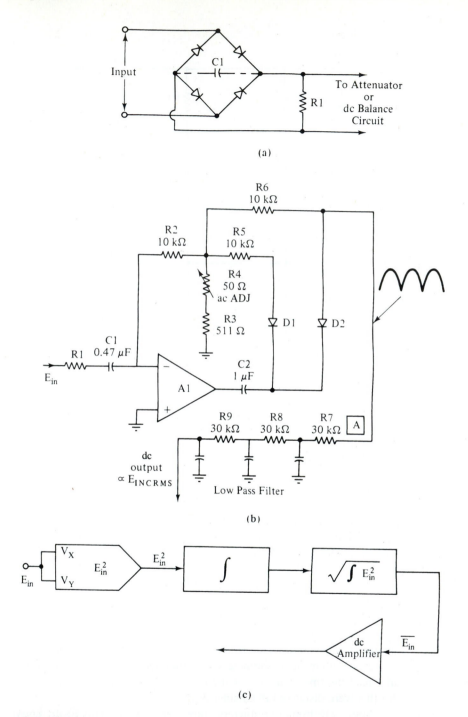

**FIGURE 7–4**
(a) Simple ac full-wave rectifier, (b) rms-averaging circuit, (c) computing-type rms converter.

Most such circuits operate to 10 or 20 kHz, while some operate to over 1 mHz. Recall the meaning of the term **rms** from Chapter 3:

$$E_{rms} = \sqrt{\frac{1}{T} \int_0^T (E_{in})^2 \, dt} \qquad (7\text{--}2)$$

The circuit in Fig. 7–4(c) is basically an analog computer that solves this equation. The first stage is an analog multiplier connected as a **squarer** circuit; that is, the $V_x$ and $V_y$ terminals are connected together, making the output proportional to $(E_{in})^2$. The second stage integrates the output of the squarer, and the third stage takes the square root of the integrator output. The output of the square root stage is proportional to the **true rms** value of the input waveform. In modern meters (digital and analog) this function is implemented in an integrated circuit.

## 7–6     ELECTRONIC OHMMETERS

Ohm's law tells us that resistance in ohms can be computed from the current in amperes and from the voltage:

$$R = \frac{E}{I} \qquad (7\text{--}3)$$

If the current through an unknown resistor is held *constant*, then the value of the voltage drop across the resistance will give us the data needed to calculate the resistance value. If that current is a power-of-10 submultiple of 1 A, (e.g., 1 mA, 100 μA, etc.), then the voltage drop will be numerically equal to the resistance. Only repositioning of the decimal point is then needed to read out directly in ohms or kilohms. Figure 7–5 shows such an ohmmeter circuit.

The constant current source (CCS) holds the current through $R_x$ at a constant 1 mA (i.e., 0.001 A). By Ohm's law, then,

| $R$ ($\Omega$) | $E$ (V) |
|---|---|
| 100 | 0.1 |
| 1000 | 1.0 |
| 10,000 | 10.0 |

The gain of the dc amplifier and the output current of the CCS are set by the range selector switch so that full-scale resistances from millohms to megohms can be accommodated.

The low current levels used in this type of ohmmeter circuit confer both advantages and disadvantages in practical situations. Ordinary ohmmeters using passive dc components and meter movement use a 1.5-V dc battery (or even larger), and these produce a relatively high current in low resistances. That type of ohmmeter will forward-bias pn

**FIGURE 7–5**
Electronic ohmmeter circuit.

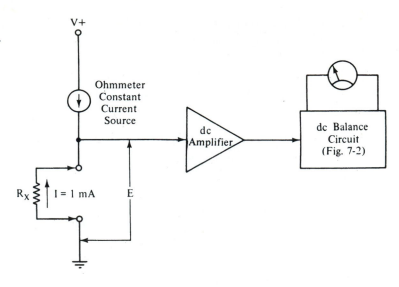

junctions, so it will result in large errors when used in solid-state circuits. But the low voltage levels used in the circuit of Fig. 7–5 usually will not forward-bias pn junctions, so it may be used in solid-state circuits.

The inability to forward-bias pn junctions is also a disadvantage, because forward and reverse resistance readings across the junctions often serve as a quick test for diodes and transistors. Some EVM manufacturers include a high-power ohmmeter scale so that such tests can be made. In meters such as the Simpson 314 shown in Fig. 7–1, high- and low-power ohmmeters are provided at different settings of the function switch, whereas in other products a pushbutton switch is provided marked "diode" or with the following symbol:

## 7–7      DIGITAL VOLTMETERS

A digital voltmeter uses an analog-to-digital converter (ADC) to convert the dc input, or the output of an ac converter circuit, to a **binary coded decimal (BCD)** digital word that is used to drive a digital display device. Most digital voltmeters (DVMs) or digital multimeters (DMMs) use an ADC circuit called the **dual-slope integrator,** which we will now discuss in detail. Other ADC circuits are used in certain other instrumentation applications, and these are further discussed in Chapter 14.

**7–8**

## DUAL-SLOPE INTEGRATION

The block diagram of a dual-slope integrator is shown in Fig. 7–6(a), and associated waveforms are shown in Fig. 7–6(b).

The heart of the circuit is the operational amplifier **integrator** (see Chapter 12) consisting of operational amplifier $A1$, plus $R1$ and $C1$, plus a **voltage comparator** ($A2$). The output of the comparator will remain LOW if the integrator output is zero, and HIGH if the integrator output is more than a few millivolts above ground potential.

At the beginning of the conversion cycle, the control logic section momentarily closes electronic switch $S2$ so that the charge on capacitor $C1$ goes to zero; it also ensures that $S1$ is set to position A.

If $S1$ is in position A, the integrator input is connected to the input voltage source, causing the voltage at the integrator output to begin rising [see time $t_0$ in Fig. 7–6(b)]. As soon as $E_A$ rises a few millivolts, the comparator output snaps HIGH, enabling the gate to pass clock pulses to the digital counter (Chapter 6) section. The counter is allowed to overflow [$t_1$ in Fig. 7–6(b)], and the output *carry* pulse from the counter is used to tell the control logic section to switch $S1$ to position B. This action connects the integrator input to a *precision reference* voltage source. The polarity of the input current created by the reference is such that it begins to *discharge* the integrator capacitor at a *constant* rate. The counter, meanwhile, has continued to increment, passing through 0000 at time $t_1$ and continuing to accumulate clock pulses until $E_A$ is back down to zero.

The value of $E_A$ at time $t_1$ was proportional to the value of $E_{in}$. At the same instant, the count was 0000. Since the counter continues to increment as the integrator discharges ($t_1$ to $t_2$), the *count* at the instant the gate is closed (when $E_A = 0$) is also proportional to $E_{in}$. By correct scaling, the count will be numerically the same as the potential applied to the input. To recapitulate:

1. At time $t_0$ switch $S2$ is closed briefly to dump any residual charge in $C1$, and $S1$ is set to position A.
2. The integrator begins to charge due to current $E_{in}/R1$, so $E_A$ begins rising from zero.
3. As soon as $E_A$ is greater than zero, the output of the comparator goes HIGH, enabling the main gate to pass clock pulses into the counter.
4. The counter increments until it overflows at time $t_1$, and the overflow pulse causes $S1$ to switch to position B, applying the reference voltage to the input of the integrator. At this instant, the count is 0000.
5. From times $t_1$ to $t_2$ the integrator *discharges* under the influence of current $-E_{ref}/R1$; meanwhile, the counter continues to increment.
6. At time $t_2$ the comparator shuts off the flow of clock pulses through the gate. The count accumulated between $t_1$ and $t_2$ represents the input voltage.
7. The control logic section issues a *strobe* pulse at time $t_2$ to update the counter's digital display.

The dual-slope integrator is one of the slower types of ADC circuits, but it offers several advantages when used in DVM circuits: relative immunity to noise riding on the input voltage, relative immunity to error due to inaccuracy or long-term drift in the clock

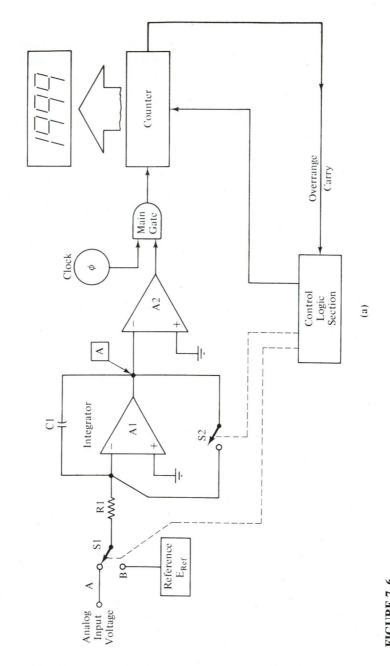

**FIGURE 7-6**

(a) Dual-slope integrator for electronic voltmeters.

**FIGURE 7–6,** *continued*
(b) Timing of the dual-slope
integrator.

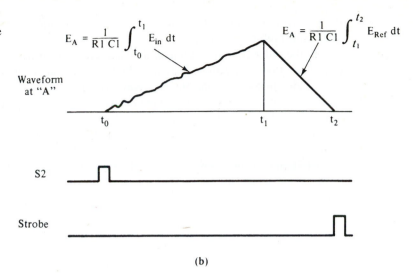

$$E_A = \frac{1}{R1\,C1} \int_{t_0}^{t_1} E_{in}\, dt$$

$$E_A = \frac{1}{R1\,C1} \int_{t_1}^{t_2} E_{Ref}\, dt$$

Waveform
at "A"

$t_0$  $t_1$  $t_2$

S2

Strobe

**(b)**

frequency, and so on. The 10-to-40-ms conversion cycle time poses no disadvantage in most DVM applications.

## 7–9    WHAT DOES THE DUAL-SLOPE DVM MEASURE?

A dc digital voltmeter tends to read the *average* input voltage over the period of integration. This feature makes the DVM particularly useful for measuring noisy signals. It also tends to average the pulsations from a rectifier-type ac-to-dc converter, provided that $t_2 - t_0$ is greater than about 18 ms.

The accuracy of a digital voltmeter, or a DMM based on a DVM, can be quite good. In fact, it can exceed the accuracy of analog multimeters. You will see terms such as "2½-digit," "3½-digit," and "4½-digit" in the advertisements for DVM/DMM products, and you will wonder what is meant by "½ digit." This term refers to the fact that the *most significant* digit can be only a "0" or a "1," while all other digits can be anything between "0" and "9." Such terminology indicates that the meter can read 100% overrange from its basic range. For example, a 3½-digit DVM will read 0–1999 mV, while its basic range is only 0–999 mV. If this range is exceeded, then the "1" lights up; otherwise it remains darkened.

## 7–10    STEPPER-TYPE DVMs

Some of the original DVM instruments used a **stepper** technique instead of an integrator, and certain modern instruments have adapted the technique to electronic methods. In both cases a voltage comparator is used to compare the input voltage with a **staircase** (i.e., quasiramp) reference voltage.

In the older type of instrument the staircase voltage was generated by a pulsed stepper relay that selected successively higher taps on a resistor voltage divider. The same pulse that moved the stepper relay also increments a digital events counter, so if each step represents an equal voltage change, then the counter reading will indicate the reference voltage value. For example, each step could represent 10 mV, so an input voltage of 1 V would be reached when the stepper had incremented 100 steps (100 steps $\times$ 10 mV/step = 1 V).

Modern versions of this circuit replace the slow, noisy, stepper relay with a precision **digital-to-analog converter (DAC)** (Fig. 7–7) that does essentially the same job (see Chapter 14). The operation of the circuit is as follows:

1. At the beginning of the conversion cycle, $E_0 = 0$, and $E_{in}$ is some potential between 0 and 1 V dc.
2. Since $E_0 \neq E_{in}$, the comparator output is HIGH, and this enables the gate to pass clock pulses to the counter.
3. DAC output voltage $E_0$ increases 1 mV per step until $E_0 = E_{in}$.

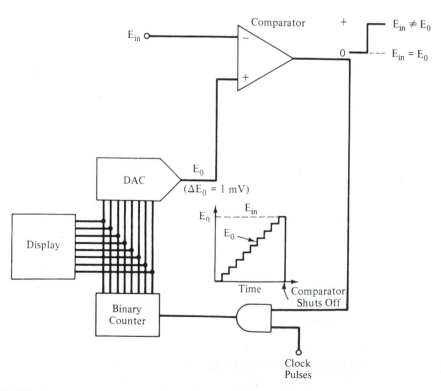

**FIGURE 7–7**
Stepper-type digital voltmeter circuit.

**4.** When $E_0 = E_{in}$, the comparator output drops LOW, turning off the gate. The output of the counter is now numerically equal to $E_{in}$ in millivolts, and this value is displayed on the digital output indicator.

Many modern digital voltmeters rely not simply on an IC integrating A/D converter but rather on a microcomputer that uses such an A/D converter as the input source. These newer instruments are capable of sensing waveforms and correcting the value displayed according to the waveform characteristics. A dc reading, for example, is valid only if the input signal is unvarying. In real circuits, however, situations often arise where the signal varies slowly. The value displayed on older digital meters would be a sample taken as a snapshot but would not represent a true picture. If the microcomputer implementation is used, however, several successive snapshots can be statistically averaged and displayed.

Similarly, the rms value of an ac waveform is valid on some instruments only when a sine wave ac signal is being measured. Use of an rms-to-dc converter IC device helps immensely, but a "sanity check" of the waveform by the microcomputer can make the measurement more meaningful. This is especially true where a high-frequency component rides on a lower-frequency ac signal or where amplitude modulation (AM) is present.

## 7–11 COMMERCIAL DIGITAL MULTIMETERS

The digital multimeter was, at one time, too expensive for most users. These meters were found in certain military, scientific, and commercial laboratories, and only a few companies, notably Fluke and Hewlett-Packard, produced the instruments. But today, integrated circuit technology has allowed the production of DVM and DMM instruments at previously unheard-of prices. One model based on a popular LSI chip sells for less than $50 in kit form, and the prices go up from there to several thousands of dollars. Do not expect a lot of accuracy or special features on a low-cost model, although very good performance is available in the medium-price range.

Figure 7–8 shows DMM products representative of a larger number of instruments. The model shown in Fig. 7–8(a), the Hewlett-Packard model 3476A, is a portable instrument designed for service technicians, field engineers, and so forth. It is battery powered, although it is able to operate from ac mains power if the charger cord is plugged in.

The H-P 3476A will measure ac volts from 3.3 mV to 700 $V_{rms}$, dc volts from 100 $\mu$V to 1000 V, direct current from 100 $\mu$A to 1.1 A, alternating current from 3.3 mA to 1.1 A, and resistance from 1 $\Omega$ to 11 M$\Omega$.

This particular model features **autoranging** operation, which means that the circuit will automatically adjust to the correct voltage, current, or resistance range. Other models by other manufacturers, but in the same price class, offer pushbutton selection of the ranges.

A pushbutton portable DMM is shown in Fig. 7–8(b). This type of instrument is based on an LSI chip and is intended for field service applications.

The Tektronix model DM-502 plug-in DMM is shown in Fig. 7–8(c). This instrument is part of the Tektronix versatile line of mainframe plug-in instruments that are

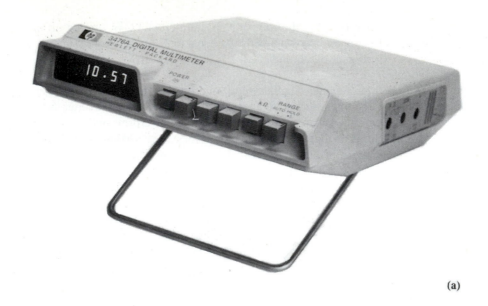

(a)

(b)

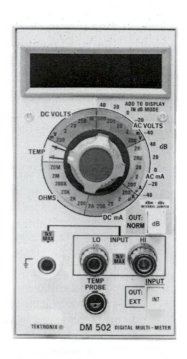

(c)

**FIGURE 7–8**

(a) Portable digital multimeter, (b) hand-held autoranging digital multimeter, (c) mainframe electronic digital multimeter [(a) courtesy of Hewlett-Packard, (b) courtesy of B&K Dynascan Corporation, (c) courtesy of Tektronix, Inc.].

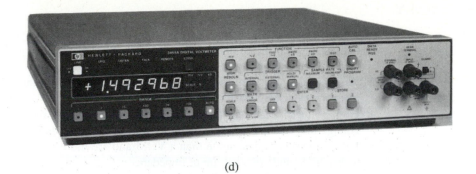

(d)

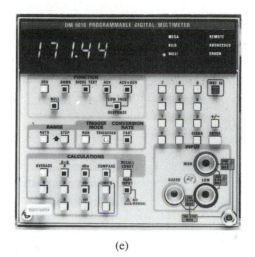

(e)

**FIGURE 7–8,** *continued*
(d) Laboratory-grade electronic digital multimeter, (e) programmable digital multimeter [(d) courtesy of Hewlett-Packard, (e) courtesy of Tektronix, Inc.].

designed to be installed in a portable or bench model instrumentation mainframe/power supply. This line allows the user to equip a laboratory bench or field service engineer with the needed instruments in a conveniently packaged form.

At the high end of the price range are precision instruments such as the Hewlett-Packard model 3455A shown in Fig. 7–8(d). This instrument is a very accurate model used mostly in precision laboratories. The 3455A is able to perform certain product and ratio measurements in addition to straight voltage determinations.

A newer category of DMM is the programmable Tektronix model DM-5010 shown in Fig. 7–8(e). This instrument is not just a high-performance DMM; it is also capable of working in a general purpose interface bus (GPIB) IEEE-488 system.

## SUMMARY

1. Analog DMMs use a dc balance circuit to drive the meter movement.
2. Digital DMMs use an analog-to-digital converter to produce a binary coded decimal "word" that is proportional to the applied input voltage.
3. Resistance is measured by passing a current through the unknown resistance and then measuring the voltage drop across the resistor.
4. High voltage ranges are accommodated through the use of an attenuator, while low voltage ranges are accommodated through the use of an amplifier.
5. EVMs can be more accurate than a VOM and are more sensitive. EVMs tend to have a very high, and constant, input impedance.

## RECAPITULATION

Now go back and try to answer the questions at the beginning of the chapter. When you are finished, answer the questions and work the problems given below. Place a mark beside each problem or question that you cannot answer, and then go back to the text and reread the appropriate sections.

## QUESTIONS

1. How does an analog EVM differ from a VOM?
2. Draw a simple dc meter balance circuit and describe its operation in your own words.
3. Name two types of analog-to-digital converters used in DVM designs.
4. What technique is used to measure resistance in many EVMs?
5. What is the difference between high- and low-power resistance circuits? Describe at least one advantage of each type.
6. What is a "3½-digit" voltmeter? What is meant by "½-digit"?
7. What is the resolution of a voltmeter described as a "4½-digit" instrument if the basic range is 0–9999 mV?
8. Can a 2½-digit DVM measure 1 mV if the basic range is 0–99 V?

## PROBLEMS

1. A 100-μA constant current is passed through an unknown resistance, and a voltmeter connected in parallel with the meter measures 888 mV. Calculate the resistance.
2. A 10-μA constant current is passed through a 100-kΩ resistor. What would a voltmeter connected across the resistor read? Assume an infinite input impedance on the voltmeter.
3. A DVM measures direct current by passing the current through a low-value fixed resistor and then measuring the voltage drop across the resistor with a dc voltmeter. Find the current if the voltage drop across a 1-Ω resistor is 200 mV.

# 8

# The Oscilloscope

## OBJECTIVES

1. To learn the principles of operation behind the cathode ray oscilloscope.
2. To learn the different types of oscilloscopes.
3. To learn how to use and operate the oscilloscope.
4. To learn how to interpret oscilloscope waveforms.

8–2

## SELF-EVALUATION

Before studying the material in this chapter, try to answer the questions given below. These questions test your knowledge of the subject matter. If you cannot answer a particular question, then look for the answer as you read the text.

1. Name two types of CRT deflection systems.
2. Draw a block diagram of a simple free-running sweep CRO.
3. Define in your own words (a) **triggered sweep** and (b) **delayed sweep.**
4. What are **Lissajous figures?** How are they produced?

8–3

## THE CATHODE RAY OSCILLOSCOPE

The cathode ray oscilloscope (also called CRO, oscillograph, oscilloscope, or simply 'scope) has been called the single most important instrument in electronics. It can be used to measure voltage (ac or dc) and frequency and to view the **wave shape** of the input signal.

The CRO is much like a television set in that it produces a picture of the shape of an input signal on a light-emitting phosphor screen. The heart of the CRO is a special vacuum tube called the **cathode ray tube (CRT).**

## 8–4        CATHODE RAY TUBES

The operation of an oscilloscope depends upon the CRT, an example of which is shown in Fig. 8–1(a). The principal components of the CRT are the fluorescent phosphor screen, the vertical and horizontal deflection plates, the accelerating electrode, the focus electrode, and an electron gun that forms an electron beam. Some CRT models also have a postdeflection accelerating electrode (sometimes called a **second anode**).

The **electron gun** is a specially constructed heated cathode that gives off electrons by thermionic emission. The cathode is surrounded by a negatively charged cylindrical control electrode that has a hole in one end. Electrons also carry a negative charge and thus are repelled from the walls of the cylinder and will stream out through the hole toward a less negatively charged **accelerator electrode.** In some CRTs this electrode is called anode 2, but beware of similar terminology when the postdeflection accelerator is introduced.

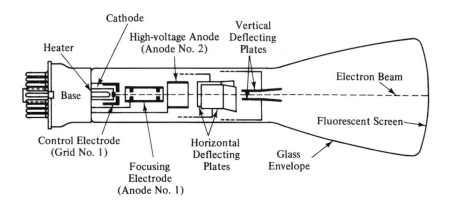

(a)

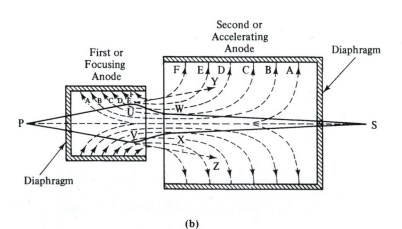

(b)

**FIGURE 8–1**
(a) Construction of the cathode ray tube, (b) focus electrodes.

The effect of the accelerator anode is to increase the velocity of the electrons, thereby increasing their kinetic energy. When the electrons strike the phosphor screen, they give up their kinetic energy in the form of **light energy.**

The beam width emitted from the electron gun is too broad for practical applications, so some focusing is required. This job is performed by the first anode, also called the **focus electrode.** A more detailed illustration of the focus system is given in Fig. 8–1(b). Note that both first and second anodes work together to form, in effect, an electron lens not dissimilar in function to an optical lens. Electric field lines exist between points on each of these electrodes. The electrons of the beam are affected by this field in such a way that they are forced to converge on a distant point external to the focus assembly. The distance between the convergence point and the second anode is determined primarily by the potential existing between the two anodes. It is usually the practice of oscilloscope designers to hold the high voltage applied to the second anode constant, and then to use a potentiometer labeled *focus* to vary the potential applied to the first anode. In practice, the operator adjusts the focus control for the sharpest dot on the CRT viewing screen, as shown in Figs. 8–1(c) and 8–1(d).

The basic CRT power supply circuit is shown in Fig. 8–2. Note that this example is but one of several alternative designs. Other models will probably use a modified version of this configuration.

A −1500-V dc power supply is connected to a resistor voltage divider network consisting of $R1$ through $R3$. Potentiometer $R3$ varies the potential applied to the control electrode, designated as grid 1. This control serves the function of **intensity control.**

Potentiometer $R2$ serves as a **focus control,** and it varies the potential applied to the focus electrode, designated as anode 1. The potential applied to the accelerator electrode is kept constant.

It is interesting to note that the names given to the structures inside a CRT are referenced to structures of an ordinary vacuum tube, for example, **anode, grid,** and so on. In Fig. 8–1(a) we label certain structures as anodes, yet the schematic representation in Fig. 8–2 clearly shows them as *grids*. This seeming contradiction may tend to confuse you,

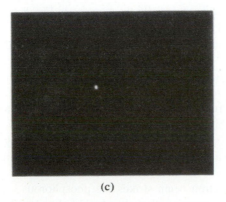

(c)

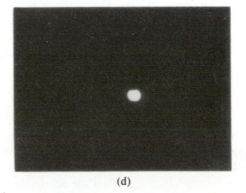

(d)

**FIGURE 8–1,** *continued*
(c) In-focus beam, (d) out-of-focus beam.

**FIGURE 8–2**

Basic dc power supply circuit for CRT.

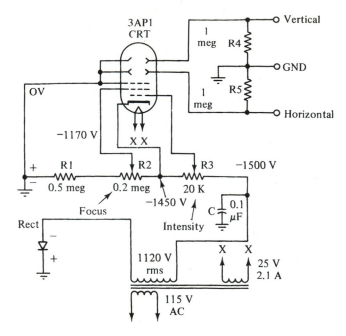

unless you break the habit of using the vacuum tube as your frame of reference. The use of grid terminology here is merely a graphic device to simplify the circuit diagrams. If the grid symbols were not used, we would be required either to draw a picture of the symbols, which would be clumsy, or to invent a totally new symbology.

It is in the treatment of the second anode, or accelerator, potential that many oscilloscope models differ. In the example of Fig. 8–2 the second anode is at a potential of zero volts; that is, it is **grounded.** Of course, being grounded, it is at a potential that is *more positive* than the other electrodes. In other models the second anode potential may be slightly positive, that is, 0 to +150 V, while in others it may be at a positive potential of several thousand volts. These situations are shown in Figs. 8–3(a) and 8–3(b), respectively. The circuit of Fig. 8–3(a) is essentially the same as in Fig. 8–2, except that a slight positive bias is applied to grid 3.

In Fig. 8–3(b), on the other hand, another electrode is used past the deflection plates, closer to the phosphor screen, and is called the **postaccelerator electrode.** The use of a postaccelerator improves the linearity of the CRT; that is, the spot remains the same in a greater area of the viewing screen.

The postaccelerator electrode is usually at a potential greater than 1 kV in low-cost oscilloscopes, 3 to 7 kV in higher-priced 'scopes, and up to 25 kV in color television receivers. Note that most CRTs using a postaccelerator electrode bring the connection out of the glass envelope through a special high-voltage nipple, rather than through the socket with the other connections.

When the accelerated electron beam strikes the phosphorous screen, the electron's kinetic energy is given up. This energy is first absorbed by the phosphor atoms but is then

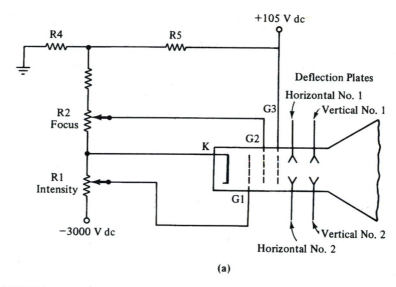

**(a)**

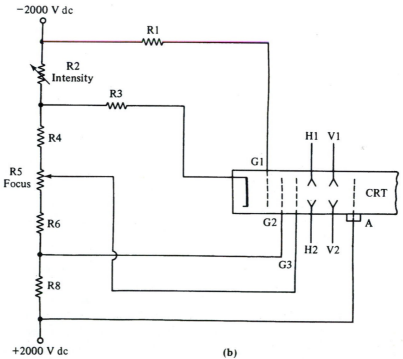

**(b)**

**FIGURE 8–3**
(a) Situation with second anode slightly positive, (b) use of the postaccelerator electrode.

released as the electrons of those atoms return to ground state. Because of the law of conservation of energy, this action causes the phosphor atoms to release light energy.

Some phosphors give up all absorbed energy rapidly, so the light emitted will diminish rapidly. Other CRTs use a different phosphor in which the emission of light is prolonged. The measure of the length of time a trace remains on the screen before fading is the **persistence** of the phosphor. Relatively short persistence phosphors are used in laboratory oscilloscopes, while longer-persistence types are used in medical electronics, where the signals occur at low-frequency rates.

## 8–5    DEFLECTION SYSTEMS

The **deflection system** of an oscilloscope moves the electron beam right and left or up and down. There are two basic types of deflection systems: **magnetic** and **electrostatic.**

A magnetic system has no internal deflection plates, but an external electromagnet surrounds the glass neck of the CRT. There are actually two separate electromagnets, one each for vertical and horizontal deflection. These two electromagnetic coils are constructed inside a single assembly called a **deflection yoke.** The yoke can create the deflection because magnetic fields will bend the path of the electron beam emitted by the gun assembly.

The inductance of the yoke coils limits magnetic deflection to low-frequency applications. A TV receiver, for example, creates a lighted area on the screen called a **raster** by sweeping the vertical at approximately 60 Hz and the horizontal at 15,734 Hz. Medical oscilloscopes deal with very slow signals (up to either 100 Hz or 1000 Hz, depending upon the application), so they often use magnetic deflection in order to take advantage of the shorter CRT length that is possible. But all service, all engineering laboratory, and most scientific oscilloscopes use electrostatic deflection.

The CRT shown in Fig. 8–1(a) is an electrostatic type; note the vertical and horizontal deflection plates located between the second anode and the fluorescent screen. This type of CRT uses the electrostatic field between two plates to deflect the electron beam.

Figures 8–4 and 8–5 show the position of the beam on the viewing screen as a function of the potentials applied to the deflection plates. When the potential on all four plates is the same, or zero as in Figs. 8–4(a) and 8–5(a), then the beam will be centered on the screen.

For this discussion let's assume the following notation for the four plates:

$$V_u = \text{the upper deflection plate (vertical)}$$
$$V_L = \text{the lower deflection plate (vertical)}$$
$$H_L = \text{the left deflection plate (horizontal)}$$
$$H_R = \text{the right deflection plate (horizontal)}$$

With these designations in mind, let's consider the following situations:

1. Figures 8–4(b) and 8–5(b): $H_R$ is more positive than $H_L$; $V_u$ and $V_L$ are at 0 V. The beam moves to the right.

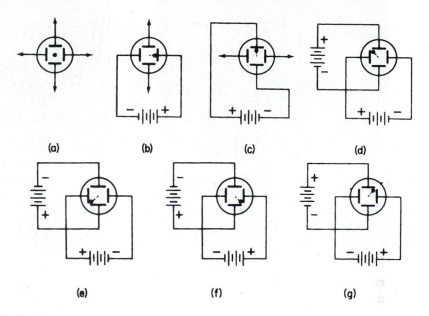

**FIGURE 8–4**
Position of the beam with different voltage situations at the deflection plates.

2. Figures 8–4(c) and 8–5(c): $V_u$ is more positive than $V_L$; $H_L$ and $H_R$ are at 0 V. The beam moves upwards.
3. Figures 8–4(d) and 8–5(d): $V_u$ is more positive than $V_L$; $H_L$ is more positive than $H_R$. The beam moves diagonally to the upper left-hand corner of the screen.
4. Figures 8–4(e) and 8–5(e): $V_L$ is more positive than $V_u$; $H_L$ is more positive than $H_R$. The beam moves diagonally to the lower left-hand corner of the screen.
5. Figures 8–4(f) and 8–5(f): $V_L$ is more positive than $V_u$; $H_R$ is more positive than $H_L$. The beam moves diagonally to the lower right-hand corner of the screen.
6. Figures 8–4(g) and 8–5(g): $V_u$ is more positive than $V_L$; $H_R$ is more positive than $H_L$. The beam moves diagonally to the upper right-hand corner.

8–6     **THE X-Y OSCILLOSCOPE**

An X-Y oscilloscope [Fig. 8–6(a)] allows the operator to position the beam anywhere on the CRT screen by controlling the potentials applied to the vertical and horizontal deflection plates. Although it is possible to use the deflection plates directly (and in some cases it is actually done), in most oscilloscopes vertical and horizontal amplifiers are used to drive the deflection plates. This arrangement allows us to display very low amplitude signals.

Note that most service and laboratory oscilloscopes sweep the horizontal plates with a time base signal [Fig. 8–6(b)] and so are "Y-time" (Y-T) designs. Many of these

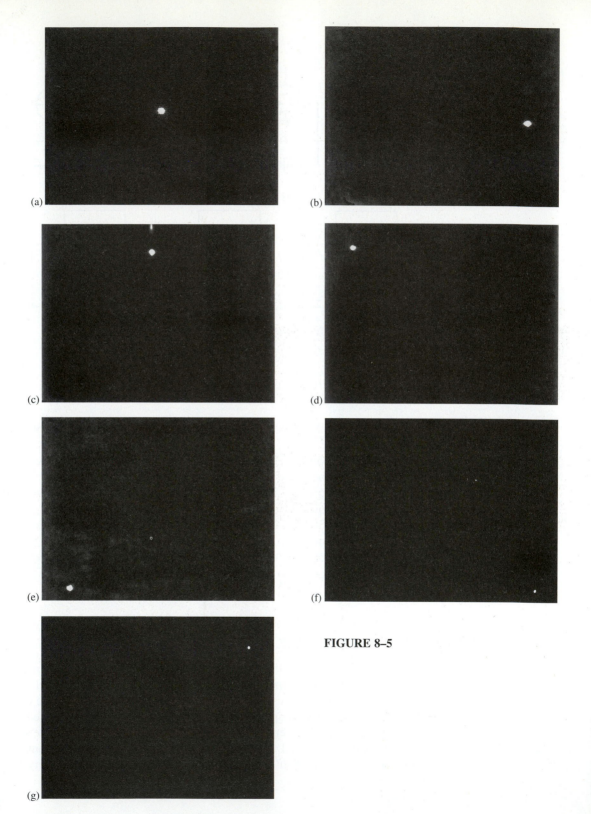

(a)

(b)

(c)

(d)

(e)

(f)

(g)

**FIGURE 8–5**

174

(a)

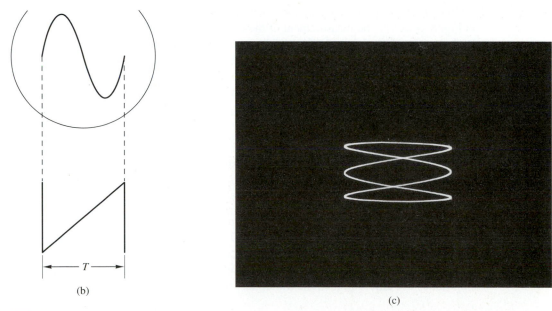

(b)

(c)

**FIGURE 8–6**

(a) X-Y display oscilloscope, (b) time base signal, (c) Lissajous pattern [(a) courtesy of Hewlett-Packard.

models can be used in the X-Y mode because a selector switch allows the user to apply either the internal time base signal or an external signal to the horizontal deflection plates.

If you apply input signals to both horizontal and vertical deflection plates of an X-Y oscilloscope, it forms a **vector pattern** that allows you to discern the relationship between the the two signals. Such diagrams are called **Lissajous patterns** [see Fig. 8–6(c)]. Color-TV repair shops, cable-TV operators, and TV broadcast studios often use X-Y 'scopes to analyze phase-modulated color signals, as do medical instrument manu-facturers (e.g., **vectorcardiographs,** an instrument that displays a heart graphic derived from the ECG).

You can measure audio frequencies with the Lissajous pattern. The ratio of the num-ber of loops along the vertical and horizontal sides of the pattern tells you the relationship between frequencies applied to the vertical and horizontal inputs [Fig. 8–6(c)].

The X-Y oscilloscope is used in a wide variety of applications, especially in the fields of computer terminal display or scientific and medical instrumentation. Often Tek-tronix or Hewlett-Packard X-Y oscilloscopes will be found as the readout display device in equipment produced by other manufacturers who find it less costly to buy an "O.E.M." (original equipment manufacturer) model than to design and build their own model.

Figures 8–7 and 8–8 show two applications for the X-Y oscilloscope. In Fig. 8–7 we see how to measure the relative *phase* between two signals of the same frequency, while in Fig. 8–8 we see the measurement of the frequency of two harmonically related signals. Both cases are examples of Lissajous figures.

Note in Fig. 8–7 that all phase differences other than 0°, 90°, 270°, and 360° produce elliptical patterns. The tilt direction of the ellipse limits the possible phase difference between $V_x$ and $V_y$ to two quadrants, while the "fatness" of the ellipse gives us the actual phase. Using the alphabetical notation of Fig. 8–7, we can calculate the phase angle $\theta$ from the expression

$$\theta = \arcsin\left(\frac{a}{b}\right) \tag{8–1}$$

There are two special cases in Fig. 8–7: 0° and 90°. When both vertical and hori-zontal plates are connected to the same signal or to in-phase signals of the same fre-quency, then the phase difference is 0° or 360° (the same actual point). This phase differ-ence produces a straight line on the CRT screen. If both signals are of the same amplitude, then the line will be at a 45° angle from the vertical or horizontal lines. But if the signal amplitudes are different, then the tilt angle will be

$$\theta_t = \arctan\left[\frac{E_v}{E_h}\right] \tag{8–2}$$

where   $\theta_t$ = the tilt angle in degrees
   $E_v$ = the vertical potential
   $E_h$ = the horizontal potential

**FIGURE 8–7**

Lissajous patterns on an X-Y oscilloscope with signals of same frequency but differing phases.

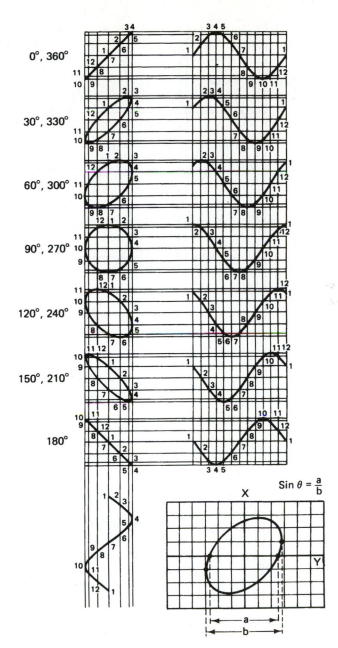

0°, 360°

30°, 330°

60°, 300°

90°, 270°

120°, 240°

150°, 210°

180°

$Sin \ \theta = \dfrac{a}{b}$

X

Y

a

b

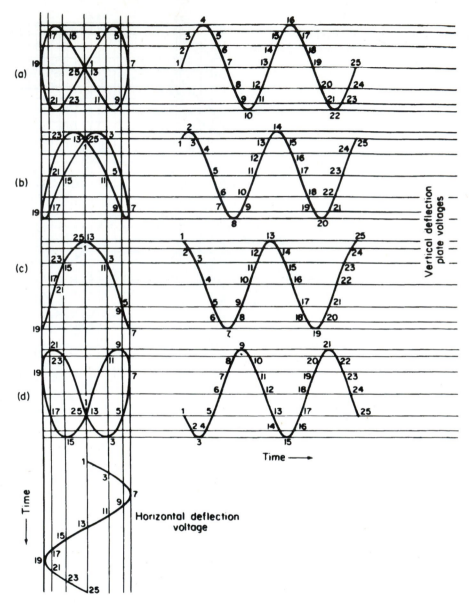

**FIGURE 8–8**
Lissajous figures for 2:1 frequency ratio.

178

■ **Example 8–1**

Calculate the tilt angle if in-phase 60-Hz ac signals are applied to the deflection plates so that $E_v$ is 38 V and $E_h$ is 14 V peak.

*Solution*

$$\theta_t = \arctan(E_v/E_h) \qquad (8\text{–}2)$$
$$= \arctan(38/14)$$
$$= \arctan(2.71) = \mathbf{69.8°}$$

The Lissajous patterns of Fig. 8–8 result from applying harmonically related signals to the vertical and horizontal deflection plates. In the example shown in Fig. 8–8, the vertical frequency is twice the horizontal frequency. Note that the *phase difference* determines the actual *shape* of the Lissajous pattern, while the *number of loops* is determined by the *frequency ratio.* The two frequencies are related by

$$\frac{F_v}{F_h} = \frac{N_h}{N_v} \qquad (8\text{–}3)$$

where    $F_v$ = the frequency applied to the vertical plates
$F_h$ = the frequency applied to the horizontal plates
$N_v$ = the number of loops observed along the vertical edge
$N_h$ = the number of loops observed along the horizontal edge

---

■ **Example 8–2**

In Fig. 8–8(a) the vertical signal is precisely 1 kHz. Find the frequency of the horizontal signal.

*Solution*

$$F_v/F_h = N_h/N_v \qquad (8\text{–}3)$$

Thus,

$$F_h = F_v N_v/N_h$$
$$= (1000 \text{ Hz})(1 \text{ loop})/(2 \text{ loops})$$
$$= 1000 \text{ Hz}/2 = \mathbf{500 \ Hz}$$

This technique is limited by the necessity of accurately knowing one of the frequencies, and it must be an integer multiple or submultiple of the unknown. In practical applications a variable, calibrated, signal source is adjusted until the pattern "locks in," and then the frequency is read from the generator dial.

---

## 8–7 THE Y-T OSCILLOSCOPE

The instrument that many people identify as the **basic oscilloscope** is the Y-time, or Y-T, oscilloscope shown in Fig. 8–9(a). The vertical channel is the same as in the X-Y system: push-pull output vertical signal amplifiers driving the deflection plates. The horizontal amplifier, however, connects not to an external signal source but to an internal time base signal.

The time base generator creates a sawtooth waveform that deflects the beam horizontally. The CRT deflection plate potentials are arranged so that the beam is positioned on the left side of the screen when the sawtooth ramp voltage is zero. The beam is pulled to the right as the ramp voltage rises. If everything is adjusted properly, then the ramp will reach maximum just as the beam disappears off the right edge of the screen [Fig. 8–9(b)].

The sawtooth waveform drops rapidly back to zero once the ramp portion is completed, snapping the beam back to the left edge of the CRT screen. This action would cause a **retrace line** to be printed on the CRT screen. To overcome this problem, a

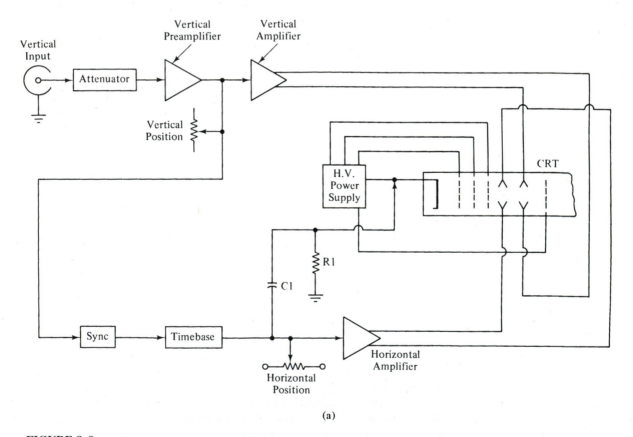

(a)

**FIGURE 8–9**
(a) Block diagram for a Y-time (Y-T) oscilloscope.

**FIGURE 8–9,** *continued*
(b) Relationship of vertical to horizontal signals produces the pattern on the screen.

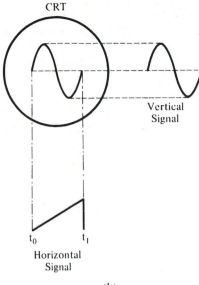

CRT

Vertical
Signal

$t_0$      $t_1$

Horizontal
Signal

**(b)**

**retrace blanking pulse** is generated that extinguishes the electron beam. This turns off the beam, eliminating the retrace line on the screen.

In low-cost oscilloscopes the time base is said to be **free running,** although the time base oscillator may, in fact, be synchronized to the vertical amplifier signal. Unless the time base is so synchronized, the waveform marches across the screen and remains unstable. Synchronization means that the time base signal sweeps across the screen in a time that is equal to an integer number of vertical waveform periods. The vertical waveform will then appear locked on the CRT screen.

The vertical section consists of a wideband preamplifier and power amplifier combination that drives the CRT vertical deflection plates. The vertical amplifier has a high gain, so large signals must be passed through an **attenuator** or, in low-cost instruments, a **vertical gain** control.

The free-running oscilloscope is a low-cost instrument, but it is relatively simple to operate.

**8–8**  **TRIGGERED SWEEP**

The low-cost free-running oscilloscope provides only limited usefulness today. The **triggered sweep oscilloscope,** on the other hand, is considered far more versatile and is the standard of the industry. Tektronix, Inc., and others now offer low- to medium-priced triggered sweep models that have supplanted the free-running models in all but the lowest-cost applications.

Figure 8–10 shows the block diagram of a triggered sweep model, at least those stages that differ from free-running models. In a triggered sweep model, the horizontal

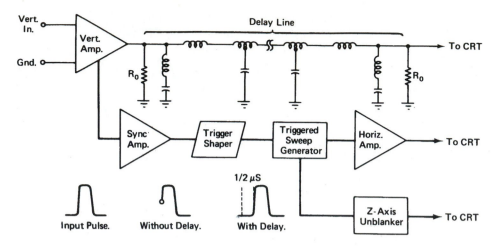

**FIGURE 8–10**

Block diagram of a triggered sweep oscilloscope.

sweep is turned off, with the CRT beam remaining at the left-hand side of the CRT screen, until a vertical signal is encountered.

The trigger circuit usually consists of a Schmitt trigger and a monostable (i.e., one-shot) multivibrator. The Schmitt trigger is a circuit that will produce either high or low output levels, depending upon whether the input signal is above or below a present threshold level. When the amplitude of the vertical amplifier signal exceeds this threshold, then the trigger output changes level, and this action causes the one-shot to produce a pulse that unlocks the sweep for one excursion across the CRT screen. As long as signals exist in the vertical amplifier and these have a level that is capable of retriggering the sweep, then the sweep will have the appearance of being free running. The difference, however, is that the sweep begins at the same point on the vertical waveform each time.

All electronic circuits have a certain propagation time, that is, the length of time delay, or phase shift, between the input and the output. The trigger circuit is no exception; it takes time for these circuits to operate. If the vertical signal were applied directly to the CRT deflection plates, then part of the vertical waveform leading edge would be lost; the sweep would not begin until several nanoseconds into the waveform. To overcome this problem, manufacturers place a **delay line** between the output of the vertical amplifier and the CRT. The delay lines in most oscilloscopes have a fixed delay time, but in a few expensive models the delay line is made variable to accommodate signals of various rise times.

This delay line should not be confused with **delayed sweep,** or **delayed triggering,** found in some models. The delayed trigger feature creates a small time delay between the beginning of a waveform and the initiation of a sweep cycle. The delayed sweep feature allows viewing of a small feature, or section, of a larger waveform—for example, a small oscillation, or some ringing, that is riding on a larger waveform. We may, for example, see several cycles of a 1-mHz oscillation riding on the peak of a 10-kHz sine wave [see Fig. 8–11(a)]. The period of a 1-mHz signal is 1 μs, while that of a 10-kHz signal is 100 μs. To view the 1-mHz signal, set the horizontal sweep to show several cycles of the

**FIGURE 8–11**
(a) Input signal with a higher-frequency oscillation on its peak.
(b) Using delayed sweep allows us to see the high-frequency signal.

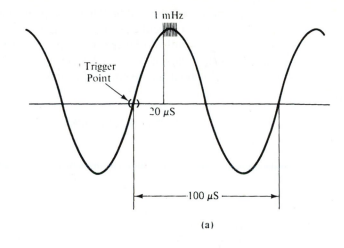

(a)

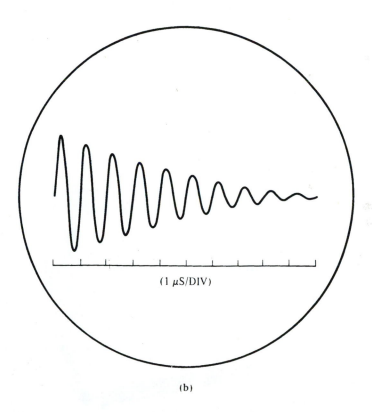

(b)

183

1-mHz waveform, and then trigger the sweep on the 10-kHz signal. Observe in Fig. 8–11(a) that the 1-mHz oscillation begins at approximately 20 μs past the trigger point. If the *trigger delay* control is adjusted for 20 μs, and if the sweep set for 1 μs/div, we will see only the 1-mHz signal on the CRT screen. The 10-kHz signal will be suppressed. More will be said about this feature in Section 8.10 (Oscilloscope Controls).

## 8–9     OSCILLOSCOPE SPECIFICATIONS

Several parameters are important to the user of oscilloscopes. How well the user understands these parameters may determine how useful an instrument is in any given situation.

### 8–9–1     Sensitivity

The **vertical sensitivity,** or **vertical deflection factor,** is a measure of how much deflection will be given for a specified input signal. The CRT screen has a grid pattern called a **graticule** (see Figs. 8–12(a) and 8–12(b). The divisions of the graticule are

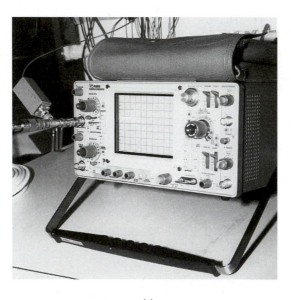

(a)

(b)

**FIGURE 8–12**
Portable triggered sweep oscilloscopes.

between 0.75 and 1.3 cm apart, with most being exactly 1 cm apart. The vertical attenuator(s) are calibrated in terms of **volts per division;** that is, the setting indicated will require the stated amount of voltage to deflect the beam one graticule division in the vertical direction.

The vertical sensitivity of an oscilloscope is the *smallest* deflection factor on the attenuator knob. If, for example, the most sensitive position of the vertical attenuator is 2 mV/div, then the vertical sensitivity of that oscilloscope is 2 mV/div.

## 8–9–2        Bandwidth

The bandwidth of an oscilloscope is also important, but it comes in for a lot of, shall we say, "creative spec writing," in advertisements. A typical oscilloscope may have a frequency response from dc to some specified upper frequency limit (e.g., 500 kHz, 15 mHz, 35 mHz, etc.). Most commonly, reputable oscilloscope manufacturers specify the range of frequencies over which the vertical amplifier gain (and hence attenuator calibration) is within ±3 dB of the proper gain. Beware of oscilloscopes that have a frequency response specified to some other point (e.g., ±3.5 dB). This is not the same, and comparisons are difficult to make. Fortunately, most advertisers of oscilloscopes are reputable.

Also beware of those models that offer a response specification of "3 dB down at 10 mHz," or some other frequency. This would lead you to believe that the bandwidth curve is flat from dc to some high frequency and that the model falls off to its rated frequency only 3 dB. You may easily believe that this is a model with a 10-mHz bandwidth. But it could also be true that the gain rolls off much faster than implied by the specification, and that the manufacturer "tweaked" the response to improve the gain at 10 mHz, in order to validate the specification. A reputable manufacturer guarantees the specification across the entire band.

Often more important than basic frequency response is the **rise time** of the vertical amplifiers. This specification gives us an indication of how well the oscilloscope will reproduce pulse waveforms.

## 8–9–3        Rise Time

The rise time of a pulse is usually defined as the time required for its leading edge to rise from 10% of its final amplitude to 90%.

An oscilloscope used to view a pulse needs a vertical rise time that is equal to or faster than the pulse rise time. If the oscilloscope manufacturer does not publish the rise time specification, then you may calculate it from the frequency response by using Equation (8–4). Alternatively, you may also use the same equation to calculate the required bandwidth if you know the pulse rise time:

$$F = \frac{0.35}{t_r} \tag{8–4}$$

where    $F$ = the ±3-dB frequency response in hertz (Hz)
           $t_r$ = the pulse rise time in seconds (s)

■   **Example 8–3**

What is the minimum frequency response required of an oscilloscope that is to reproduce without distortion a pulse that has a 15-ns ($1.5 \times 10^{-8}$ s) rise time?

*Solution*

$$F = 0.35/t_r$$
$$= 0.35/1.5 \times 10^{-8} \text{ s}$$
$$= 2.3 \times 10^7 \text{ Hz} = \textbf{23 mHz}$$

(8–4)

**8–9–4**     **Horizontal Sweep Time**

The low-priced free-running oscilloscope uses a horizontal section that is calibrated in sweep frequency. Better models, including all of those used in engineering, calibrate the horizontal section in units of **time per division.** Since these better models have ten horizontal graticule divisions, the sweep time across the CRT screen is

$$\frac{0.2 \text{ μs}}{\text{div}} \times 10 \text{ div} = 2 \text{ μs}$$

Using this information, we may determine the approximate frequency or period of displayed waveforms. For example, a trace shows that each complete pulse has a *period* of 2.3 divisions, which in units of time converts to

$$2.3 \text{ div} \times \frac{0.2 \text{ μs}}{\text{div}} = 0.46 \text{ μs}$$

The frequency in hertz is the reciprocal of the period in seconds, so we determine that the frequency of the displayed waveform is

$$F_{(Hz)} = \left[ \frac{1}{0.46 \text{ μs} \times \dfrac{1}{10^6 \text{ μs}}} \right]$$

$$= \frac{10^6}{0.46 \text{ s}}$$

$$= 2.17 \times 10^6 \text{ Hz} = 2.17 \text{ mHz}$$

The oscilloscope is uniquely suited to waveform analysis, giving not merely shape information but also frequency, period, amplitude, and rise time.

**8–9–5**     **Dual-beam Models**

The popularity of the dual-beam oscilloscope has grown immensely in the past decade or so. The name "dual beam," however, refers not to two electron beams but to two *traces* on the CRT screen. The creation of the two traces from a single electron beam is done by switching.

The oscilloscopes in Fig. 8–12 are both dual-beam models, so they allow the user to view and compare two harmonically related waveforms. Note that these are not like older two-channel models that allowed you to view *either* channel A *or* channel B but not *both* at the same time.

Figure 8–13(a) shows how a dual-beam oscilloscope can be made. A single vertical output amplifier is driving the vertical deflection plates of the CRT. But there are two pre-amplifiers, one each for channels A and B.

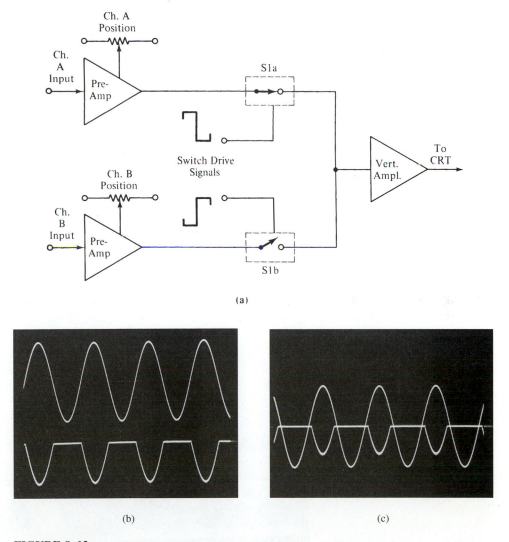

(a)

(b)                                                          (c)

**FIGURE 8–13**
(a) Electronic switching allows two traces on the same CRT. (b) and (c) Two separate signals can be viewed simultaneously against the same time base.

A dual CMOS electronic switch is used to switch the input of the vertical amplifier between the outputs of the two vertical preamplifiers. These switches are driven *out of phase* with each other, so that only one is connected to the vertical amplifier at any given time.

There are two separate vertical inputs on dual-trace 'scopes, so two separate signals can be viewed simultaneously against the same time base [Figs. 8–13(b) and 8–13(c)]. The two position controls can be adjusted to either separate [Fig. 8–13(b)] or superimpose [Fig. 8–13(c)] the two input signals. Dual-trace capability is extremely useful for examining the time relationship between two signals or for comparing the signals with each other on a point-by-point basis. Another application is in troubleshooting. For example, dual-trace CROs can be used to examine the input and output signals of a circuit to determine which is malfunctioning.

## 8–10    OSCILLOSCOPE CONTROLS

Although different oscilloscope models and manufacturers have somewhat different sets of controls, a certain commonality makes it possible to generalize the controls found on most 'scopes.

### 8–10–1    Controls Group

The controls group—some of which may appear on the back of the 'scope, as shown in Fig. 8–14—includes the **on-off switch** (power), **intensity, focus, astigmatism, illumination, trace rotation, beam finder,** and internal **calibration** signals. These controls operate as follows:

**On-Off/Power.**    The power control turns the instrument on and off; it is the ac mains power switch (or the battery switch in portable models). In some 'scopes the on/off switch is part of another control (usually intensity), similar to the on/off switch on a radio being

**FIGURE 8–14**
Controls group on back of oscilloscope.

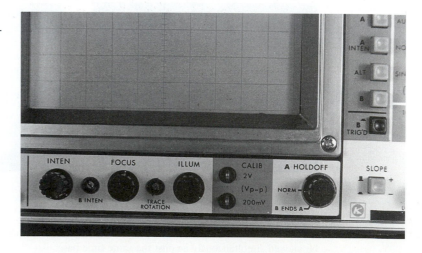

ganged to the volume control. In other 'scopes, perhaps on most modern models, the on/off switch is a separate entity. Although on older models it may have been a toggle or rotary switch, on most 'scopes today it is a pushbutton switch.

**Intensity.**   The intensity control controls the brightness of the CRT beam. As a general rule, keep the intensity control just high enough to see the entire waveform comfortably. If the same waveform is expected to remain on the CRT screen for a long time (or if there is no waveform), keep the intensity low in order to prevent burning of the CRT phosphor.

**Focus.**   This control adjusts the size of the electron beam spot on the CRT screen. It is often interactive with the astigmatism control.

**Astigmatism.**   This control adjusts the "roundness" of the CRT spot and is often interactive with the focus control. A good way to adjust the astigmatism control is to set it for a uniform line thickness when the CRT is swept horizontally (but no signal is present in the vertical channel). On many 'scopes the astigmatism control is a screwdriver adjustment available on either the front or the rear panel, while the focus control is an adjustable knob on the front panel.

**Illumination.**   The illumination control adjusts a lamp that lights up the graticule lines inscribed on the CRT screen. At the lowest settings of this control, no light appears on the graticule. You must use care with this control when photographing the CRT screen, as the graticule lighting can overexpose the ASA 3000 film typically used in 'scope cameras.

**Trace Rotation.**   Also called **trace align** on some models, this screwdriver adjustment compensates for the effects of local magnetic fields on the CRT trace. This control is adjusted so that the CRT beam is horizontal with respect to the graticule.

**Beam Finder.**   This control disconnects the horizontal and vertical inputs so that the beam collapses to a spot of higher intensity than the original trace. It helps the operator locate the beam, which may be off the screen at times.

**Calibration.**   The calibration control provides a standard signal to calibrate the vertical amplifier controls. Typically, such signals are 400 or 1000 Hz, either 1 or 2 V peak-to-peak. In some cases it is a sine wave, in others a square wave. In the model shown in Fig. 8–14, the 1000-Hz square wave has both 2-V and 200-mV levels.

8–10–2      **Vertical Group**

The vertical group controls the vertical position of the beam and the amplification factor and input selection of the oscilloscope. Controls in this group (Fig. 8–15) include **input, input selector, position, step attenuator, vernier (variable) attenuator, ground, 5×
magnification, channel-2 polarity,** and **vertical mode.** Also shown in Fig. 8–15 is an **internal trigger selector (INT TRIG),** which is technically part of the trigger control group (its purpose is to select the channel that supplies the triggering signal).

**FIGURE 8–15**
Vertical control group on oscilloscope.

**Input Connector.** The input connector is the point at which signal is applied from the external world to the oscilloscope vertical amplifier. In the model shown in Fig. 8–15, the input connector is a coaxial BNC-style chassis-mounted connector. This form of connector is the standard on modern oscilloscopes. On older 'scopes the input connector might be an SO-239 female UHF coaxial connector or even a pair of 5-way binding posts spaced on 0.75-in centers. Because most modern probes and 'scope accessories are now BNC equipped, owners of older models usually opt to buy either an SO-239-to-BNC or a post-to-BNC adapter (as appropriate).

**Input Selector.** This switch is marked "AC-GND-DC" and is used to select the coupling of the input connector to the input of the vertical amplifier. The "DC" setting means that the connector is dc (direct current) coupled to the amplifier input; the "AC" setting means that a blocking capacitor is in series with the connector center conductor (hence only ac signals will pass; dc is blocked). In the "GND" setting the input of the vertical amplifier is shorted to ground, and the input connector center conductor is open circuited ("GND" does NOT ground the input connector, which would short-circuit the signal source!).

**Position.**    The position control moves the electron beam up and down on the CRT face. On a dual-trace 'scope the two vertical position controls are normally used to prevent overlapping of the two traces (and resultant confusion). Otherwise, the control can be used to position the trace precisely over the graticule markings for amplitude measurements. It is common practice, for example, to set the AC-GND-DC input selector to GND and then use the position control to set the straight line trace over either the bottom or the center graticule line, which then becomes the 0-V reference point.

**Step Attenuator.**    The sensitivity of the oscilloscope amplifier is a measure of its gain and is expressed in terms of the voltage required to deflect the CRT beam a specified amount, that is, V/div (or V/cm). The step attenuator (see close-up in Fig. 8–16) is a resistor/capacitor voltage divider that allows the instrument to accommodate higher potentials that would otherwise over-deflect the CRT beam. Each position of the step attenuator is calibrated in volts or millivolts per division (V/div or mV/div). To make the actual peak-to-peak voltage measure of the input signal, note how many CRT screen divisions the signal occupies, and then multiply that figure by the sensitivity factor in V/div. For example, suppose a sine wave signal occupies 5.6 divisions peak-to-peak when the step attenuator is set to 0.2 V/div. The peak-to-peak voltage is (5.6 div) × (0.2 V/div) = 1.12 V (i.e., 1120 mV).

**Vernier Attenuator.**    This variable control is concentric to (in the center of) the step attenuator and allows continuously variable adjustment of the sensitivity factor (hence it also traces vertical size). The calibration of the step attenuator is valid only when the vernier attenuator is in the CAL'D (also called CAL) position, which on most 'scopes is detented for easy location. When the vernier control is not in the CAL'D position, a red UNCAL lamp on the front panel warns the operator that the step attenuator settings are not to be trusted.

**FIGURE 8–16**
Close-up view of step attenuator.

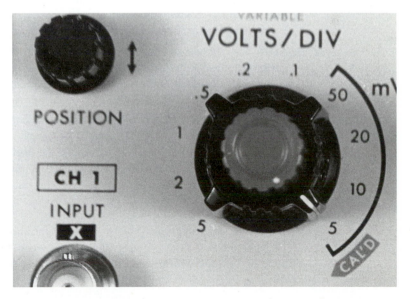

**Ground.**   This ground jack is connected to the chassis ground at the input of the vertical amplifiers. It can be used to provide a proper "star" ground in order to eliminate (or prevent) ground loop errors.

**5× Mag.**   The 5× magnification control increases the sensitivity factor by five times, which means that all of the V/div and mV/div calibrations must be divided by 5. For example, when the VOLTS/DIV knob is in the 50 mV/div position and the 5× MAG button is pressed, a 5× MAG light turns on to warn the operator, and the sensitivity increases fivefold (10 mV/div in this example). This feature is especially useful when you are dealing with low-level signals that are ordinarily below the threshold of the normal settings, so it effectively doubles the number of available sensitivity factors.

**Channel-2 Polarity.**   When pressed, the polarity control inverts the channel-2 vertical signal. If this control is left unoperated, the polarity of the signal on the screen from channel-2 is normal. The polarity control allows us to have a pseudodifferential input on a single-ended 'scope (see ADD control below).

**Vertical Mode.**   This control forms a subgroup that includes the following: CH1 and CH2; ALT and CHOP; ADD; and X-Y. These controls are described as follows:

> **CH1, CH2.** This control selects the single-channel mode. When "CH1" is pressed, the 'scope operates as a single-channel model and displays only the channel-1 signal. When "CH2" is pressed, only the channel-2 signal is examined.
>
> **ALT, CHOP.** These are dual-channel modes. There is only one electron beam in the CRT, and it must be shared between the channels. In the ALT mode the channel-2 trace does not start until the channel-1 sweep is finished. In other words, the 'scope alternates between the two signals. In the CHOP mode, the electron beam is switched back and forth rapidly between channel 1 and channel 2. The input signal must have a frequency that is very much less than the chopping frequency.
>
> **ADD.** The signals are combined into one, with the resultant amplitude being the algebraic sum of the two channels (CH1 + CH2). If the CH2 polarity control is pressed, then the inputs become pseudodifferential, and the summation is CH1 − CH2.
>
> **X-Y.** In this mode the internal oscilloscope time base is disconnected, and the instrument becomes a vectorscope. Channel 1 becomes a horizontal (X) input, while channel 2 is the vertical (Y) input. In this mode the oscilloscope can be used for modulation measurements, color-TV Lissajous patterns, and so forth.

## 8–10–3        Horizontal Group

The horizontal control group [Fig. 8–17(a)] determines the horizontal deflection and sweep characteristics. These controls consist of **sweep time, sweep vernier, horizontal position, 10× magnification,** and **sweep mode.** These are as follows:

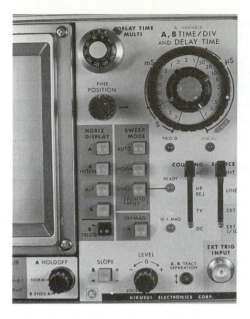

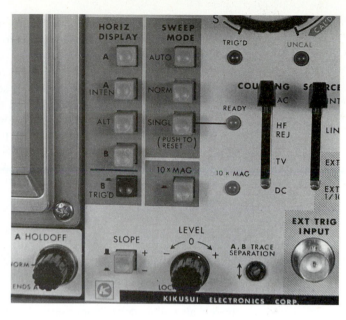

(a)                                                                                          (b)

**FIGURE 8–17**
(a) Horizontal and trigger control group on oscilloscope, (b) close-up of trigger controls.

**Sweep Time.**    This is the main horizontal timing control and is used to determine the amount of time required per division to sweep the beam across the CRT face left to right. The calibration of this control is in units of time/division (s/div, ms/div, or µs/div). From this control the period of a signal and the number of divisions occupied by one cycle of the signal can be determined. For example, if exactly one cycle of a sine wave occupies 6.2 divisions of the horizontal graticule, and if the switch setting is 2 ms/div, then the period of the signal is (6.2 div) × (2 ms/div) = 12.4 ms. Because frequency is the reciprocal of period, we can calculate the frequency as $F = 1/T = 1/0.0124$ s $= 80.65$ Hz.

**Sweep Vernier.**    This is a continuously variable time control that allows us to interpolate between step time settings. The step time settings are accurate only when the vernier is in the CAL'D position. The vernier is ganged to the step attenuator and is thus concentric to and in the center of that control.

**Horizontal Position.**    The horizontal (or "fine") position control moves the trace left and right on the CRT screen. Like the equivalent vertical control, the horizontal position control is used to place key features right over graticule points for purposes of precision measurement.

**10× Mag.**   The 10× magnification control speeds up the sweep by 10 times. For example, if the time/div sweep control were set to 10 ms/div, then the 10× MAG control would force it to become 1 ms/div.

**Sweep Mode.**   The sweep mode control selects automatic (AUTO), normal (NORM), and single sweep (SINGLE) submodes. In the AUTO mode the sweep will periodically retrigger even if no signal is present in the vertical amplifier. The NORM mode requires a vertical signal to begin sweeping the CRT, and the screen will remain blank otherwise. In the SINGLE mode the CRT beam will sweep only once. Two means of operation are noted. If there is a periodic signal present, pressing the button in the AUTO position will force one sweep to take place. If the NORM mode is selected, then the SINGLE button will reset the circuit, which will sweep only after a valid input signal is received.

**8–10–4**     ### Trigger Group

The triggered sweep oscilloscope is considerably more useful than the old untriggered forms. The triggered sweep 'scope will not allow the CRT beam to sweep across the screen unless a signal is in the vertical amplifier to trigger the sweep generator. Some models also allow delay time triggered sweep function. That is, the sweep will not actually begin until some preset time after the triggering event occurs. The trigger group of controls is also shown in Fig. 8–17(a), along with the horizontal controls, and closer up in Fig. 8–17(b). The controls include **trigger level ("level"), slope, source, coupling, external trigger input,** and a **horizontal display** selector. In addition, some models [such as in Fig. 8–17(a)] also have a **time delay vernier** control. Keep in mind that the sweep mode, which we discussed under "Horizontal Group," is also part of the trigger group [Fig. 8–17(b)], as is the CH1-CH2-NORM switch shown along with the vertical group controls.

**Trigger Level.**   The trigger level control determines what minimum amplitude vertical signal is required to trigger the horizontal sweep and where on the waveform sweep begins [Fig. 8–18(a)]. Figures 8–18(b) and 8–18(c) show the effect of this control. Both traces were taken moments apart with the same signal input to the same oscilloscope. The only difference in these displays was the setting of the LEVEL control [see Fig. 8–17(b)]. The range of the control runs from negative, through zero, to positive voltage values.

**Slope.**   This control determines whether the trigger occurs on a negative-going or a positive-going edge of the input waveform. In Figs. 8–18(a) and 8–18(b) the slope control is set to the "+" position, so the triggering occurs on the positive-going edge of the sine wave. In Fig. 8–18(c), on the other hand, we see exactly the same signal with the level control set as it was in Fig. 8–18(b) and the slope control changed to "−." Note that the triggering now occurs on the negative-going slope of the waveform.

**Source.**   The source control selects the source of the signal applied to the triggering circuits. The selections are INT, LINE, EXT, and EXT/10. The INT is the internal selection, which means that the source is selected by the CH1/CH2/NORM switch in the vertical

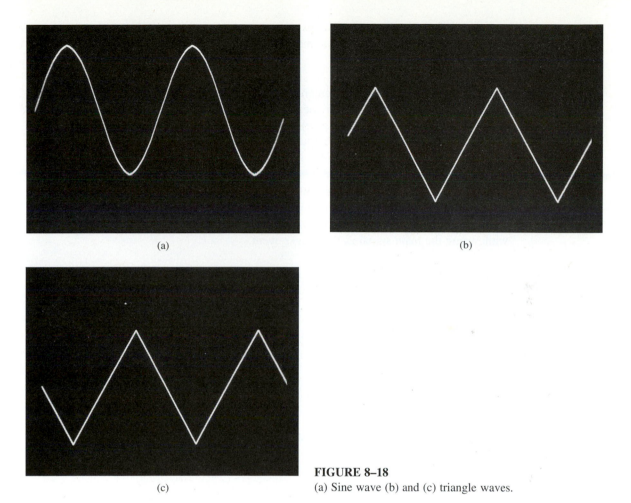

(a)

(b)

(c)

**FIGURE 8–18**

(a) Sine wave (b) and (c) triangle waves.

section. For example, with the source control in INT and the other switch in CH1 posi-tion, the signal in the CH1 vertical amplifier will cause triggering. The LINE selection means that the 60-Hz ac line will cause triggering, a feature useful in some measure-ments. The EXT means that the signal applied to the external trigger input (EXT TRIG INPUT) will trigger the sweep circuits. Again, some useful measurements are possible with this feature. The EXT/10 is the same as EXT, but a 10:1 attenuator is in place.

**Coupling.**    The coupling control allows us to tailor the triggering. It has selections of AC, HF REJ, TV, and DC. The AC and DC are self-explanatory and are similar to the same markings on the vertical selector switch. The HF REJ uses a low-pass filter at the input of the trigger circuit that rejects high frequencies. This system will, for example, allow us to trigger on the modulation of a modulated RF carrier while ignoring the RF signal. Some 'scopes also have an LF REF, which is similar except that the filter is a high-pass filter. The TV selection allows us to sync to the horizontal/vertical frequencies used in television sweep systems. Some models have separate TV VER and TV HOR selections.

**External Trigger Input.**    This control has an input connector that provides the trigger circuit with an external signal for special-purpose syncing and triggering. Some 'scopes also have a TRIG GATE (trigger gate) function on this connector in certain switch selections (or a separate TRIG GATE output). That feature produces a short-duration pulse for synchronizing external circuits to the sweep system.

**Time Delay.**    The time delay control allows us to program a short delay between the triggering event selected according to the LEVEL and SLOPE controls and the actual onset of the sweep. Using this control, we can view small segments of the waveform while using the main signal as the trigger event.

**Horizontal Display.**    The horizontal display is a switch bank [Fig. 8–17(b)] that allows certain submodes. Not all 'scopes have this feature, even though it is very useful. Figure 8–19 shows the operation of certain features of this selector. When button A is pressed, the 'scope operates as any triggered sweep 'scope operates. But in the A INTEN mode, we see a trace such as in Fig. 8–19(a). Note the segment of the waveform that is intensified. The position of this intensified segment is a function of the time delay control, while the length of the intensified portion is a function of the delay time control that is concentric with the time/div control. We can use this mode to designate a small segment of the waveform for a closer look. When the B switch is pressed, that portion of the waveform is displayed, as in Fig. 8–19(b). A slightly different function is shown in Fig. 8–19(c), which is the trace that results when the ALT button is pressed. In this case we see both the main waveform and the time-delayed "close-up" portion. A screwdriver "A-B Trace Separation" control allows us to either separate or superimpose these waveforms.

**8–11**          **COMMON OSCILLOSCOPE PROBES**

Figure 8–20(a) shows the most basic form of input probe for oscilloscopes. Here we see a length of shielded cable, usually coaxial cable, with a BNC (banana plugs or PL-259 on older instruments) on one end and a pair of alligator clips on the other end. This method works well for signals with frequencies from dc up to a certain point, and for some readers this probe set is all that is required. But there is a problem that must be recognized. The cable has capacitance on the order of 20 pF per foot. The input impedance of a typical oscilloscope is a 1-M$\Omega$ resistance shunted with a 20-pF capacitance. If the cable is 3 ft long, then it has a capacitance of $3 \times 20$ pF, or 60 pF, which when added to the natural input capacitance results in 80 pF shunting 1 M$\Omega$. The RC network thus created has a low-pass filter characteristic that rolls off −6 dB/octave above the following −3 dB frequency:

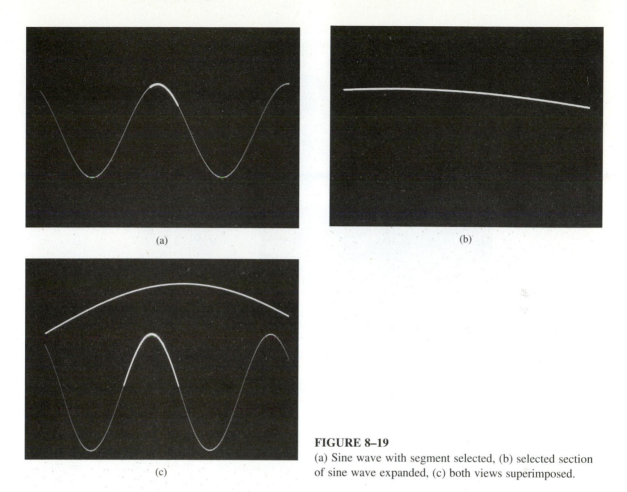

(a)

(b)

(c)

**FIGURE 8–19**
(a) Sine wave with segment selected, (b) selected section
of sine wave expanded, (c) both views superimposed.

$$F_{(Hz)} = \frac{1}{2\pi RC}$$

$$= \frac{1}{(2)(3.14)(10^{-6}\ \Omega)(8 \times 10^{11}\ F)}$$

$$= \frac{1}{0.0005}\ Hz = 1990\ Hz$$

This probe will load down any high-frequency circuit that it is used to measure, and thus it is not the best solution. Also, the fundamental frequency need not be anywhere near the cutoff frequency for there to be problems. Nonsinusoidal signals are made up of a collection of sine waves consisting of a fundamental plus harmonics. Thus, a fast rise time, 100-Hz square wave is made of a 100-Hz sine wave plus even harmonics up to the

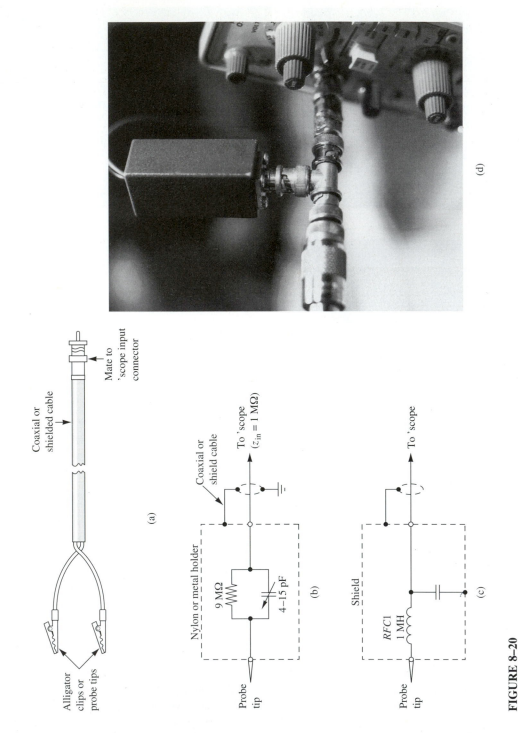

**FIGURE 8–20**

(a) Basic form of input probe for oscilloscopes, (b) circuit for low-capacitance probe, (c) circuit for RF-blocking probe, (d) TDR adapter.

**198**

zillionth or so. The low-pass filter effects of the probe in Fig. 8–20(a) will roll off the higher harmonics and round off the shoulders of the square wave.

The answer to the frequency response problem is to use a low-capacitance probe, two examples of which are shown in Figs. 8–20(b) and 8–20(c). The probe in Fig. 8–20(b) is the standard 10 :1 ratio probe. The output signal of this probe is 1/10 the input signal (if the input impedance of the oscilloscope is 1 MΩ). If the resistors used are precision types, then the scale factor on the vertical attenuator of the 'scope is multiplied by 10. For example, when the vertical attenuator is set to 0.5 V/cm, the actual scale factor is 5 V/cm.

In all three types of low-capacitance probe, the capacitor is adjusted to flatten the frequency response. In most cases a fast rise time, 1000-Hz square wave is applied to the input of the probe when it is connected to the 'scope. Adjust the capacitance to show as square a square wave on the screen of the 'scope as possible.

Another problem is the matter of isolation from external fields. The classical problem is looking at a waveform in the presence of an interfering electromagnetic field. The classical approach to this problem is insertion of an RF choke in series with the 'scope probe. Figure 8–20(c) shows a probe that can be used on medical electrosurgery machine and radio transmitter measurements. The 1-millihenry (1-mH) RF choke suppresses the RF that is present on the probe when it is in the presence of a radio field.

A problem with the probe in Fig. 8–20(c) is self-resonance. All RF chokes, indeed all inductors, have a certain amount of capacitance between windings and a stray capacitance to ground. These capacitances interact with the inductance of the coil to make either series or parallel resonances (or both!) . . . and that spells trouble in some cases.

A different kind of oscilloscope input device is shown in Fig. 8–20(d). Certain RF and computer measurements require special adapter devices to make the oscilloscope work. This particular adapter is used in time domain reflectometry, a method for "doping out" coaxial cable transmission lines such as those used to interconnect the elements of the receiver and antenna system of the cardiac telemetry system in a hospital.

## 8–12    MAKING MEASUREMENTS ON THE OSCILLOSCOPE

The standard oscilloscope display (Fig. 8–21) is calibrated in two axes: vertical and horizontal. The horizontal axis is calibrated in units of time, typically time per division (most modern 'scopes use the centimeter as the division). The vertical axis is the amplitude (usually voltage) of the applied signal. The graticule shown in Fig. 8–21 has two forms of divisions on the screen. The major divisions are each 1 cm, while the minor divisions are 0.2 cm each. These minor divisions are inscribed only on the center vertical and horizontal axes. In the example of Fig. 8–21, the vertical displacement is approximately 4.4 divisions, the horizontal 3.25 divisions.

## 8–13    STORAGE OSCILLOSCOPES

The input signal must be periodic, that is, it must repeat itself at fixed intervals, before the pattern will remain stable long enough to be analyzed. But single events often pass too rapidly to be viewed on an oscilloscope unless it is equipped with a *storage* feature. There

**FIGURE 8–21**
Standard oscilloscope display.

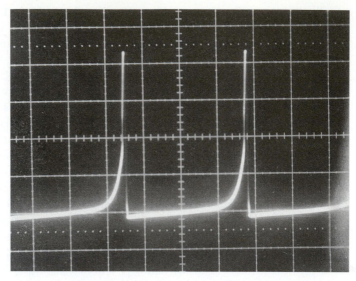

are now two basic types of storage oscilloscopes, one type using a special CRT and the other using digital memory techniques. Figure 8–22 shows the block diagram of the type that requires a special storage CRT.

A storage oscilloscope retains the image of the waveform on the screen for a period of time before it fades out. In the class of instruments using special CRT designs, there are several subclasses that operate on slightly different principles. Figure 8–23 shows three types of storage CRTs.

The two types of CRTs shown in Figs. 8–23(a) and 8–23(b) depend for operation on a phosphor screen in which individual particles are insulated from each other. In Fig. 8–23(a) the phosphors lie in the same plane to form target dots, while the other type [Fig. 8–23(b)] uses layers of scattered particles of increasing weights. A special **flood gun** in the CRT emits high-energy electrons that excite the phosphors. The phosphor particles struck by these electrons take on a charge of 150 to 200 V, but unenergized particles remain at 0 V. When electrons from the main electron gun have preenergized certain phosphors, forming an image on the CRT screen, these phosphors attract more flood gun electrons. The image is thereby retained by the continuous bombardment of flood gun electrons. Erasure of the screen is accomplished by grounding the phosphor screen, thus returning all phosphor particles to a 0-V potential.

The other type of special CRT is the wire-mesh variety shown in Fig. 8–23(c). The flood gun charges one mesh so that no further electrons can pass through, although the **write** or **main gun** electrons will pass through if sufficiently energetic.

A **split-screen** storage oscilloscope allows the operator to store waveforms in either the top or the bottom half of the screen. If both halves are turned on, then the device will store traces appearing anywhere on the screen.

Some storage oscilloscopes have a feature called **variable persistence,** which allows the operator to vary the length of time that the image will remain on the screen.

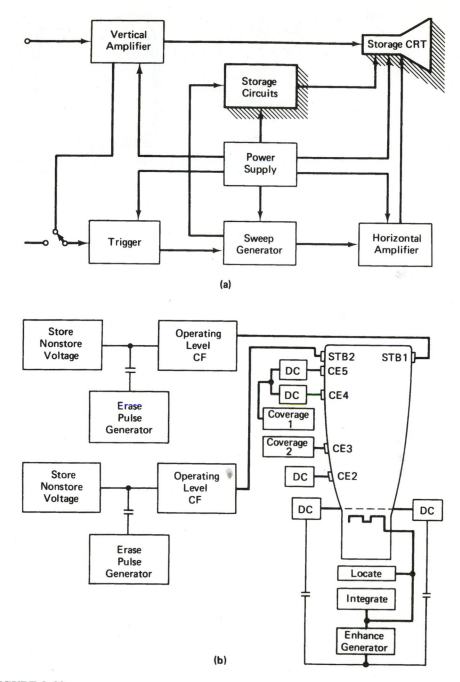

**FIGURE 8–22**
Functions of typical storage-type oscilloscopes: (a) Block diagram of typical storage scope, (b) subdivision of the storage circuits.

**FIGURE 8–23**
Three types of storage CRTs.

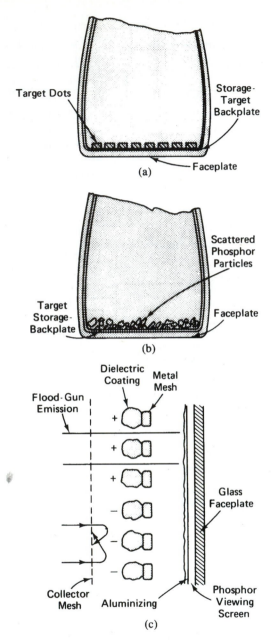

(a)

(b)

(c)

## 8–14   DIGITAL STORAGE

A digital storage oscilloscope (see Fig. 8–24) uses an analog-to-digital converter to **digitize** the input waveform. That means it will sample the waveform at many points and then convert the instantaneous amplitude at each point to a binary number value proportional to the amplitude. These binary numbers are then stored in memory. A digital-to-analog converter at the output of the memory circuit reconverts the binary words to analog voltages capable of driving the CRT vertical deflection system. The memory is scanned many times per second, so the CRT screen is constantly being "refreshed" by the data stored in memory before it can fade out.

The digital storage technique is used in some high-frequency oscilloscopes designed as "transient catchers," but the technique is very expensive for this frequency range. A sampling rule called **Nyquist's theorem** requires a sampling rate for any given waveform of *twice* the highest-frequency Fourier series component making up the waveform. Consider a short, irregularly shaped pulse. It may contain Fourier series components up to 10 mHz or so. To accurately sample that pulse would require $2 \times 10^7$ samples per second. Even if the pulse is very short, a lot of memory is used, and memory is expensive—as are ADC and DAC circuits that will track that fast.

The digital storage technique does, however, find extensive use in medical and physiological oscilloscopes, such as that shown in Fig. 8–25(a). The sweep time is typically 25 mm/s, so it takes approximately 4 s to sweep the CRT; thus, you would normally expect to find the beam fading on the left side of the screen before the trace reaches the right side. However, because of the digital storage capability, this 'scope is a **nonfade** model. The waveform on the screen is refreshed continuously, until a new waveform overwrites the old data. Many of these models have a **freeze** control that turns off the ADC so that no new data are written into memory. This action leaves the old waveform on the screen to be refreshed indefinitely. A standard CRT storage oscilloscope made by Tektronix is shown in Fig. 8–25(b).

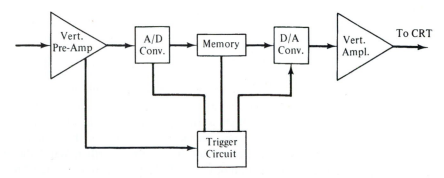

**FIGURE 8–24**
Digital storage oscilloscope.

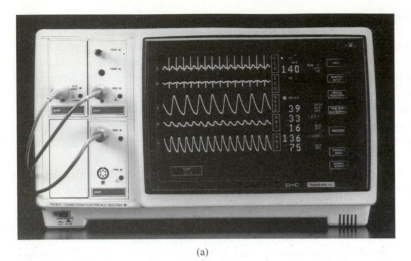

(a)                                              (b)

**FIGURE 8–25**
(a) Medical oscilloscope, (b) CRT storage oscilloscope [(a) courtesy of Space Labs, (b) courtesy of Tektronix, Inc.].

## 8–15   MEDICAL OSCILLOSCOPES

In the sections to follow we will take a closer look at oscilloscopes designed specifically for medical, biological, and physiological applications. Medical oscilloscopes used in ECG and similar applications are often calibrated along the horizontal axis in units of distance and time, rather than just time. For example, the standard external lead ECG, recorded on a strip chart recorder, is usually calibrated at 25 millimeters per second (25 mm/s). Other instruments will be calibrated at 50 mm/s or 100 mm/s. Some multipurpose instruments are available with all three calibrations.

The graticule on many medical 'scopes is inscribed with millimeter marks, and the major divisions represent 5 mm each. This calibration makes it easy to measure ECG feature parameters.

A newer form of oscilloscope now on the market is intended for use with computer graphics displays. These 'scopes are often calibrated in the units being measured, for example, distance or some other physical parameter. The computer will inscribe the CRT screen with a calibrating scale or will directly label the features as to calibration.

Three types of oscilloscopes are typically found in the medical world. First, there are the straight analog **bouncing ball** models. The nickname "bouncing ball" comes from the fact that the beam of light seems to bounce as the ECG waveform parades across the 'scope face. These 'scopes are medical variants of the straight nonstorage oscilloscope discussed earlier, so we will not discuss them further. While the medical models tend to have magnetic deflection of the CRT electron beam (via a CRT neck yoke), they are otherwise very similar.

The phosphor on the CRT screen is also a bit different. The regular 'scope will probably use a P1 or P4 phosphor. These phosphor mixtures are relatively fast and thus fade immediately after the electron beam passes a point. This feature is desirable for the labora-tory user, but for slow waveforms such as the ECG, arterial pressure, or EEG, the rapidly fading light beam does not allow examination of the entire waveform. For these reasons, a P7 or similar "long-persistence" phosphor is usually selected by the 'scope designer.

The standard long-persistence phosphors have an interesting property. The light emit-ted is a mixture of yellow, green, and violet-blue colors. The standard unfiltered display looks "violet-ish" to most people (especially under fluorescent lighting). However, if a filter is used over the CRT screen, then either green or yellow colors are emitted, but these colors are a bit dimmer (especially the yellow) than the unfiltered versions.

The second type of 'scope used in medical applications is the **analog storage** 'scope. These instruments use trickery inside the CRT to store the signal on the face of the CRT for a long period of time. These instruments are used in older vectorcardiogram 'scopes and other applications, but like the straight bouncing ball displays, they are all but obso-lete. The modern approach, however, is to use a **digital storage** oscilloscope—the so-called **nonfade** 'scopes seen in medicine, discussed in the following section.

**8–15–1**   **Nonfade Medical Oscilloscopes**

The traditional form of oscilloscope uses a beam of electrons to sweep the screen, writ-ing the analog waveform as it is deflected. Even with long-persistence phosphors, how-ever, the trace vanishes from the CRT shortly after it is written onto the screen. This type of CRT is usually called a bouncing ball display in the jargon. To the medical personnel using the bouncing ball display, it is very difficult to evaluate waveform anomalies because the trace fades too rapidly.

The analog storage oscilloscope is a partial solution to the problem for some research applications, but it is generally not suited to monitoring and other clinical applications. A problem with the analog storage 'scope is that the trace tends to "bloom" out and become fuzzy a few minutes after it is taken—or if circuit conditions are not exactly right.

The solution to these problems is a special variant of the digital storage oscilloscope, also called the **nonfade oscilloscope** in medical instrument jargon. Analog CRT storage systems are not used in most medical 'scopes because, at the low frequencies involved, the digital types offer a better display at competitive prices and the display is more per-manent. Also, the digital type of nonfade display does not bloom when the display is either old or erased. While most earlier nonfade 'scopes used discrete digital logic (as do many today), modern computer techniques (which are all but ubiquitous in medicine now) offer an easy approach to the nonfade display.

Two different formats of nonfade 'scopes are commonly used: **parade** or **erase** bar (see Fig. 8–26). In both forms the waveform may march across the screen from right to left rather than from left to right.

Figure 8–26(a) shows the parade type of display. The newest data, which are being written in real time, appear in the upper right-hand corner of the screen. The light beam bounces up and down at a fixed horizontal point in response to the vertical waveform; it does not move along the time base as it does in regular analog 'scopes. The waveform is nicknamed the "parade" display because the oldest data march across the screen, seem-ingly leading a parade of waves.

**FIGURE 8–26**
(a) Parade display, (b) erase bar format.

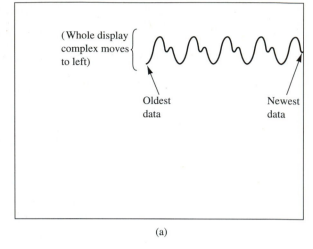

(Whole display complex moves to left)

Oldest data

Newest data

(a)

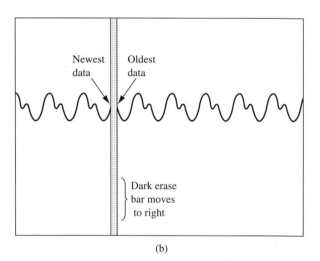

Newest data

Oldest data

Dark erase bar moves to right

(b)

With the erase bar format [Fig. 8–26(b)], the beam of light travels left to right (it is not stationary as on parade models). There is an erase bar (i.e., a dark region) traveling ahead of the beam that obliterates the oldest data so that the new data can be written onto the CRT screen.

There are two basic forms of digital circuits for nonfade oscilloscopes. Figure 8–27 shows a complex form that uses a regular computer memory, as might be found in a personal computer. The signal is applied to the input amplifier and is scaled to match the dynamic range of the analog-to-digital (A/D) converter. The A/D converter is used to convert the analog voltage to an equivalent binary word. For example, in a unipolar system with a 0-to-+10-V range, the binary word 00000000 might represent 0 V, while 11111111 might represent +9.96 V (it is not possible to get exactly to +10 V if zero is to be represented properly).

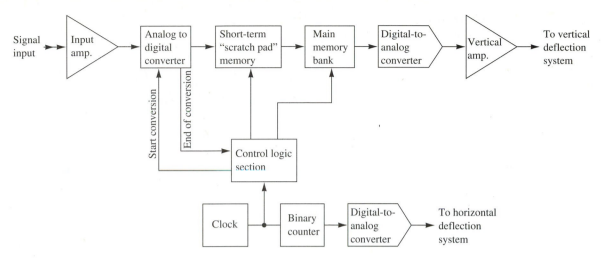

**FIGURE 8–27**
Complex oscilloscope using computer memory.

The output of the A/D is stored first in a short-term "scratch pad" memory. The contents of the scratch pad memory are periodically dumped to the main display memory. The rate of transfer from the scratch pad to the main memory is faster than the eye can see (typically 40 to 80 times per second). The memory is scanned via a signal from the control logic section, and then it is output in sequence to a digital-to-analog converter (DAC) that forms the signal applied to the oscilloscope vertical amplifier.

The horizontal sweep signal should be a sawtooth waveform. This signal is formed by a DAC that is driven by a binary counter. As the clock causes the counter to increment through its range, the output of the DAC ramps upwards. When the counter overflows, however, the DAC output snaps back to zero so the process can start anew. The type of nonfade 'scope depicted in Fig. 8–27 can store an immense number of waveform data. In fact, it can also transfer those data to a computer or to long-term magnetic media such as a floppy disk, a tape, or a main disk. However, in simpler instruments a different form of nonfade 'scope might be used.

Computers have added a new dimension to the medical instruments industry, and medical oscilloscopes have fared well with these developments. Figure 8–28 shows a form of graphics display for a computer that can be used for medical applications. (VGA video monitors are used extensively, although some special format graphics protocols are also used.) The screen of the CRT is broken into a matrix of tiny square or rectangular zones called **picture elements,** or **pixels.** Each storage location in the digital memory represents one pixel on the CRT screen. The pixel can be lighted to an intensity defined by the protocol (6 to 32 shades of gray). It is lighted by the electron beam as its raster scans the CRT surface.

Figure 8–29 shows typical medical monitor 'scopes used in intensive care units, emergency rooms, and like places. These 'scopes typically have at least a built-in ECG amplifier (a special-purpose vertical amplifier), as well as carrier amplifiers for blood pressure monitoring. Some also have temperature, EEG, and other functions.

**FIGURE 8–28**
Magnified segments show composition of trace.

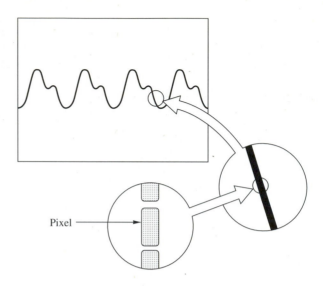

Pixel

The 'scope in Fig. 8–29(a) is hanging from a wall bracket in an intensive care unit. The pattern on the screen is the **built-in self-test** pattern that comes on the screen either on operator manual initiation or automatically at turn-on. This type of monitor is used in a computer-based monitoring system. The medical monitor in Fig. 8–29(b) is a portable unit designed for transporting patients. Note that it also contains a strip chart recorder for making a permanent record. This type of monitor 'scope can be mounted on an IV pole attached to the hospital "gurney" used to transport the patient. The model in Fig. 8–29(b) has ECG, arterial blood pressure, and a pulse meter readout from an electro-optical sensor used on the finger. The version in Fig. 8–29(c) is a slave 'scope used to simultaneously monitor several patients at the same time from a central location.

The monitors of Fig. 8–29, like certain other displays, use a "touch screen" method for the selector "switches." Figure 8–30 shows how most such 'scopes operate. Figure 8–30(a) shows a layout similar to those of Fig. 8–29, but in schematic form. Figure 8–30(b) shows how the finger sensors are arranged. Positioned along the edges of the display [Fig. 8–30(b)] are a series of infrared sources [light-emitting diodes (LEDs) operating in the IR region] and infrared detectors. The IR light is invisible to the naked eye and thus is not seen by the operator. The function labels are either painted onto the CRT by the computer [as in Figs. 8–29 and 8–30(a)] or affixed to the edge. When the operator touches the screen over any label, his or her finger interrupts one vertical and one horizontal beam, causing a unique pattern. For example, suppose all detector outputs are at binary LOW when the IR reaches the detector. The $X$ outputs $X1$-$X2$ are L-L, and the $Y$ outputs $Y1$-$Y2$-$Y3$-$Y4$-$Y5$-$Y6$ are L-L-L-L-L-L. Then someone touches the SELF-TEST label. In this case, $X1$ and $Y1$ are both interrupted and their respective outputs go HIGH. Thus, the patterns become H-L and H-L-L-L-L-L. The internal computer recognizes this as an operator command to branch to the SELF-TEST software stored in program memory.

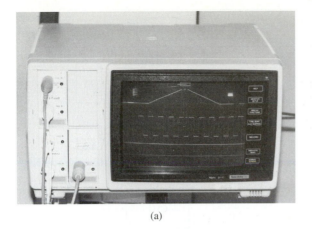

(a)

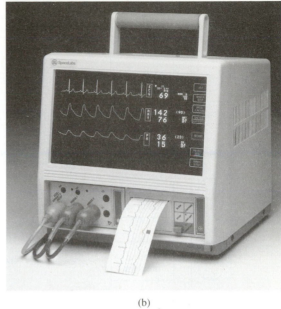

(b)

(c)

**FIGURE 8–29**

Typical medical monitor 'scopes. (a) Multifunction bedside monitor, (b) portable monitor, (c) central monitor [(b) and (c) courtesy of Space Labs].

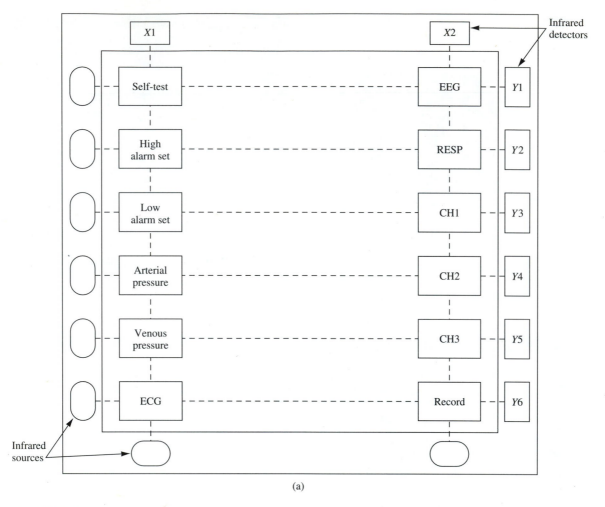

**FIGURE 8–30**
(a) Schematic layout of touch screen of oscilloscope.

**8–16          DELTA-TIME (Δ-TIME) OSCILLOSCOPES**

Some oscilloscopes are equipped with a Δ-time feature that allows the measurement of the time interval between two points on the applied waveform. The model shown in Fig. 8–31(a) uses an internal electronic period counter to measure time intervals. The numerical readout is in a small window on the oscilloscope front panel. Markers on the CRT screen, in the form of enhanced trace brightness, allow the user to select the two points on the input waveform between which the measurement is made.

In the second example, shown in Fig. 8–31(b), the measurement is made by a digital voltmeter mounted on top of the unit. A voltage proportional to the interval is available on the back panel.

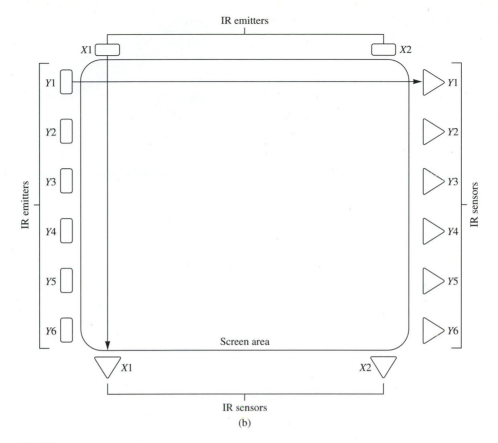

**FIGURE 8–30,** *continued*
(b) Arrangement of finger sensors on touch screen.

## 8–17     OSCILLOSCOPE CAMERAS

The oscilloscope is capable of measuring very rapid repetitive or transient waveforms, but the waveform is viewed only temporarily. Repetitive waveforms exist only as long as all of the instruments that measure the waveforms and the circuit that produces them are operating, whereas transient waveforms exist only for a fleeting moment unless a storage oscilloscope is used. But even storage images fade in an hour or so.

Also, in the oscilloscope there is no provision for making a hard copy—that is, permanent records of the waveforms. Such records document the performance of circuits and so are deemed very important.

The solution to this problem is to make a photograph of the waveform on the oscilloscope screen. Although a few life sciences researchers and clinicians in medicine use a special 35-mm camera—equipped with a close-up lens, 100-ft magazine, and motor drive—the vast majority of oscilloscope photography is done on Polaroid® film, which gives the advantage of an instant print.

**FIGURE 8–31**
(a) Digital oscilloscope (courtesy of
Hewlett-Packard).

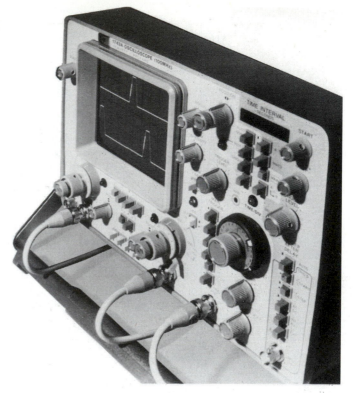

**(a)**

A number of different camera options are available. The Polaroid Corporation, for example, offers their model CR-10 hand-held camera. Both Tektronix and Hewlett-Packard manufacture various models that mount directly on the oscilloscope (see Fig. 8–32). Some of these cameras are manually operated, and others are electrically operated. The model shown in Fig. 8–32 includes a viewing hood so that the operator can see the waveform on the CRT screen before he takes the picture.

Although special oscilloscope films are used, most cameras take the standard Polaroid film pack, so they will accept any black-and-white film that has an ASA speed rating of 3000 (e.g., Polaroid types 667, etc.).

An important parameter in considering any oscilloscope photographic system is the **writing speed,** which is usually expressed in centimeters per microsecond. This parameter measures the effectiveness of the entire system and is determined by factors such as the camera lens, CRT phosphor, beam spot size, intensity, and so on.

**Postfogging** of the film can result in as much as 2× or 3× increase in writing speed. This technique exposes the film to a light source after it has been exposed to the CRT trace. On some oscilloscopes this is accomplished by turning on the CRT storage flood gun, whereas in others the graticule illumination is used. In the H-P model 197A shown in Fig. 8–32, an ultraviolet light source inside the camera is used.

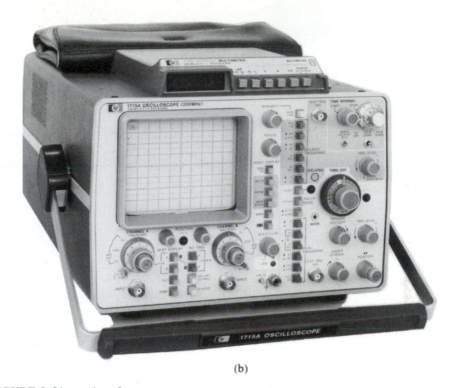

(b)

**FIGURE 8–31,** *continued*
(b) Add-on electronic voltmeter creates a digital oscilloscope (courtesy of Hewlett-Packard).

## 8–18    SAMPLING OSCILLOSCOPES

The vertical amplifier bandwidths of even the costliest oscilloscopes extend to only 250 mHz or so. Similarly, sweep speeds go up only to 0.1 µs/div. When attempting to view pulse waveforms in the over-500-mHz region, you must use a **sampling oscilloscope.**

A simple view of sampling is given in Fig. 8–33(a). The input signal is too fast (i.e., its frequency is too high) to be displayed directly on the CRT. But if the waveform is repetitive, then we may take amplitude samples from different pulses in the wave train and display them on a CRT screen. After several sweeps, the storage oscilloscope will have averaged a sufficient number of samples that a good representation of the waveform will appear on the CRT screen. If the sampling rate is within the range that the oscilloscope can handle, then the sampling effectively extends the range of operation well into the gigahertz region.

## 8–19    Z-AXIS MODULATION

The intensity, or brightness, of the CRT trace is dependent upon the strength of the electron beam and the kinetic energy imparted to each electron in the beam by the accelera-

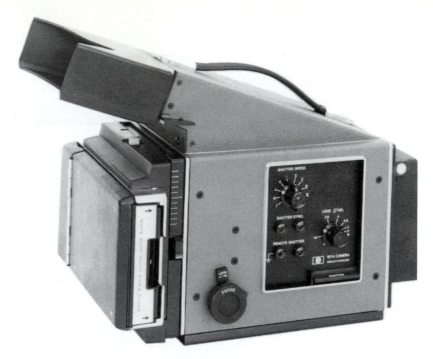

**FIGURE 8–32**
Oscilloscope camera (courtesy of Hewlett-Packard).

**FIGURE 8–33**
(a) How a sampling oscilloscope operates, (b) reconstructed waveform.

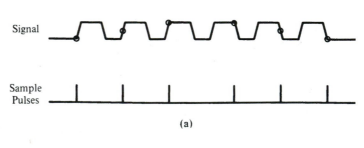

(a)

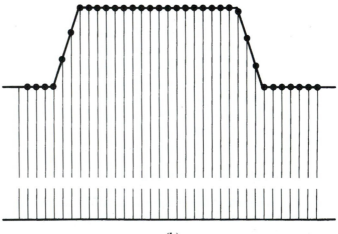

(b)

tor electrodes. We can vary the strength of the electron beam relatively easily, so by this method we obtain control over intensity.

The X-axis of the CRT is the horizontal channel, and the Y-axis is the vertical channel. It is common practice to refer to the intensity modulation channel as the "Z-axis." Many oscilloscope models contain circuitry to vary the intensity through an external jack that is usually labeled **Z-axis, Z-input,** or **intensity modulation.**

Figure 8–34 shows two popular methods for accomplishing Z-axis, or intensity, modulation. In Fig. 8–34(a) the CRT cathode is held constant by a bias supply, and the Z-axis signal is applied to a control grid. In Fig. 8–34(b), on the other hand, we have exactly the opposite situation: the control grid bias is constant, and the Z-axis signal is applied to the cathode.

In most oscilloscopes using a Z-axis input, it is necessary merely to reduce the intensity (using the normal front panel *intensity* control) so that the beam is just barely extinguished. If this is done when the Z-axis signal is zero, then applying a signal will brighten the CRT trace.

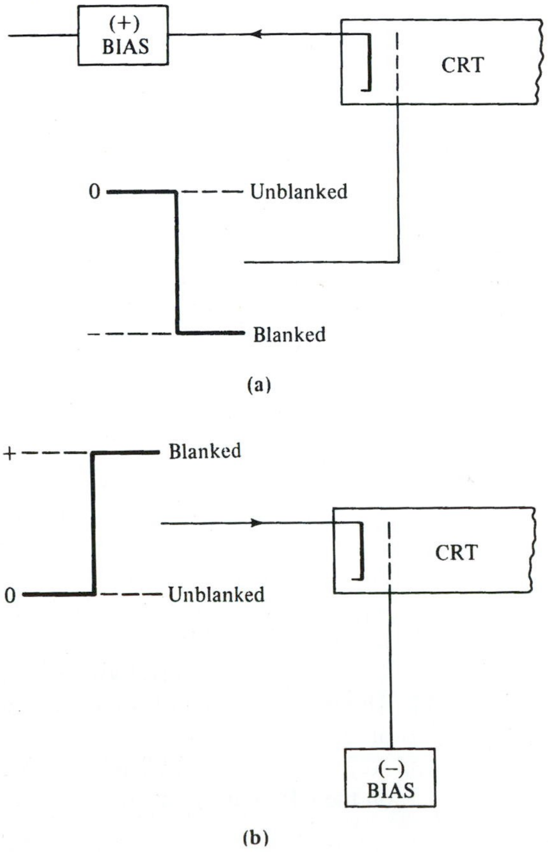

**FIGURE 8–34**
(a) Grid blanking, (b) cathode blanking.

## SUMMARY

1. The CRT oscilloscope allows evaluation of a signal's wave shape and measurement of its frequency, period, duration, phase, and other factors.
2. Two types of deflection are used: magnetic and electrostatic. Most laboratory and service oscilloscopes use the electrostatic because it is capable of higher-frequency operation.
3. In a Y-T oscilloscope the horizontal axis is connected to a time base (sawtooth) signal, and the vertical input (i.e., Y-axis) is driven by the input signal.
4. The vertical section is calibrated in volts/division, while the horizontal is calibrated in time/division.
5. A storage oscilloscope will retain the trace for up to an hour or until it is intentionally erased.

## RECAPITULATION

Now go back and try to answer the questions at the beginning of the chapter. When you are finished, answer the questions and work the problems given below. Place a mark beside each problem or question that you cannot answer, and then go back to the text and reread the appropriate section.

## QUESTIONS

1. In addition to wave shape, it is also possible to measure the following parameters of a signal displayed on an oscilloscope: _____, _____, _____, and duration.
2. Name the principal components of an electrostatic cathode ray tube.
3. List two different types of CRTs, classified by the deflection method used.
4. Describe the construction of an electron gun.
5. What purpose is served by the acceleration electrode in a CRT?
6. Light on the CRT screen is created by _____ bombardment of a _____ surface.
7. Which two elements in a CRT are responsible for *focusing* the beam?
8. Many oscilloscopes have a positively charged electrode located between the deflection plates and the screen. This _____ _____ electrode is kept at a potential of several kilovolts.
9. The electrode in Question 8 is known to improve _____ .
10. An X-Y oscilloscope can be used to produce _____ patterns that determine the relative frequency between the vertical and horizontal frequencies.
11. In a Y-T oscilloscope the horizontal deflection system is driven by a _____ waveform.
12. The time base controls on a free-running oscilloscope are usually calibrated in units of _____.
13. A _____ blanking pulse is used in a Y-T oscilloscope to eliminate the line on the CRT during retrace.

14. A _____ sweep oscilloscope holds off the beginning of a sweep until the input signal begins.

15. A quality oscilloscope will usually have a vertical attenuator that is calibrated in _____ / _____ and a time base calibrated in _____ / _____.

16. Describe the purpose of the delay line in the vertical amplifier of a triggered-sweep oscilloscope.

17. Describe in your own words what is meant by "delayed sweep."

18. What advantages are conferred by delayed sweep?

19. Describe in your own words the difference between **focus** and **astigmatism.**

20. Define **vertical sensitivity** and **vertical deflection factor** in your own words.

21. Define **vertical bandwidth.** List two ways this specification is sometimes abused in advertising.

22. Define the rise time of a pulse.

23. Electronic _____ is used to create a dual-beam oscilloscope, even though the CRT has but one electron gun.

24. Define ALT and CHOP modes as they apply to dual-beam oscilloscopes.

25. Three types of storage CRT are _____, _____, and _____.

26. Describe how a digital storage oscilloscope works. Use a diagram if necessary.

27. A **flood gun** is used in _____ CRTs.

28. What is meant by "variable persistence"?

29. Describe the action of an 5× horizontal magnifier.

30. The **lf reject** button in the trigger circuit _____ frequencies below a specified point.

31. Describe the differences between the **normal** and **auto** triggering modes.

32. Most oscilloscope photography is done with a Polaroid film with an ASA rating of _____.

33. What factors affect the photographic writing speed? What units are generally used to measure writing speed?

34. Describe the technique of postfogging and what it does.

35. Gigahertz-range signals may be viewed on a _____ oscilloscope.

## PROBLEMS

1. In Fig. 8–35(a) find the frequency of the waveform if the time base is set to 0.5 μs/div.

2. In Fig. 8–35(a) find the rms amplitude of the waveform if the vertical attenuator is set to 2 V/div.

3. In Fig. 8–35(b) find the period of the waveform if the time base is 10 ms/div.

4. In Fig. 8–35(b) find
   a) the amplitude if the vertical attenuator is set to 1 V/div and
   b) the slope of the leading edge of the waveform.

5. In Fig. 8–35(c) find the frequency if the time base is set to 2 μs/div.

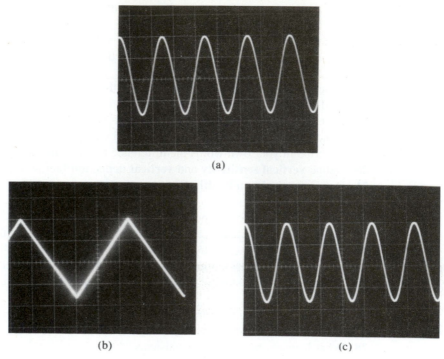

(a)

(b)                                        (c)

**FIGURE 8–35**
Waveforms, for Problems 1–6.

**FIGURE 8–36**
Signals for Problems 8 and 9.

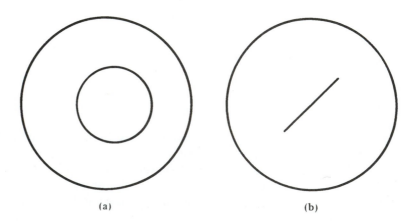

(a)                                        (b)

**FIGURE 8–37**
Lissajous pattern for Problem 10.

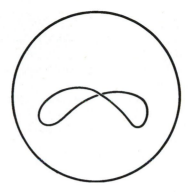

6. In Fig. 8–35(c) find the amplitude if the vertical attenuator is set to 5 V/div.
7. A 1-kHz sine wave exactly fits onto a 10-cm-wide CRT graticule. What is the setting of the time base knob in time/div units?
8. In Fig. 8–36(a) the Lissajous pattern shows a _____ degree phase differ-ence between vertical and horizontal signals.
9. What is the phase angle of the signals in Fig. 8–36(b)?
10. The Lissajous pattern in Fig. 8–37 is produced when a 41-kHz signal is applied to the vertical channel and a signal of unknown frequency is applied to the horizontal input. Find the frequency of the unknown signal.
11. Find the bandwidth in hertz required of an oscilloscope vertical amplifier if the rise time of the input pulse is 26 ns.
12. Consider Fig. 8–38. A 3.58-mHz burst signal eight cycles in duration is on the "back porch" of an 11-µs pulse. If a delayed sweep oscilloscope were used to view this sig-nal, without the main pulse, we would set the delay time to _____. Which of the following time base speeds is most suited to viewing the burst: 1 µs/div, 2 µs/div, 5 µs/div, or 1 ms/div?

**FIGURE 8–38**
Signal for Problem 12.

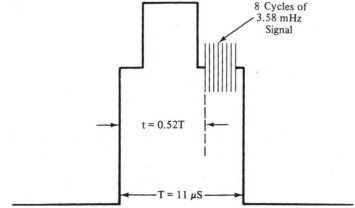

8 Cycles of
3.58 mHz
Signal

$t = 0.52T$

$T = 11\ \mu S$

# BIBLIOGRAPHY

Brown, John M., and Carr, Joseph J. *Introduction to Biomedical Equipment and Technology.* Englewood Cliffs, NJ: Prentice-Hall, 1993.

Cameron, Derek. *Advanced Oscilloscope Handbook.* Reston, VA: Reston Publishing Co./Prentice-Hall, 1977.

Carr, Joseph J. *Biomedical Equipment: Use, Maintenance, and Management.* Englewood Cliff, NJ: Prentice-Hall, 1992.

Herrick, Clyde. *Oscilloscope Handbook.* Reston, VA: Reston Publishing Co./Prentice-Hall, 1974.

# 9

# Signal Generators

9–1 **OBJECTIVES**

1. To learn the basic types of signal generators.
2. To learn the types of signals available.
3. To become familiar with certain commercial models.
4. To learn appropriate applications for each type of signal generator.

9–2 **SELF-EVALUATION**

Before studying the material in this chapter, try to answer the questions given below. These questions test your knowledge of the subject. If you cannot answer a particular question, place a check mark beside it, and then look for the answer as you read the text.

1. Name two types of oscillator circuits commonly used in audio generators.
2. Give two different methods for creating square waves from a sine wave source.
3. What type of stage is needed to create triangle waves from square waves?
4. State the standard output impedances for (a) RF signal generators and (b) audio signal generators.

9–3 **SIGNAL GENERATORS**

A **signal generator** is an instrument that provides a controlled output waveform or signal for use in testing or aligning, or in measurements on other circuits or equipment. Like most other electronic devices made in the past 20 years or so, the modern signal generator is considerably more sophisticated than its predecessors. Yet, at the same time, some recent quality products are little more than solid-state versions of designs that originated prior to World War II. The effect of solid-state technology has been to update traditional designs and to prompt even newer designs and techniques. Computer-based signal generators both control and generate analog waveforms. Regardless of the techniques used in the design, however, solid-state circuits have allowed manufacturers to make packages smaller; what was once several cubic feet of cabinet (and could heat a small room) is now in a box small enough to get lost on a crowded workbench.

It used to be easy to discern the different classes of signal generators. The classes were relatively rigid, with only a few exceptions, so one knew exactly what performance and capability to expect in any given product. But today, performance and specifications overlap to such an extent that classification becomes almost hopelessly difficult. There are very often several instruments in any given laboratory or shop that will do any given job. With this situation in mind, let's examine some of the traditional classes, even while caution reminds us of the many exceptions.

### 9–3–1        Audio Generators

This class of instruments traditionally covers the range 20 Hz through 20 kHz, although a few models produce signals up to 100 kHz. Audio generators always produce pure sine waves, and most also produce square waves. An audio signal generator typically uses either output metering or a calibrated output attenuator, and it produces controllable signal levels down to approximately −40 or −50 VU.

### 9–3–2        Function Generators

These generators typically cover at least the same frequency range as audio signal generators, but most modern designs have extended frequency ranges. Some, for example, operate as low as 0.001 Hz, while others operate to 11 mHz. One is known to operate to 80 mHz. A very common frequency range is 0.01 Hz to 2 mHz.

One difference between a function generator and an audio generator is in the number of output waveforms that are produced. The audio signal generator produces only sine waves and square waves, while almost all function generators produce these basic waveforms plus triangle waves. Some function generators also produce sawtooth, pulse, and nonsymmetrical square waves.

The output attenuator forms the other difference between these two types of signal generators, although this is not a universal situation. Most audio generators use a precisely calibrated or metered output control capable of producing known output amplitudes, even at very low levels. The function generator, however, typically has a coarse, uncalibrated, output level control. Although the specifications sheet will list the output range as 0 to 10 V rms, it is very difficult to obtain precise amplitude levels, especially at levels less than a few hundred millivolts. Both types of signal generators, however, will typically be rated to have a 600-$\Omega$ output impedance (the standard for audio circuits).

It is also quite common to find that the total harmonic distortion (THD) specification for audio generators is very much lower than for function generators, especially if high-fidelity audio equipment testing is the intended application.

### 9–3–3        RF Generators

An RF signal generator produces output frequencies in the "over 30 kHz" region. Most high-quality RF signal generators have a precision output attenuator that allows setting of

**TABLE 9–1**
RF frequency range designations

| Designation | Frequency Range |
|---|---|
| Audio* | 20–20,000 Hz |
| Extremely low frequency (ELF) | 10 Hz–10 kHz |
| Very low frequency (VLF) | 10–100 kHz |
| Low frequency (LF) | 100–500 kHz |
| Medium wave (MW) | 500–3000 kHz |
| High frequency (HF) | 3000–30,000 kHz |
| Very high frequency (VHF) | 30–300 mHz |
| Ultra high frequencies (UHF) | 300–1000 mHz |
| Microwaves | 1000–30,000 mHz |
| Millimeter waves | >30,000 mHz† |

\* Audio waves are actually *acoustical*, but since they cover the same frequency range, the word *audio* is used to describe electrical oscillations in the 20 to 20,000-Hz range. When an electrical oscillation, of whatever frequency, is radiated into space as an electromagnetic wave, it is *radio*. The 13-kHz Navy submarine communications stations, therefore, are producing RF signals, even though most frequency spectrum charts show 13 kHz as "audio." The audio frequencies are shown for comparison.

† Some call frequencies over 10,000 MHz "SHF" for super high frequencies, while others call them UHF for unbelievably high frequencies.

output levels from under 1 to 100,000 μV. RF signal generators typically have a standard 50-Ω output impedance. Few RF signal generators cover the entire spectrum from "over 30 kHz to daylight." Some will, however, cover extremely wide ranges, with 1 to 400 mHz being very easy to obtain in one instrument.

Table 9–1 shows the various RF ranges and the generally accepted names for each range. Note that most signal generators will actually cover overlapping portions of several of these ranges.

RF signal generators operating in the over 1-GHz (i.e., 1000-mHz) region are generally called **microwave signal generators,** while those over 30 GHz are millimeter wave generators.

**9–4**      **AUDIO GENERATORS**

The audio oscillator is used typically in tests on high-fidelity, public address, and other audio equipment. These generators typically produce sine waves and symmetrical square waves at frequencies between 20 Hz and 20 kHz. Most use a 600-Ω output impedance and produce output levels from −40 dBmW to +4 dBmW.

Two methods of frequency selection are typically used in audio signal generators: **continuous** and **step.** On the continuous type of dial you must turn a knob to the desired frequency. Many such audio generators have a scale that reads 20 to 200 (or, alternatively, 2 to 20), and a **range** selector switch determines whether the output frequencies will be 20 to 200 Hz, 200 to 2000 Hz, or 2000 to 20,000 Hz. In a step-tuned generator these controls are replaced by a rotary or pushbutton switch bank. As many as four decade switches might be used, although three is a more common number. These will be marked 0 through 100, 0 through 10, and 0.1 through 1.0 in decade steps. A **multiplier** switch determines whether the actual frequency will be ×1, ×10, ×100, or ×1000 the frequency indicated on the selector switches. Consider, for example, a generator in which the tens switch is set to 60, the units switch to 3, and the fraction switch to 0.4, and the multiplier is set to ×100. The output frequency will be $(60 + 3 + 0.4) \times 100 = (63.4)(100) = 6340$ Hz.

### 9–4–1   Audio Oscillator Circuits

Although many different oscillator circuits can be made to function in the audio range, only a few are truly practical for use in audio signal generators. An LC oscillator, for example, can be made to work in the audio range, but certain mechanical problems make them unusable; the LC components become excessively large. With the exception of the heterodyne technique, most audio oscillators used in audio signal generators tend to be RC designs.

An RC **phase shift oscillator** is shown in Fig. 9–1. In any oscillator the criteria for oscillation are (1) a 360° (0°) phase shift and (2) gain greater than unity around the feedback loop. An active element such as a transistor or an operational amplifier provides the needed gain. In Fig. 9–1 a JFET is used as the active element. Since $Q1$ is operated in the common-source mode, it provides a 180° phase shift between input and output. But to satisfy the criteria for oscillation, an additional 180° phase shift at the desired frequency must be provided by the feedback path. The shift is provided by three 60° RC networks: $R1C1$, $R2C2$, and $R3C3$. In some cases, $R1$ through $R3$ are potentiometers so that continuously variable frequency tuning is available. Or, alternatively, these resistors might be

**FIGURE 9–1**
RC phase shift oscillator.

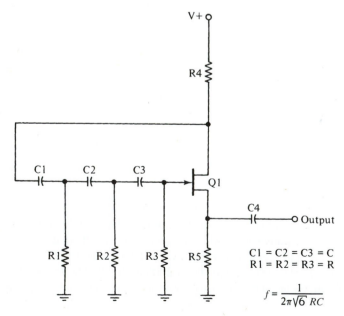

$$C1 = C2 = C3 = C$$
$$R1 = R2 = R3 = R$$

$$f = \frac{1}{2\pi\sqrt{6}\ RC}$$

part of a decade resistor assembly to provide step tuning of the oscillator frequency. Capacitors $C1$ through $C3$ are part of the range switch, a different value being used for each range of frequency coverage. The oscillator's frequency of oscillation is given by

$$F_0 = \frac{1}{2\pi\sqrt{6}RC} \tag{9-1}$$

where  $F_0$ = the operating frequency in *hertz* (Hz)
$R$ = the resistance in ohms ($\Omega$)
$C$ = the capacitance in *farads* (F)

An example of a Wien bridge oscillator is shown in Fig. 9–2. A differential amplifier $A1$ is connected across the output nodes of an RC Wien bridge consisting of four legs: $R1$, $I1$, $R3C1$, and $R4C2$. At all times, $E_0$ and $E1$ are in phase with each other and are 180° out of phase with $E2$. The feedback in this loop is degenerative, and therefore stable, at all frequencies except $F_0$. Frequency $F_0$ is also described by Equation (9–1). Voltage $E3$ is out of phase with $E_0$, but at frequency $F_0$, voltages $E3$ and $E_0$ are in phase; that is, $\theta = 360°$. Since $E3$ is applied to the noninverting input of the amplifier, this circuit will oscillate.

Lamp $I1$ is used to stabilize the output amplitude and prevent saturation of the amplifier device. This job is accomplished by the changing resistance of the lamp as $E_0$

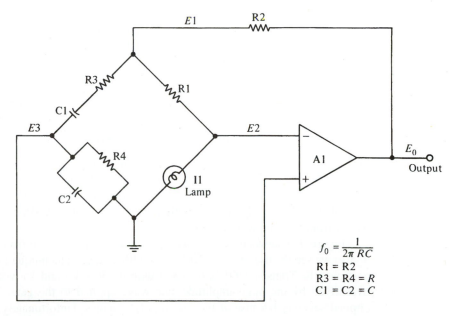

**FIGURE 9–2**
Wien bridge oscillator.

increases, causing more current to flow in the lamp filament. In most cases, $I1$ is operated at current levels just below incandescence.

The Wien bridge oscillator is used in many audio signal generators because of its relatively low distortion. Although $R3/R4$ may be made variable by using either a potentiometer or a decade switch and resistor bank, it is the latter that is most popular.

A twin-T oscillator circuit is shown in Fig. 9–3(a). Transistor $Q1$ provides 180° of the required 360° phase shift, while the remaining 180° is provided by a twin-T RC network. There are two qualitative ways to describe this circuit and its operation. There is a 180° phase shift through the network at frequency $F_0$ [Equation (9–1)]; hence, the input of the amplifier sees an in-phase feedback signal. Some other textbooks view the circuit in quite a different way. They point out that the twin-T is a notch filter; that is, it severely attenuates $F_0$ but passes all other frequencies removed from $F_0$. Because of this fact, all frequencies except $F_0$ are given a tremendous amount of negative feedback, so the circuit gain at those frequencies is limited. But $F_0$ receives very little negative feedback because only a small amount of the signal at this frequency passes the feedback network.

There are two variations on the twin-T design that tend to make the circuit more stable. These circuits, collectively called **bridged-T oscillators,** are shown in Figs. 9–3(b) and 9–3(c). The resistor-bridge is shown in Fig. 9–3(b), while the capacitor-bridge is shown in Fig. 9–3(c).

The last oscillator circuit that we will consider is the **heterodyne** design, shown in Fig. 9–4. This signal generator has fallen into disuse in recent years but was once very common.

In the heterodyne system a variable-frequency oscillator is mixed with a fixed-frequency oscillator to produce a difference frequency ($F2 - F1$). Most heterodyne audio signal generators used LC oscillators operating in the 50- to 500-kHz region, although a few were known that used higher-frequency oscillators. In general, though, lower-frequency LC oscillators produce more stable signals than do high-frequency oscillators. The fixed oscillator might be a fixed-frequency LC oscillator, but in most cases it was a crystal oscillator.

## 9–4–2     Generating Square Waves from Sine Waves

Square waves produced by an audio signal generator must be symmetrical, not only in the left-right sense, but across the 0-V baseline as well. The square waves can be produced by an astable multivibrator that is frequency locked to the sine wave oscillator, but this is difficult to achieve reliably in practice. It is, then, considered best to derive the square waves from the sine waves directly.

Figure 9–5 shows two methods for deriving a square wave from a sine wave. In Fig. 9–5(a) we see the use of an overdriven amplifier stage, a technique that is popular in low-cost models. Transistor $Q1$ is operated with the low $V-$ and $V+$ voltages, so it is easily overdriven by the large-amplitude sine wave applied to the gate. The output signal is clipped severely because of the low supply voltages. Unfortunately, the rise time of the square waves produced in this manner is not as good as might be required in some applications.

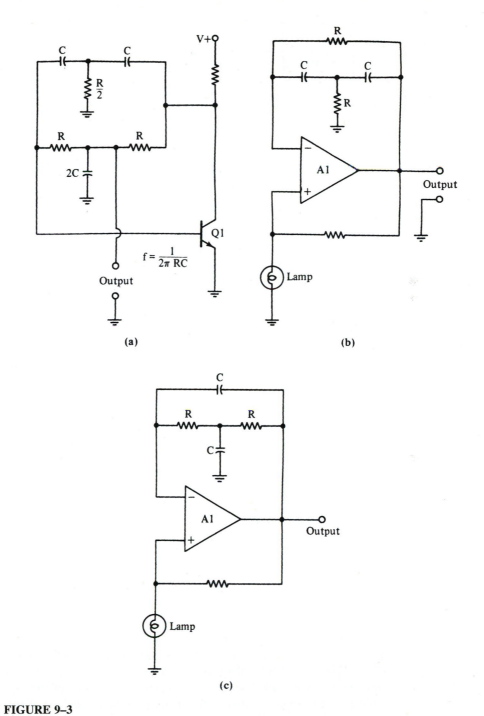

**FIGURE 9–3**

(a) Twin-T oscillator, (b) bridged-T oscillator (resistance-type), (c) bridged-T oscillator (capacitor-type).

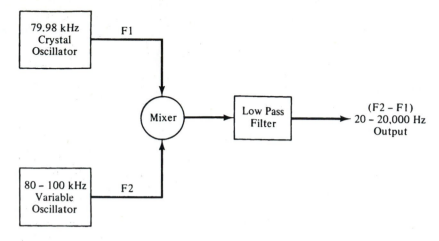

**FIGURE 9–4**
Block diagram for a heterodyne-type audio generator.

**FIGURE 9–5**

(a) Overdriven amplifier used as a clipper, (b) use of a comparator as a square wave generator.

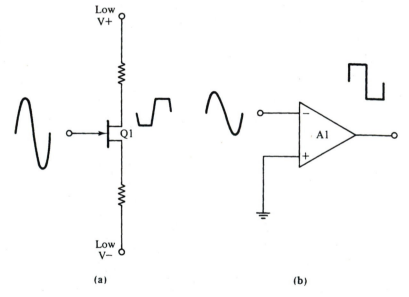

(a)                                    (b)

A somewhat superior approach is the circuit of Fig. 9–5(b). Here we have an operational amplifier operated as a voltage comparator. The lack of feedback causes the amplifier gain to be very high (see Chapter 12). Since the noninverting input is grounded (i.e., at a potential of 0 V), the output saturates any time the input potential is more than a few millivolts greater than zero. On negative excursions of the input sine wave, the amplifier output snaps positive, whereas on positive excursions of the input signal, the output snaps negative. The bipolar power supply accounts for the baseline symmetry of the output waveform.

A high-frequency comparator will produce output square waves with very good rise times. But even this circuit can be improved by the addition of a small amount of positive feedback to speed up the output transition time, thereby improving rise time.

## 9–4–3      Output Circuits

The audio signal generator must be able to deliver as much as 10 V rms into a 600-$\Omega$ load. It must also be capable of reducing the output level to a few millivolts without changing the generator output impedance. Additionally, the output signal will be symmetrical about zero, so the output range will be ±10 V rms, or ±14 V peak. The output stage of such a signal generator will usually consist of a power amplifier operated from bipolar power supplies (see Fig. 9–6). This amplifier will be required to deliver several hundred milliwatts at very low total harmonic distortion. In any event, a buffer stage between the oscillator and the output load is always desirable so that the operation of the oscillator is not affected by changes in the load external to the signal generator.

Figure 9–7 shows the block diagram of a typical audio signal generator. The audio oscillator section will use one of the circuits given earlier. A power amplifier stage provides buffering between the load and the oscillator, and it develops the output signal amplitude.

The ac voltmeter at the output is strictly optional, but in some models it is used with a **level control** to set precisely the input signal to the attenuator. Not all quality audio signal generators use this feature, so the lack of an ac output meter is not, in itself, indicative of quality. In some models an audio digital frequency counter is used ahead of the attenuator to provide digital display of the output frequency.

Typical attenuator circuits are shown in Fig. 9–8. The circuit of Fig. 9–8(a) is used in lower-grade instruments and may take the form of a potentiometer or a

**FIGURE 9–6**
Complementary output stage.

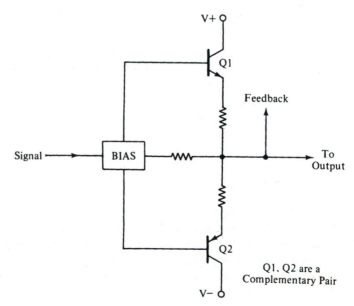

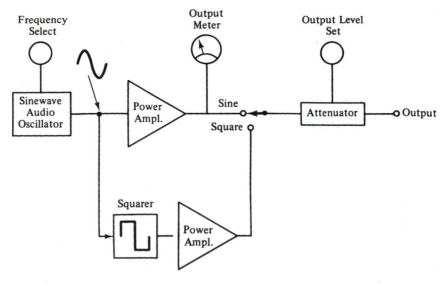

**FIGURE 9–7**
Block diagram of an audio signal generator.

switched resistor bank (as shown). This circuit is not often used in high-quality audio signal generators because the output impedance of the instrument is not constant at all output level settings. In addition, whatever load is connected to the output affects the amplitude of the output signal severely, and each position is affected differently. This problem will lead to significant output voltage errors even if standard loads are used, so beware.

An unbalanced ladder network attenuator is shown in Fig. 9–8(b). This network and several popular variations are used in most commercial audio signal generators. It provides a constant 600-Ω output impedance and therefore also eliminates the output voltage error of the previous design.

A square wave is created from the sine wave in a squarer circuit. In some models the square wave section has its own output terminals, while in others a **function selector** switch selects the output waveform. The output calibration is for the sine wave function and has little meaning for square waves. The square wave amplitude is usually measured in peak-to-peak volts.

**9–4–4**   ### Audio Signal Generator Examples

Figure 9–9 shows two examples of audio signal generators. The instrument shown in Fig. 9–9(a) is a step-tuned audio signal generator intended for testing high-fidelity equipment. Three rows of pushbutton frequency selectors are provided: 0 to 0.9, 0 to 9, and 0 to 90. Multiplier switches (×1, ×10, ×100, and ×1000) are used to set the overall range at which the other switches operate. A **frequency vernier** knob is used to provide a small amount

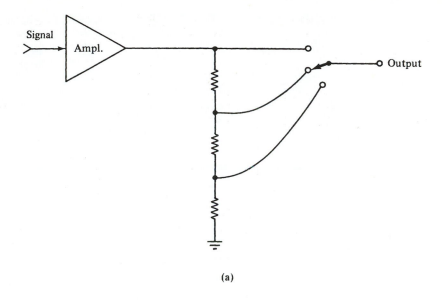

(a)

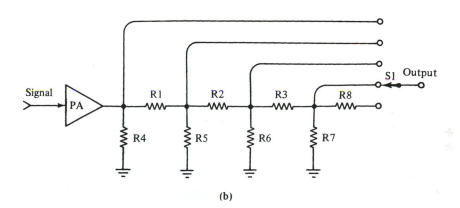

(b)

**FIGURE 9–8**

(a) Simple attenuator, (b) constant-impedance output attenuator.

of variation between steps so that precise frequencies can be set. A coarse output control sets the level of the signal appearing at the output jack. This example is also equipped with a **sync output** that provides a short-duration pulse at the start of each cycle in order to synchronize the sweep of an external oscilloscope (or other instrument). This output is considered essential for some test purposes.

A continuously tuned audio signal generator is shown in Fig. 9–9(b). Like the previous example, it has multiplier switches and a frequency vernier, but the specific frequency is set by the large, continuously variable, frequency control dial. This instrument has two

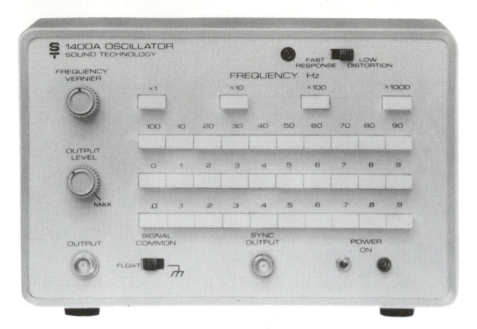

(a)

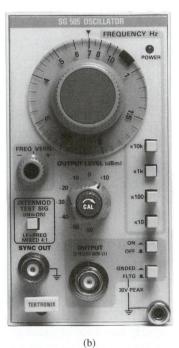

(b)

**FIGURE 9–9**

(a) Step-tuned audio signal generator, (b) continuously tuned audio signal generator [(a) courtesy of Sound Technology, Inc., (b) courtesy of Tektronix, Inc.].

concentric knobs for output level setting. The larger, switch-stepped, coarse level knob sets the overall range of the output signal level; it is calibrated in decibels relative to one milliwatt (1 mW) from −60 dBm through +10 dBm. A vernier output level control knob in the center of the stepped knob sets the level from zero to the maximum level established by the coarse control setting.

Another feature of this signal generator is that it will provide the two-tone test signal, mixed in the standard 4 : 1 ratio, needed to do intermodulation distortion (IMD) tests. This test measures the linearity of a circuit by noting how much heterodyning (mixing) takes place at the output of the test circuit between two independent input signals. In a perfectly linear circuit no frequency mixing takes place.

## 9–5     FUNCTION GENERATORS

We have already mentioned that function generators and audio generators have much in common, so in this section we will consider only those points in which the typical function generator is different from the audio generator. Keep in mind, however, that definitions are not fixed and that any manufacturer may offer instruments that blend attributes of several classes of instruments. The world is not so crisp as textbook writers would have you believe.

Figure 9–10(a) shows a simplified schematic of a typical sine-square-triangle function generator. In most of these circuits a square wave multivibrator originates the signal. Most of the sine wave oscillators discussed in the previous section have problems with amplitude and frequency stability unless properly compensated by the designer. But certain types of square wave generator circuits have significantly better performance in this respect and use simpler (and hence less costly) circuits.

In Fig. 9–10(a) amplifier $A1$ is used as a voltage comparator, while amplifier $A2$ is used as an integrator. When the circuit is first turned on, the output of amplifier $A1$ (i.e., point A) will snap high, creating current $I2$. The voltage at point A will also cause the feedback capacitor in the integrator to begin charging, thereby causing a voltage at point B to begin rising. Voltage $E_B$ creates current $I1$. When $E_B$ rises to the point where $I1$ is equal to $I2$, then $E_A$ snaps back to zero, making $I2$ also zero. But $E_B$ is still at the same level, so now $I1$ is greater than $I2$. This makes the amplifier output pass through zero to become high in the negative direction. With $E_A$ now negative, the integrator capacitors will begin to discharge, so $E_B$ reduces toward zero. The waveforms at points A and B are shown in Fig. 9–10(b). The upper trace shows the waveform at point A (the square wave), while the lower trace shows the waveform at point B (the triangle).

A sine wave can also be generated from a square wave, because the square wave is a composite of a fundamental sine wave and an infinite number of odd harmonics. A low-pass filter is shown in Fig. 9–10(a); it removes the harmonics, leaving only the sine wave fundamental. Figure 9–10(c) shows the result of using a low-pass filter on a square wave at point A. Although it is not apparent from the waveforms in Fig. 9–10(c), the filtering causes a substantial loss of amplitude, so an amplifier stage [e.g., $A4$ in Fig. 9–10(a)] must follow the filter section.

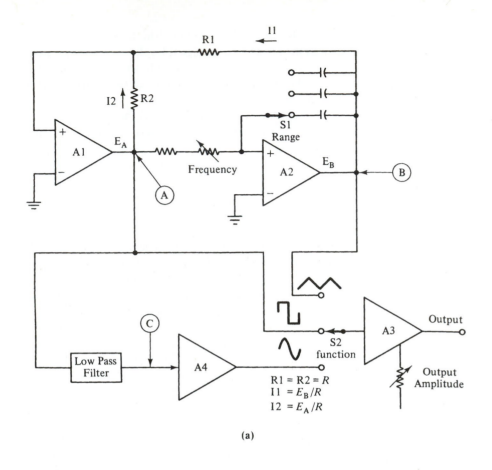

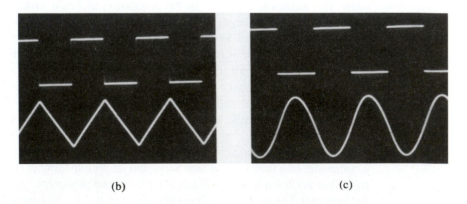

**FIGURE 9–10**

(a) Block diagram of a function generator, (b) waveforms at points A and B in (a), (c) waveforms at points A and C in (a).

**9–5–1**          **Sawtooth Generators**

A sawtooth waveform (Fig. 9–11) consists of a ramp leading edge, followed by a rapid drop back to zero on the trailing edge. The sawtooth is used to drive the horizontal plates of an oscilloscope and the voltage controlled oscillator (VCO) in a sweep signal generator.

Figure 9–12(a) shows one traditional sawtooth circuit: the **unijunction transistor (UJT)** relaxation oscillator. The UJT remains essentially turned off until voltage $E$ exceeds a threshold point, at which time the UJT breaks down the base-1/emitter junction. Current $I$ charges capacitor $C$ until the voltage across $C$ exceeds the UJT breakover voltage. The low resistance across the $B1$-$E$ junction rapidly discharges the capacitor.

**FIGURE 9–11**
Sawtooth waveform.

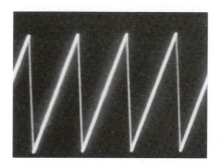

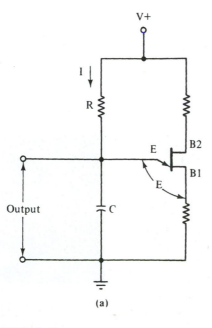

(a)

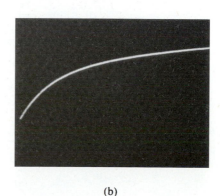

(b)

**FIGURE 9–12**
(a) UJT semi-sawtooth oscillator, (b) waveform.

Unfortunately, the simple circuit of Fig. 9–12(a) produces a poor sawtooth wave-form. The leading edge of a sawtooth must be a linear ramp, not an exponentially rising, capacitor-charging waveform such as that shown in Fig. 9–12(b). We could linearize this waveform by using only the first 10% or 15% of the RC time constant, but this is difficult in UJT circuits. Another solution to the problem is to use a **constant current source (CCS)** to charge the capacitor. [Fig. 9–13(a)]. A CCS may be a special diode-connected JFET configuration, or it may be a simple bipolar transistor circuit. The CCS keeps the capacitor charge current $I$ constant, so the output waveform will be more linear [Fig. 9–13(b)].

A circuit in which it is practical to use the first 10% to 15% of the capacitor charge curve is shown in Fig. 9–14. This circuit finds use in several modern function generators. It uses the standard Miller integrator that is used to create triangle waves, but in this case a CMOS electronic switch is connected across the feedback capacitor $C$. A voltage com-parator is used to turn the switch on and off. When $E_0$ rises to a point equal to $E1$, then the comparator output snaps high, turning on the switch, which rapidly discharges $C$. When $E_0$ drops back to zero, the comparator opens the switch, allowing the procedure to repeat itself. The input to the integrator is a constant reference voltage, and this keeps both $I1$ and $I2$ also constant.

## 9–5–2        Function Generator Examples

Several examples of typical function generators are shown in Fig. 9–15. The example shown in Fig. 9–15(a) is a step-tuned instrument that will produce sine wave, square

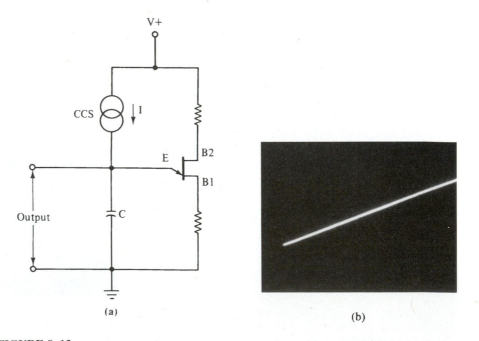

**FIGURE 9–13**
(a) Improved version of UJT circuit, (b) waveform.

**FIGURE 9–14**

Improved sawtooth generator based on a Miller integrator.

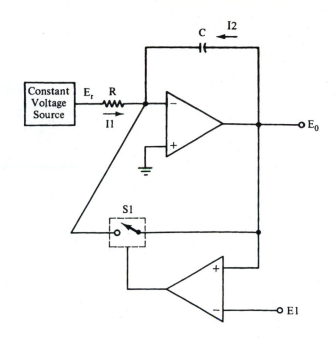

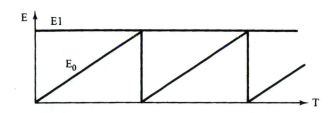

wave, and triangle wave signals at frequencies from less than 0.1 Hz up to 1 MHz, depending on the settings of the dial and the multiplier switches. This instrument provides two forms of output, one at the standard RF signal circuit impedance of 50 $\Omega$ and the other at the standard audio and function generator output impedance of 600 $\Omega$. The 600-$\Omega$ output allows switchable selection of either internal or external 600-$\Omega$ load. Like many other function generators, this model has the ability to provide a dc offset to the output waveform and, for the non–sine wave functions, to provide a symmetry control. The modulation controls are not seen on all function generators and indicate that this instrument can provide AM and FM (sweep) modulation of the frequency using a variety of internal waveforms or an external waveform. Sweep/FM produces a frequency that varies over a short time over a specified range; it is used to test the frequency response of electronic circuits and devices.

The instrument shown in Fig. 9–15(b) is a plug-in model that is intended to be racked with other instruments in a measurement system. It will provide function generator waveforms up to 40 MHz, so it is quite a bit wider in frequency range than other instruments.

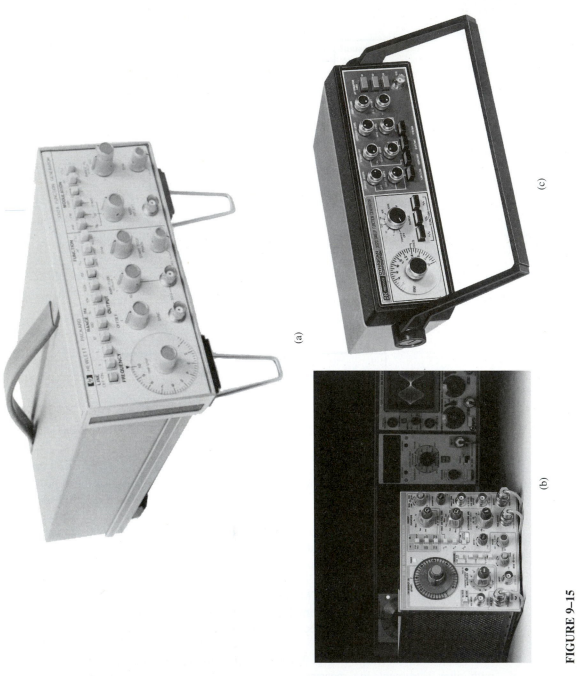

(a)

(b)

(c)

**FIGURE 9–15**
Function generators. (a) Step-tuned, (b) plug-in [(a) courtesy of Hewlett-Packard, (b) courtesy of Tektronix, Inc.] (c) Low-cost function generator with many features of more expensive models [(c) courtesy of B&K/Dynascan].

It is a sweep generator in that it can produce a swept frequency output. The output frequency will vary either with an internal sawtooth waveform or in response to an external waveform provided for some test purpose. The external trigger control allows the signal generator to initiate the sweep of an oscilloscope or other instrument. Another feature is that this sweep function generator also permits an external source to trigger the sweep frequency run. Both trigger mechanisms are needed for many sweep tests of circuits and other instruments.

The instrument shown in Fig. 9–15(c) is intended for service shop use, and while it is low cost, it also provides many of the features of the more expensive models.

| 9–6 | ## PULSE GENERATORS |
|---|---|

A pulse differs from a square wave in that it needs neither baseline nor left-right symmetry. Figure 9–16(a) shows one popular approach to constructing a pulse generator. A monostable multivibrator (i.e. one-shot) follows a square wave oscillator. The pulse repetition rate is set by the square wave frequency. The one-shot triggers on the leading edge of the square wave and produces one output pulse for each input cycle. The duration of each output pulse is set by the on-time of the one-shot, and it may be very short or may approach the period of the square wave. The relationship between the waveforms is shown in Fig. 9–16(b).

**FIGURE 9–16**
(a) Block diagram of a pulse generator, (b) waveforms.

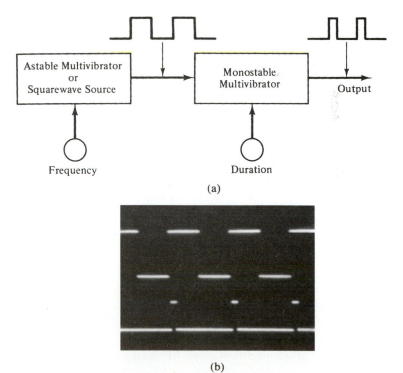

(a)

(b)

A typical pulse generator will allow the user to select the repetition rate, duration, and amplitude, and also the number of pulses to be output in a given burst.

## 9–7     RF GENERATORS

An RF signal generator uses an oscillator to produce ac (usually sine wave) signals in the region above 20 kHz. Most RF signal generators use LC oscillators as the signal source. Most RF signal generators also produce either AM or FM (or both) modulation of the output signal so that receivers and other circuits may be tested or aligned.

The most basic RF signal generator configuration is shown in Fig. 9–17. Here we have a band-switched, tunable oscillator and buffer driving the attenuator directly. The output meter determines the level applied to the attenuator. In most cases the meter is not calibrated (the attenuator *is*) in units, but it has a calibration point, often a red line on the scale. A level control varies the gain of the buffer, or the signal to the attenuator, to the red line. When the output is set to the red line, then the level at the attenuator input is known, so its calibrations are valid. In some older signal generators, the output frequency changes slightly when the attenuator is adjusted. But the calibrations will be accurate if the level control is then adjusted to the red line.

An automatic approach to signal generator design is taken in the circuit of Fig. 9–18. The previous design required a level control to keep the output signal amplitude constant. But in Fig. 9–18 an **automatic gain control (AGC)** circuit is used. An amplitude modulator is placed between the output of the oscillator and the input of the attenuator. The oscillator signal is the "carrier," and the output of an AGC rectifier is the modulating signal. The rectifier output is proportional to the RF signal at the modulator output, so by feedback action it keeps this level constant. Since the stage is an amplitude modulator, we

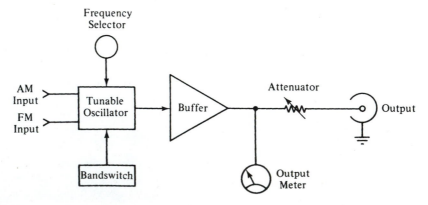

**FIGURE 9–17**
Block diagram of simple RF signal generator.

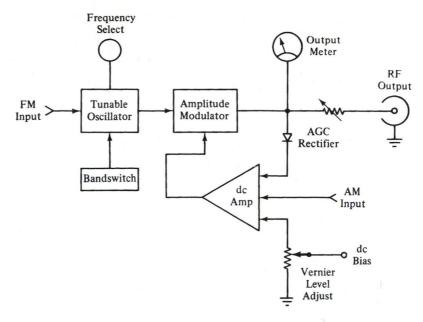

**FIGURE 9–18**
Block diagram of RF signal generator with automatic output level control.

may also apply an ac signal (usually 400 or 1000 Hz) to this stage to provide the AM modulation function.

**9–7–1        Frequency Synthesizers**

In the past decade or so a new type of RF signal generator has become popular: the **frequency synthesizer.** Various frequency synthesis schemes were tried over the years, but many suffered from various faults or design problems that made them essentially either useless or too costly. In one technique, for example, a wideband gaussian frequency spectrum is created. RF filters tuned to the various harmonics and frequency components of the spectrum pick off the components required to construct any given frequency. The "crystalplexer" synthesizers were used in some equipment. In these instruments a bank of crystal oscillators are heterodyned together to produce the desired frequency. In many citizens' band (CB) transceivers the crystalplexer synthesizer reduced the total number of crystals needed (80) to only a few. In that application the crystalplexer was economic. But in applications using a wider frequency range the technique becomes prohibitively expensive. Modern wide-range frequency synthesizers use **phase locked loop (PLL)** circuits to generate the output frequencies.

A simplified block diagram of a PLL circuit is shown in Fig. 9–19. The principal components of a PLL circuit are a **voltage controlled oscillator (VCO), divide-by-N counter, phase detector, reference oscillator, low-pass filter,** and **dc amplifier.**

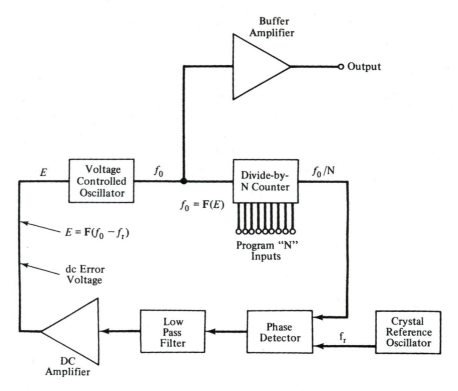

**FIGURE 9–19**
Block diagram of a PLL synthesizer.

The VCO oscillates at an RF frequency $F_0$ that is a function of dc control voltage $E$. Voltage $E$ is created by comparing the output frequency of the divide-by-$N$ counter (a subharmonic of $F_0$) to a low-frequency reference oscillator signal. A phase detector circuit makes the comparison.

The divide-by-$N$ counter is a digital counter circuit that divides the input frequency $F_0$ by an integer $N$. The integer division ratio is set by the binary word applied to the **pro-gram-N** inputs of the counter. The output of the counter will be a frequency that is equal to $F_0/N$.

The output of the $N$-counter and a low-frequency reference (i.e., 5 to 10 kHz) are applied to opposite inputs of the phase detector circuit. As long as $F_r$ is equal to $F_0/N$, then the VCO will oscillate at the frequency determined by the dc amplifier offset control and the $N$-code applied to the counter. But if the VCO drifts, then the two signals applied to the phase detector will no longer be equal, so the dc error voltage at the output of the phase detector will change in a direction and an amount needed to pull the VCO back on frequency. The PLL, then, constantly corrects changes in the VCO frequency. The stability and precision of this type of signal generator, then, are essentially those of the crystal reference oscillator (which can be quite good).

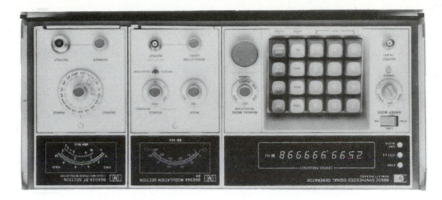

**FIGURE 9–20**

RF synthesizer-type signal generator (courtesy of Hewlett-Packard).

Deliberate frequency changes can be made by reprogramming the divide-by-$N$ counter to a different division ratio. The signal generator in Fig. 9–20 uses a numeric keyboard to enter the desired frequency. The operator enters the frequency, and the keyboard produces digital words that are interpreted as $N$-codes by the counter.

In most wide-range PLL synthesizers, several PLL circuits are used in a heterodyne system that is similar to the crystalplexer (although considerably more sophisticated). An example of such a signal generator is the Marconi Instruments model TF2020, shown in block diagram form in Fig. 9–21.

**9–7–2**          **RF Signal Generator Examples**

Several examples of RF signal generators are shown in Fig. 9–22, and these span the RF range from about 100 kHz well into the microwave region. The signal generator shown in Fig. 9–22(a) is a low-cost service or hobbyist-grade instrument that covers from around 100 kHz to 30 MHz on fundamental frequencies and up to 150 MHz on harmonics. (The latter is usually not very satisfactory for high-quality measurements but serves some purposes well.) A similar instrument intended for professional service shop applications is shown in Fig. 9–22(b). Both of these instruments use an "analog" frequency dial in which multiple bands are laid out on the dial and are selected by a pointer-and-knob system. There are also digital-dial versions of this class of instrument available in which the frequency indicator is a digital frequency counter.

The instrument shown in Fig. 9–22(c) is at the other end of the price spectrum. It is digitally tunable from 0.01 MHz (10 kHz) to 1024 MHz (1.024 GHz). The output frequency is input via a keypad. This instrument produces AM and FM modulated waveforms, both of which are controllable via the keypad entry. Modulating frequencies can be selected from either fixed standards (400 and 1000 Hz) or others in a range to 20 kHz. The main 50-$\Omega$ RF output connector is neither the BNC nor the SO-239 UHF connector

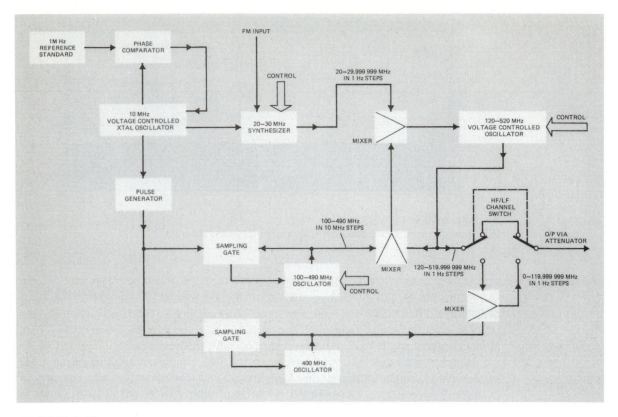

**FIGURE 9–21**
Block diagram of a wide-range, PLL-synthesized, RF signal generator (courtesy of Marconi
Instruments).

seen in many instruments, but rather the better shielded type N connector (which is common on high-quality RF instruments).

## 9–8        OTHER SIGNAL GENERATORS

The classes of signal generators discussed thus far are representative of a wide range of
instruments. Other signal generators are available, however, that do not necessarily fit
well in the various preestablished classes. Indeed, the ingenuity of manufacturers defies
classification except as a means for conveniently separating the instruments in our minds.

One class of instruments seen sometimes are RF marker generators. These devices
are usually crystal controlled and have a fixed output frequency for use as a reference. For
example, an RF marker may have standard outputs of 100 kHz, 1 MHz, and 10 MHz. It
may also have certain other standards depending on the intended use. A 3.58-MHz oscillator may be used to calibrate color TV signals, while 455-kHz and 10.7-MHz markers

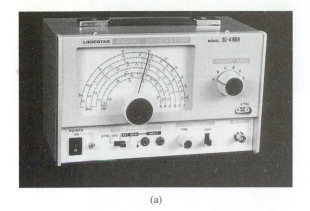

(a)                                              (b)

(c)

**FIGURE 9–22**

RF signal generators. (a) Low-cost service grade model, (b) professional service grade model, (c)
laboratory grade model [(b) courtesy of B&K/Dynascan, (c) courtesy of Marconi Instruments].

may be used to calibrate the standard intermediate frequencies (IFs) used on AM band
and FM band broadcast receivers, respectively.

Another class of instrument is the digitally programmable test oscillator [Fig.
9–23(a)]. These instruments can have extremely wide frequency ranges, although some
versions are much narrower in range. The set frequency can be programmed either

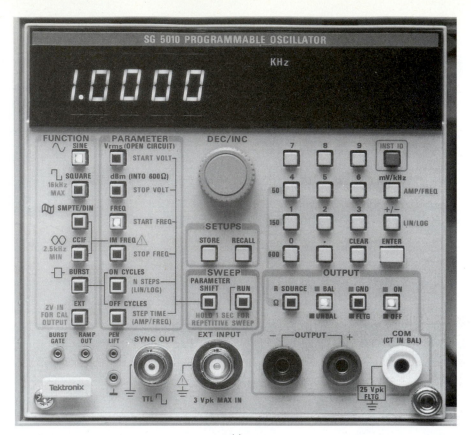

(a)

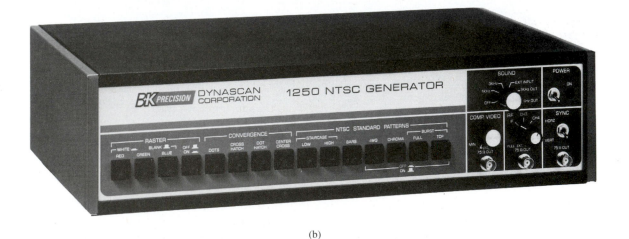

(b)

**FIGURE 9–23**

(a) Digitally programmable test oscillator, (b) NTSC TV generator [(a) courtesy of Tektronix, Inc., (b) courtesy of B&K/Dynascan].

through the front panel keypad or via a computer interface input, such as the IEEE-488 general purpose interface bus (GPIB).

Specialist industries often need specialized signal generators. For example, the instrument shown in Fig. 9–23(b) is intended for the television and cable servicing industries. It provides the standard sound, video, and composite patterns needed to troubleshoot television receiver systems. Note that the system output impedance is not 600 or 50 Ω, as in the case of audio and RF generators, respectively, but rather the 75-Ω standard impedance used in television systems.

## SUMMARY

1. Audio signal generators usually produce sine and square waves over the range of at least 20 Hz and 20 kHz. Some will operate down to 1 Hz and up to 100 kHz.
2. The function generator generally produces sine, square, and triangle waveforms at frequencies from fractional Hz to the mHz range.
3. Audio and function generators use 600-Ω output impedance, while RF generators have a 50-Ω output impedance. Proper operation of signal generators depends upon using a load matched to the output impedance.
4. RF signal generators operating over 1 GHz are usually called **microwave generators.**

## RECAPITULATION

Now go back and try to answer the questions at the beginning of the chapter. When you are finished, answer the questions and work the problems given below. Place a mark beside each problem or question that you cannot answer, and then go back to the text and reread appropriate sections.

## QUESTIONS

1. List two principal differences between an audio generator and a function generator.
2. List two types of frequency selectors used in audio generators.
3. List the typical frequency ranges of audio signal generators.
4. List the typical frequency ranges for function generators.
5. RF signal generators operate in the over-_____-kHz region.
6. List four different types of sine wave audio oscillators.
7. Draw a simple circuit for a function generator.
8. List two types of circuits for creating square waves from sine waves.
9. What type of circuit can be used to generate a sine wave from a square wave?
10. What type of circuit can be used to generate a triangle wave from a square wave?
11. A _____ audio generator uses two RF oscillators to create the audio signal.
12. List two ways to create a *linear* sawtooth waveform.
13. Some RF signal generators use a(n) _____ to keep the output amplitude constant.
14. List the principal stages of a PLL synthesizer.

## PROBLEMS

1. A standard RF signal generator is matched with its load. Calculate the voltage produced across the load if the generator attenuator is set to −10 dBm.

2. A standard audio generator is connected to a 600-Ω load. Find the rms voltage across the load if the signal generator is producing +3 VU. (*Hint:* 0 VU is 1 mV across 600 Ω, and VU is a power dB scale.)

3. A 10-to-15-mHz PLL operates from a 5-kHz reference. Find the *N*-code required to produce a VCO output frequency of 12.51 mHz when a divide-by-100 predivider is used ahead of the divide-by-*N* counter.

# 10

# Mechanical Graphics
# Chart Recorders

## 10–1    OBJECTIVES

1. To learn the principles of operation behind the PMMC galvanometer chart recorder.
2. To learn the principles behind the operation of servomechanism chart recorders.
3. To learn the causes of, and solutions for, certain types of recording problems.
4. To learn the different types of writing systems used in chart recorders.

## 10–2    SELF-EVALUATION

Before studying the material in this chapter, try to answer the questions given below. These questions test your knowledge of the subject. If you cannot answer a particular question, then look for the answer as you read the text.

1. Describe in your own words the PMMC galvanometer.
2. Describe the operation of a recording potentiometer.
3. What type of paper is used in thermal recorders?
4. What is a "deadband signal"?

## 10–3    MECHANICAL CHART RECORDERS

The term **mechanical chart recorders** refers to a broad class of devices that make a permanent paper record of analog waveforms, alphanumeric data, and graphics. To an analog data acquisition and instrumentation system, such as an ECG machine, the recorder serves the same function as a printer in a personal computer system. That is, it creates a **permanent record** ("hard copy") of the data output produced by the instrument. As you will see later in this chapter, for some modern analog recorders (which are actually digitally based) the analogy to a computer printer is more than merely metaphorical.

Mechanical chart recorders are used for a variety of reasons in a variety of applications in medicine, science, and engineering. First, they provide a means for evaluating a waveform in a non-real-time situation, perhaps long after the recorded event is over. In medicine, the physician will not always have the time to carefully consider all the implications of a recording when he or she is tending to the emergency, but afterward may want to review the record. Alternatively, another physician may be asked to evaluate a strip.

Second, the paper recorder allows a permanent legal record to be made of an event, and this record might be useful later in the event of engineering analysis, product liability lawsuits, or medical malpractice suits. As of this writing, the court system still prefers "original" records, such as a contemporary paper recording, over computer records that are still deemed too open to falsification. Thus, the analog recorder permits a legal record to be made of an event as evidence in a lawsuit. The importance of a reliable legal record cannot be overstated. The paper recording of a heart attack episode, for example, will help establish the fact that it was correctly identified and (in some cases) that it was correctly treated.

For all of the above reasons the care and feeding of mechanical recorders is a necessary topic of discussion in the context of electronic instrumentation and measurement systems. In this chapter you will learn how these machines operate, how they are maintained, and how they are abused.

## 10–4    TYPES OF MECHANICAL CHART RECORDERS

Several different categories of analog recorders are used: **recording volt-ohm-milliammeters (VOMs), strip charts, X-Y servorecorders,** and **plotters** are among the most common. We also now have **analog-digital dot matrix printing mechanisms** for analog recorders. In some cases a laser printer can also be used for making annotated analog graphical records.

The recording VOM is essentially a volt-ohm-milliammeter (see Chapter 3) integrated with a mechanical chart recorder. In these instruments a heavy-duty meter pointer is configured so that it rests above slowly moving chart paper. The paper is a special carbon-backed paper that leaves a mark when tapped or touched (care must be exercised to not "dirty" the record!). A cam on the paper motor lifts the pointer up and down so that it taps the paper, leaving a trail of dots that follow the pointer position.

The recording VOM typically uses a very slow paper drive speed, up to about 12 in per hour and as slow as 1 in per hour. This type of instrument is especially useful for recording changes and trends occurring over relatively long periods of time. A typical application for recording VOMs is recording temperatures or humidities in controlled environments such as greenhouses or scientific laboratories.

The strip chart recorder uses paper that is continuous either on a roll (like adding machine paper) or in a Z-fold pack (like personal computer dot matrix or daisy wheel printer paper). The X-Y servorecorder uses a single sheet of paper and an analog pen. The plotter is actually a sophisticated, digitally controlled version of the analog X-Y recorder. Of these, the strip chart recorder is by far the more commonly used. For this reason we will begin our study with this category of recorder. But first we must examine the principal means for writing in the strip chart recorder: the permanent magnet moving coil (PMMC) galvanometer mechanism.

**10–4–1**      **PMMC Galvanometer Movements**

The heart of any standard strip chart recorder is a device called a **permanent magnet moving coil (PMMC) galvanometer,** or "galvie" as it is called in the trade (see Fig. 10–1). The moving coil consists of a bobbin on jeweled bearings, with a lightweight coil of wire wound over it. When an electric current is applied to the coil, a weak magnetic field is set up around the coil. The signal being recorded is applied to the coil so that it causes a current to flow in the coil. Thus, the magnetic field surrounding the coil is proportional to the amplitude of the signal applied to the coil.

The permanent magnet applies a strong magnetic field across the space occupied by the moving coil. The magnetic field of the moving coil therefore interacts with the field of the permanent magnet. Recall the standard rules for magnetic field interaction: opposite poles attract and like fields repel. The reaction of the coil will be to move when a current flows. The jeweled bearings allow rotational motion only. By attaching a pen or stylus (more on this topic later) to the coil's pivot point, we produce an angular deflection proportional to the applied signal amplitude.

The pen tip is positioned over a strip of chart paper that is pulled under the pen tip at a constant speed, thereby establishing a time base. Deflection of the pen across the paper reproduces the wave shape of the signal applied to the coil.

A PMMC galvanometer with a short pen arc will write in a curvilinear manner [Fig. 10–2(a)]; the pen will scribe an arc at its tip instead of a straight line. In some applications this may be tolerable, and the user merely records on a paper that has a semicircular grid marked on it. In medical, scientific, and most engineering applications, however, this instrument is limited to such jobs as recording the time-varying temperature of rooms and machines that must maintain an internal temperature constant. In these cases the amount of deflection (e.g., a temperature) is more important than the shape of the wave. Thus, this type of pen is almost useless for recording most analog information and waveforms.

**FIGURE 10–1**

Permanent magnet moving coil (PMMC) galvanometer.

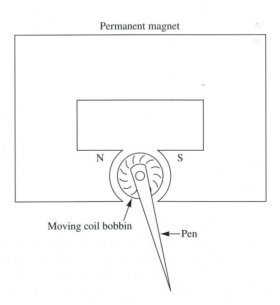

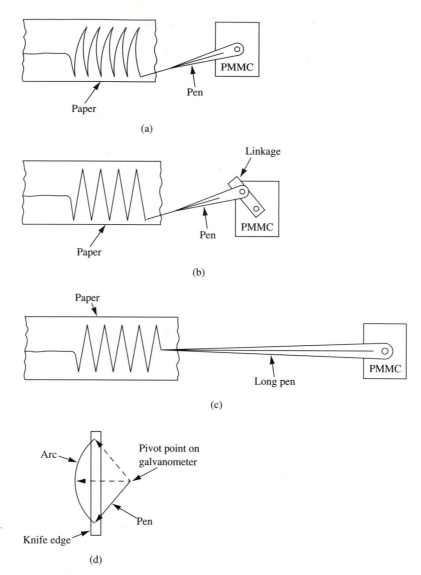

**FIGURE 10–2**
PMMC galvanometer recordings. (a) curvilinear, (b) rectilinear, (c) pseudorectilinear, (d) knife edge.

A solution to the problem of curvilinear recording is shown in Fig. 10–2(b). The PMMC mechanism is not connected directly to the pen but through a mechanical linkage that translates the curvilinear motion of the PMMC coil bobbin to a rectilinear motion needed at the pen tip. This type of mechanism is used extensively, although not as extensively as the knife edge method discussed below.

A pseudorectilinear technique is shown in Fig. 10–2(c). In this type of assembly the pen is very long compared with the arc that it must scribe (and also compared with the chart paper width). The pen tip, therefore, travels in an arc that is very small compared with its radius (i.e., the pen length). The trace, then, is nearly linear. Some curvature will exist, however, so this method is used only on some noncritical applications.

Several writing methods are used in strip chart recorders, but two are amenable to a special type of pseudorectilinear writing. Both of these writing systems use a knife edge (also called "writing edge" in some service manuals) to effectively linearize the trace. The method is shown in Fig. 10–2(d). The pen tip still travels in a curvilinear method (which allows the simplest form of PMMC pen mechanism), but the trace is very nearly linear because the knife edge is straight. This technique works in both thermal writing and direct pressure systems (as opposed to ink) because the pen, which is actually a heated stylus, can write anywhere along its own length—not just at the tip. By keeping a straight knife edge in a fixed position beneath the stylus, and by allowing the stylus's point of contact with the heat- or pressure-sensitive paper to vary, we obtain a rectilinear recording. It is the straightness of the knife edge that creates the rectilinear operation.

The thermal "knife edge" recording system is used more widely than the direct contact type. The direct method uses specially treated paper similar to the "carbonless" multicopy form paper marketed by companies such as NCR for commercial purposes. The heat-sensitive paper is paraffin coated and turns black when heated.

## 10–4–2  PMMC Writing Systems

Several different writing systems are used on PMMC recorders: **direct pressure, thermal, ink pen,** and **optical.**

Those mechanisms that use pens or styluses (i.e., direct pressure, thermal, or ink pen) tend to have relatively low "frequency response" due to the mechanical inertia of the massive pen or stylus. Most such recorders have an upper frequency limit of 100 to 200 Hz. Ink jet and optical recorders, on the other hand, have responses up to 1000 or 2000 Hz. Why is this important? All analog waveforms, such as ECG, EEG, EMG, and so forth, are made up of components of frequencies higher than the apparent "fundamental" frequency. For example, a 60-BPM ECG has a fundamental frequency of one beat per second, or 1 Hz. But an analysis of its components in a Fourier spectrum analyzer will show that it contains frequency components in its spectrum of 0.05 to 100 Hz. Similarly, a phonocardiogram may have spectrum components up to 1500 Hz. Therefore, a stylus recorder can easily handle an ECG waveform but not a phonocardiogram waveform. In other words, the type of recorder used for any given purpose is very important.

The thermal recording system uses a specially treated paper that turns black when heated. The paper is paraffin coated, and the writing instrument is an electrically heated stylus. Early models formed the stylus tip from a U-shaped electrical resistance element. More modern models use a resistance element inside a cylindrical metal stylus. In both cases a low-voltage electrical current is passed through the element, heating it almost to incandescence. The heated stylus leaves a black mark on the treated paper wherever it touches.

(a)

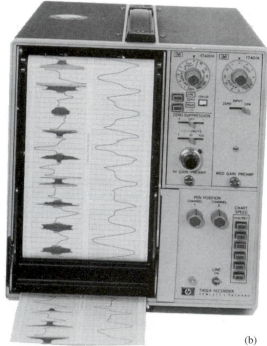

(b)

**FIGURE 10–3**
(a) Multichannel direct writing recorder, (b) two-channel recorder [(a) courtesy of Gould/Brush, (b) courtesy of Hewlett-Packard].

Ink pen systems write using a hollow pen and ink supply. In some machines the ink is relatively lightweight (i.e., it has a low viscosity), and pressure is applied manually through an atomizer type of pump (also called a "squeeze ball" pump). Other machines, such as the Gould/Brush instruments [Fig. 10–3(a)] and the Hewlett-Packard model 7402 [Fig. 10–3(b)], use a thick, high-viscosity ink. The pressure is applied through a spring-driven piston (Fig. 10–4) inside the ink supply cartridge. An ink manifold distributes the ink to the several pens that might be used in the system. Pressure is applied to the manifold by a solenoid that is energized when the machine power is turned on.

A related type of writing system, used in some high-frequency response phonocardiogram machines, is the high-velocity ink jet. In that type of recorder, ink under high pressure is fed to a nozzle mounted on the PMMC galvanometer in place of the pen. The ink jet is directed at the moving paper and, when the system is properly adjusted, produces a line that is almost as fine as that produced by thermal and ink pen recorders. Only a small amount of trace fuzziness is apparent, and this is due to ink splattering.

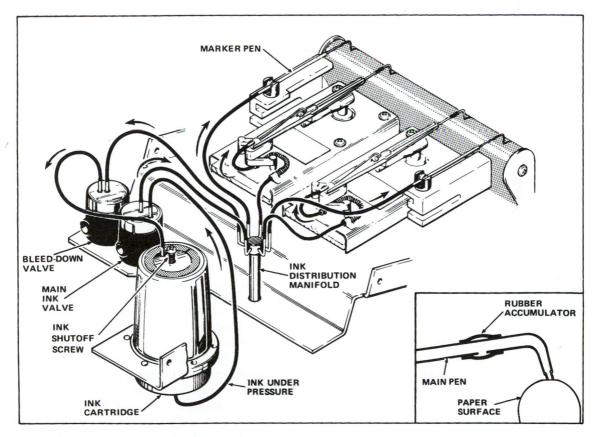

**FIGURE 10–4**
Pressurized ink system (courtesy of Hewlett-Packard).

The high-velocity ink jet recorder finds use wherever a high-frequency response is needed. The low mass of the ink jet nozzle, compared with the relatively bulky mass of the ink pen and thermal stylus, gives the machine a 1000-to-2000-Hz frequency response.

An example of an optical PMMC recorder is shown in Fig. 10–5(a), and a cathode ray tube (CRT) optical recorder is shown in Fig. 10–5(b). The PMMC optical writer uses a mirror mounted on the PMMC galvanometer to reflect the light beam from a collimated light source onto the photosensitive chart paper. In most such cases the paper is at least 6 in wide, so a greater span (or resolution) is possible. Additionally, on a multichannel recorder the time relationships between two different traces can be seen more easily because the beams can be made to cross each other—a difficult trick to accomplish with pen or stylus recorders.

The paper is exposed to an ultraviolet light source that develops the trace as the paper comes out of the recorder. The trace will fade over a long period of time or if it is exposed to a strong light (sunlight will cause damage). Therefore, for long-term storage a light-tight box is used. Alternatively, some machines allow the paper to be wet developed like photographic printing paper, after which process the image is stable.

A few models of older recorders use a CRT as shown in Fig. 10–5(b). The old VR-6 and VR-12 recorders by Electronics for Medicine are still occasionally found in use. This type of chart recorder is called a **camera recorder.**

There is no time base sweep on the CRT screen. The beam sweeps back and forth along one axis in response to amplitude variations of the input signal [examine Fig. 10–5(b) closely]. A time base is provided by the constant speed at which the photosensitive paper is drawn in front of the CRT screen.

The frequency response of the PMMC optical recorder is better than that of any other PMMC system because of the low mass of the reflecting mirror. The CRT camera has an even higher frequency response because it is limited only by the writing speed of the photosensitive paper. As a result, CRT cameras are often used for high-frequency vibration signals; for medical signals such as phonocardiograph, electromyograph, and Bundle of His (and other intracardiac ECG) recordings; and for other applications where a frequency response of several kilohertz is needed.

## 10–5  RECORDING POTENTIOMETERS AND SERVORECORDERS

A potentiometer [Fig. 10–6(a)] is a three-terminal device that acts as a variable resistor. Two ends of the potentiometer are fixed, while the middle arm is attached to a "wiper" that selects a resistance proportional to its position. The two fixed ends are connected across a reference voltage, while the variable voltage appears at the movable wiper arm. The output voltage is proportional to the applied voltage and the relative position of the wiper on the resistance element. A galvanometer (zero-center meter) connected between the wiper and an unknown voltage will register zero deflection when the known voltage from the potentiometer is equal to the unknown voltage. This condition is called the **null state.** But if these two voltages are unequal, then the PMMC galvie will deflect an amount proportional to the difference between the two potentials.

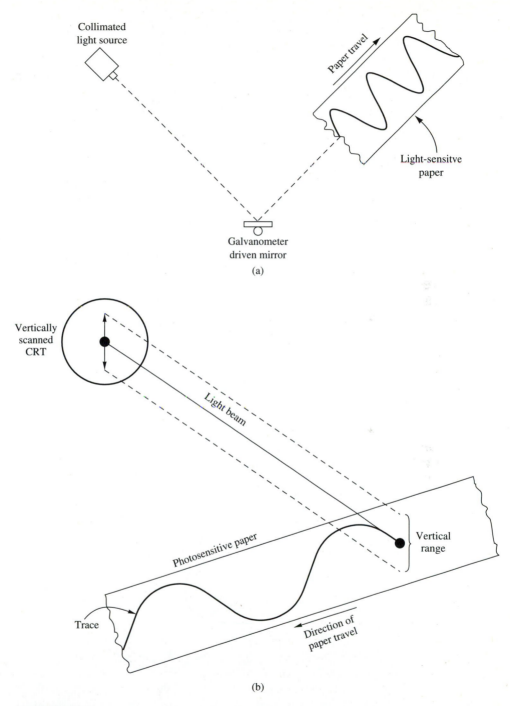

**FIGURE 10–5**

(a) Optical PMMC recorder, (b) cathode ray tube (CRT) optical recorder.

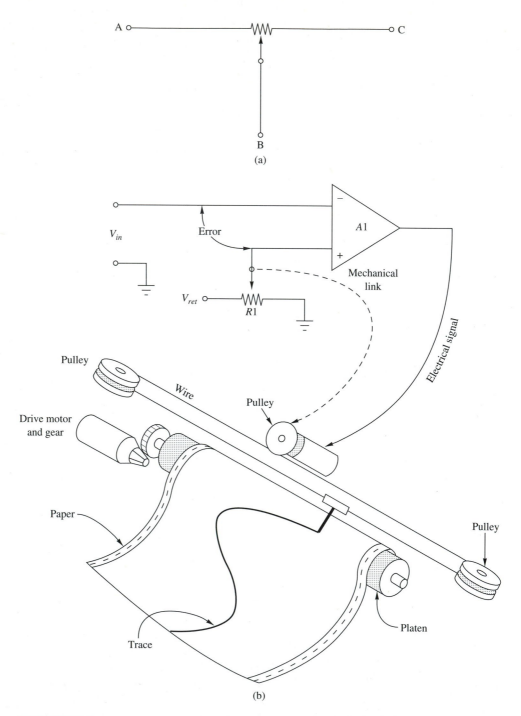

**FIGURE 10–6**

(a) Potentiometer, (b) basic dc potentiometer servorecorder system.

It is possible to build a self-nulling potentiometer. If a pen is connected to the potentiometer's wiper drive mechanism, then the applied voltage can be recorded on paper as the mechanism constantly seeks to null out the applied voltage from the signal source.

Figure 10–6(b) shows the basic dc potentiometer servorecorder system. The pen is attached to a string that is wound around two idler pulleys and a drive pulley on the shaft of a dc servomotor. The pen assembly is also linked with a potentiometer ($R1$) in such a way that the position of the wiper arm on the resistance element is proportional to the pen position.

The potentiometer element is connected across a stable, precision reference potential, $E_{ref}$, so potential $E$ will represent the position of the pen. That is, $E$ is the electrical analog of pen position. When the pen is at the left-hand side of the paper [in Fig. 10–6(b)], then $E$ is zero, and when the pen is full scale (i.e., at the right-hand side of the paper), $E$ is equal to $E_{ref}$.

The pen position is controlled by the dc servomotor, which is in turn driven by the output of the servoamplifier. This amplifier has differential inputs. The input signal $E_{in}$ is applied to one amplifier input, and the position signal $E$ is applied to the other amplifier input. The difference signal ($E_{in} - E$) represents the error between the actual pen position and the position it should take in response to the command from $E_{in}$.

If the error signal is zero, then the amplifier output is also zero, so the motor remains turned off. But if $E$ is not equal to $E_{in}$, then the amplifier sees an input signal and creates an output signal that turns on the motor.

The motor drives the pen potentiometer assembly in such a direction as to cancel the error signal. When the input signal and the pen position signal are equal, then the motor shuts off and the pen remains at rest.

A paper drive motor forms a time base because it pulls the paper underneath the pen at a fixed, constant rate. In most servorecorders a sprocket drive is used instead of the roller drive system popular in PMMC machines. The paper used in these machines (see the examples in Fig. 10–7) has holes along each margin to accept the sprocket teeth, much like personal computer printer paper.

Most high-grade servorecorders use either a stepper motor that rotates only a few degrees for each pulse applied to its winding or a continuously running motor. The stepper motor method is low in cost and can be very precise when a stable reference clock oscillator is used as the source of the electronic drive pulses.

Examples of servorecorders are shown in Fig. 10–7. The instrument in Fig. 10–7(a) is essentially a form of recording VOM using an X-Y servorecorder as the chart recorder. The instrument in Fig. 10–7(b) is a dual pen servorecorder that can chart two waveforms simultaneously. In Fig. 10–7(c) the servorecorder is part of a medical instrument used to monitor fetuses in hospital obstetrical units.

The actual potentiometer resistance element used in a servorecorder may be any of the following: a **slide wire,** a **rectilinear,** or a **rotary potentiometer.** The slide wire system (Fig. 10–8) is very common because it can be built with fewer friction losses and no mechanical linkage. A shorting bar on the pen assembly serves as the wiper element on the slide wire resistance element; it also serves to connect the shorting bar to the wire. The letters A, B, and C in Fig. 10–8 refer to the potentiometer terminals shown in Fig. 10–6(a). The shorting wire $B$ serves as the terminal for the wiper.

(a)

(b)

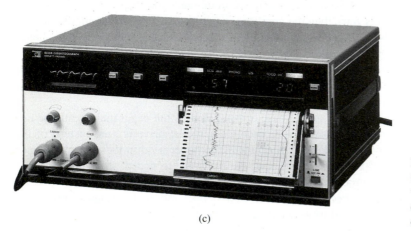

(c)

**FIGURE 10–7**
(a) Recording VOM using X-Y servorecorder as the chart recorder, (b)
dual pen servorecorder, (c) obstetrical servorecorder.

## 10–5–1     X-Y Recorders and Plotters

An **X-Y recorder** uses two servomechanism pen assemblies mounted at right angles to
one another. The vertical (or Y-plane) amplifier moves the pen vertically along the bar,
while the X-plane amplifier moves the bar back and forth across the paper. Examples are
shown in Figs. 10–9 and 10–10.

The paper itself does not move and is usually held in place by either clamps or, in
high-grade machines, a vacuum drawn by a pump connected to a hollow chamber beneath
the paper platform. Holes in the paper platform create the negative pressure needed to
keep the paper in place.

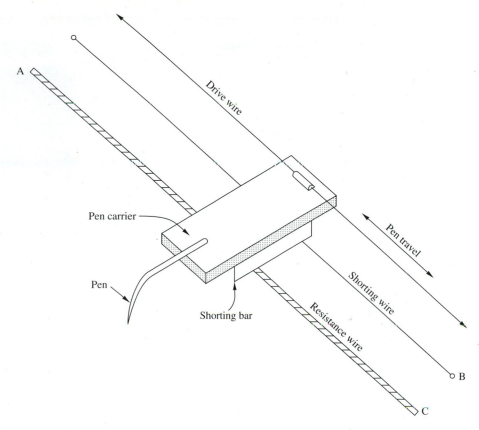

**FIGURE 10–8**
Slide wire servorecorder.

One advantage of the X-Y recorder over either PMMC or servorecorders is that almost any type of paper may be used. Most X-Y recorders are designed to accept the standard size graph papers used by scientists, engineers, and physicians.

The X-Y recorder can be made to record time-varying signals by applying a linear ramp voltage to the X-plane input. The ramp is adjusted so that it traverses the horizontal width of the paper in the desired length of time.

An X-Y plotter (Fig. 10–10) is an X-Y recorder that has an electrically operated pen-lift mechanism. A plotter can make patterns of complex shape, including alphanumerics. Many modern plotters are designed to do computer graphics, while others handle both analog and digital data.

**Digital Recorders.** A relatively new class of graphics recorder is the digital type (Fig. 10–11). The input signal is applied to an analog-to-digital (A/D) converter. The A/D device produces a binary output "data word" (of the sort used in computers) that is proportional to the signal amplitude. If a large number of successive "samples" of the ana-

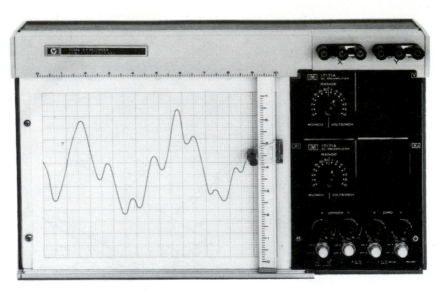

**FIGURE 10–9**
X-Y recorder.

**FIGURE 10–10**
X-Y graphics plotter.

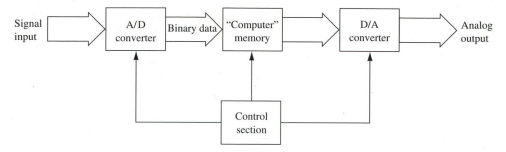

**FIGURE 10–11**
Digital graphics recorder.

log waveform are taken and stored in a computerlike memory, then the pattern of digital data in the memory will represent the time-varying analog signal.

The data in memory can be scanned and played back to either a strip chart recorder or a CRT screen for viewing. In that case, a digital-to-analog (D/A) converter is used to recreate the analog voltage signal required by the display device.

An advantage of this system is that low-frequency (which also means low-cost) paper recorders can be used to record very high frequency analog signals simply by varying the time base. For example, suppose we want to record a 1000-Hz analog signal. According to a standard engineering convention, the sampling rate of an analog signal should be at least twice the highest frequency, so we need to take $2 \times 1000 = 2000$ samples per second. These samples are stored in the computer memory. When we want to make a recording, we need only reduce the sampling rate proportionally, say, to 20/s, and play it back through a standard low-frequency machine.

This method is one means for standard slow-speed (25 mm/s) medical ECG machines to be used for higher-frequency studies. The same method is also used for producing the nonfade oscilloscopes. Normally, the waveform will fade almost immediately after it occurs. But with the digital recording oscilloscope it will remain on the screen until the memory is used up and is overwritten by newer data.

## 10–6 RECORDER PROBLEMS

The pen assembly in a chart recorder always has a certain amount of mechanical inertia because of its mass, regardless of whether PMMC or servo systems are used. Because of its mass, the pen assembly will not start moving until a certain minimum signal is applied. Figure 10–12 shows this "deadband" phenomenon. The **deadband** is the largest-amplitude signal to which the instrument will NOT respond. In most quality instruments the deadband is not more than 0.05% to 0.1% of full scale.

The deadband can cause severe distortion of the recorded trace, especially on low-amplitude signals whose level approximates the deadband itself. The solution to the problem of deadband is to slew the signal through the deadband as rapidly as possible.

Another problem is overshoot or undershoot of the pen in response to a step function (e.g., square wave). This problem also affects both forms of mechanical writers. Figure

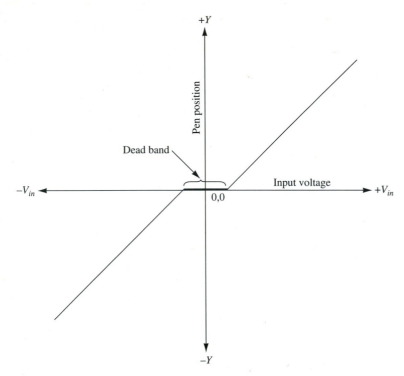

**FIGURE 10–12**
Deadband.

10–13 shows three different responses. In a properly damped recorder the pen will rise to the correct position with very little rounding of the trace. This situation is called **critical damping.**

A subcritically damped signal will overshoot the correct position and then hunt back and forth across the correct position for a few cycles before it settles on the correct point. If a square wave is applied to such a system, then the recording will appear to be "ringing."

An overcritically damped system approaches the final value very sluggishly. The pen in this type of system will appear to be very sluggish. A square wave applied to such a system will have rounded edges. Damping is an important parameter in analog medical recorders because it can distort the resultant traces and *thereby mimic pathological traces.*

The three damping factors are a mechanical analogy of frequency-selective filtering, and filters can be used to correct the damping problems. Some PMMC recorders have a **pen position transducer** inside the PMMC galvanometer housing, and servorecorders already have a position signal. The position signal can be integrated (i.e., low-pass filtered) to vary the damping factor of the pen assembly.

In the PMMC recorder the integrated position signal is fed back to the servoamplifier to be summed with the error signal.

**FIGURE 10–13**
Damping.

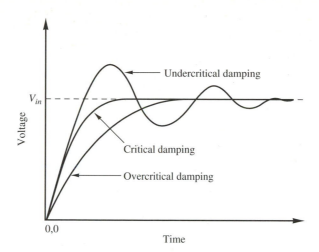

Pen assemblies can be damaged if they are allowed to strike the mechanical limits-of-travel stops. If a signal that is very large relative to the full-scale signal is applied to the input, then the pen may hit the mechanical stops and break. Some PMMC recorder amplifiers use output limiting (e.g., diode clamp circuits) to guard against such damage. Alternatively, a pair of zener voltage regulator diodes connected back-to-back across the PMMC coil are sometimes used to accomplish the same job. The zener potential of the diodes is selected so that the diodes break over and conduct current only when a voltage greater than the normal full-scale potential is applied to the amplifier input.

If a PMMC position signal is available, then it can be differentiated to produce a velocity signal (i.e., $dv/dt$), and the position signal can be differentiated twice to form an acceleration signal (i.e., $a = d^2 \, dt^2 = dv/dt$). If appropriate circuitry is used, then these signals can be used to apply a hard braking signal to the PMMC coil if either pen velocity or pen acceleration exceeds certain preset limits. The pen will strike the stops, but with considerably less force.

Most recorders have a damping control that can be adjusted when a square wave, typically 1 Hz, is applied to the amplifier input. The damping control is adjusted to provide a trace with the best "squareness."

## 10–7    MAINTENANCE OF PMMC WRITING STYLUSES AND PENS

Several common faults are found on paper chart recorders. In some cases these faults are amenable to adjustment, while in others repair is necessary. Let's separately consider the ink pen and heated stylus types of recorders. Because they find only limited use, we will not consider some of the other types (even though you need to be aware of them conceptually). In this section we will take a look at some common maintenance actions for mechanical recorders.

Figure 10–14 shows how to remove an ink blockage from an ink pen recorder that has been allowed to stand too long without being used. As a general rule, such recorders

**FIGURE 10–14**
Removing an ink blockage from an
ink pen recorder.

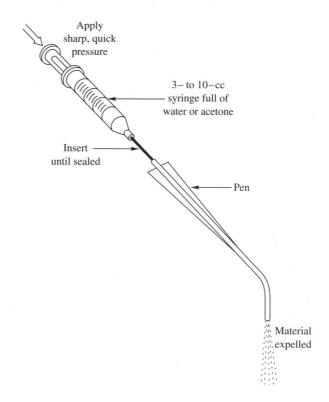

Apply
sharp, quick
pressure

3– to 10–cc
syringe full of
water or acetone

Insert
until sealed

Pen

Material
expelled

should be run for about 5 minutes or so once a week when not in regular service. Otherwise, the ink will dry up at the tip and prevent the recorder from writing. As shown in the figure, fill a 3- to 10-cc syringe (the sort used by diabetics to inject insulin will work) with water (or acetone if certain types of ink are used), and insert the needle end into the ink inlet on the rear end of the pen assembly. On most recorders the pen will have to be removed from the machine for this operation. The needle should be inserted up to the Luer lock hub in order to make a good fluid seal. Quickly, and with a single sharp motion, drive the plunger "home" so that a high-pressure jet of water or acetone is forced into the pen. The ink clot should be forced under high pressure out the other end.

This procedure usually works quite well. Some precautions are in order, however. First, always wear protective goggles over your eyes when doing this operation. Also, wear protective clothing, as ink/water will splatter everywhere, and the thick, high-viscosity ink used in these machines stains everything it touches. Second, be sure that the pen tip is aimed downward into a sink (there is still an electronics laboratory at George Washington University Medical Center with a blue spot on the ceiling tiles over the sink . . . where I failed to observe this precaution). Finally, as always when dealing with needles, be careful not to stick yourself.

Ink pen tips are designed to operate parallel to the paper surface [see Fig. 10–15(a)]. If the pen is worn, or when a new pen is installed, it is necessary to "lap" the tip in order to reestablish the parallelism [Fig. 10–15(b)]. The symptom that lapping is needed will be

**FIGURE 10–15**
(a) Correctly lapped pen tip is parallel to paper surface. (b) Improperly lapped tip is not parallel and will cause ink blobs and smearing.

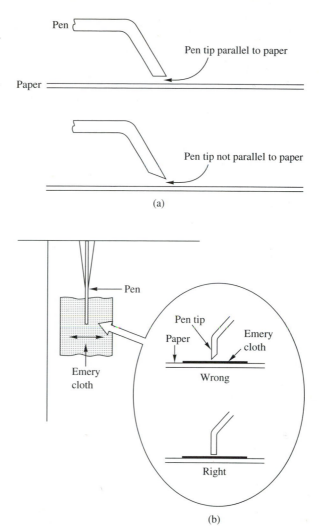

either (or both) of the following: (1) a "blob" of ink when the machine first starts recording a waveform or (2) a too-thick trace. In most such machines the ink should be dry before the paper leaves the paper platform at the drive roller end of the surface. If it is not dry, then lapping may be needed. To lap the pen, place a piece of very fine emery cloth (a sandpaper-like material available at any hardware store) under the tip. The pen tip is worked back and forth 5 to 10 times to "sand" the tip parallel to the paper.

The pressure of the stylus or pen is also important. If the pressure is not correct, then the waveform may be distorted. On medical equipment it is possible to make a normally healthy Lead-I ECG signal look like it has either a heart block or a recent myocardial infarction because of improperly adjusted stylus pressure. The manufacturer will specify a pressure in grams. These numbers vary from 1 to 10 g, depending upon the machine model (and its year of manufacture—older model machines tended to use heavier stylus pressures).

A stylus pressure gage (see Fig. 10–16) is used to lift the pen or stylus from the paper as the machine is running until the trace just disappears. The pressure reading is then made from the barrel of the gage. Suitable stylus pressure gages can be purchased from machine manufacturers' service or parts departments. Alternatively, the stylus pressure gages used for other purposes are also useful if the specified pressure is within the relatively limited range (usually 0 to 4 g of the gage). The stylus pressure adjustment is made using a screw that is usually located on the rear of the stylus (or pen) or on the assembly that holds it in place.

Figure 10–17 shows several different situations where a combination of improper pressure or (in the case of thermal writers) temperature causes improper recording. Our example shows the 1-mV calibration pulses on an ECG machine. These traces can be made by pressing the "1-MV" or "CAL" button on the front panel of the machine. The ideal shape is perfectly square, as shown in Fig. 10–17(a), but this ideal is almost never achieved in practice because of the inertia of the pen or stylus assembly. Usually we see the slightly rounded features of the pulse shown in Fig. 10–17(b). This waveform is usually acceptable. What is not acceptable, however, are the overdamped and underdamped waveforms of Figs. 10–17(c) and 10–17(d), respectively.

Some recorders have a damping control available for adjustment by a properly trained technician. This control is adjusted (usually internally to the machine) in order to compensate for problems. On ink pen machines the stylus pressure can affect the wave shape, especially if it is set to too high a value (it produces the overdamped waveform). On heated stylus machines both the stylus pressure and the heat can affect the waveform.

On some heated stylus machines the standard procedure is to set the pressure to a specified value, set the voltage applied to the stylus heating element to a specified value (usually either 5.00 or 7.00 V), and then adjust the internal damping control to produce the waveform of Fig. 10–17(b) in response to either a square wave input or successive presses of the "1-MV CAL" button.

**FIGURE 10–16**
Stylus pressure gage.

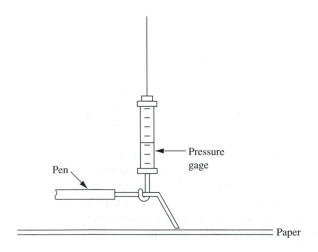

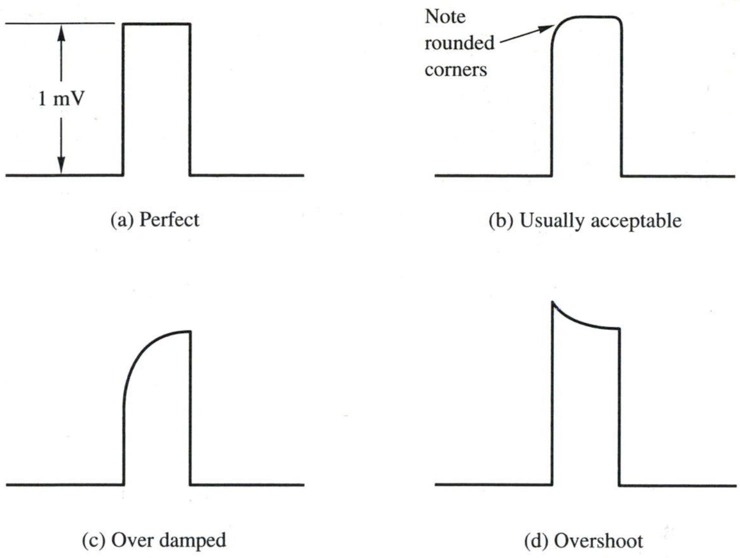

**FIGURE 10–17**
Examples of improper recording.

(a) Perfect

(b) Usually acceptable

(c) Over damped

(d) Overshoot

If the manufacturer of the machine did not provide a knob on the stylus heat control, then don't adjust it without the correct equipment (usually stylus pressure gage and voltmeter) AND the manufacturer's service manual!

I remember a time when the director of the emergency room called and snorted that he didn't really believe that the last 40 patients—only one of whom had a cardiac complaint—all had recent myocardial infarctions. The problem was that an ambitious medical student (who also held an electrical engineering degree) thought he knew how to adjust the heat—but he did not. He had brought a small screwdriver to work and had adjusted the heat to make the trace lighter "for reliability reasons." The trace was lighter, all right, but the machine would not recover from fast waveform transitions (such as an ECG QRS complex) fast enough. As a result of the extra inertia caused by insufficient melting of the paraffin on the paper, the ECG tracings all erroneously showed S-T segment anomalies.

## 10–8        DOT MATRIX ANALOG RECORDERS

The dot matrix printer is long familiar to users of computer equipment. It was developed in response to the high cost of traditional computer printers. At the time that the dot matrix printer became popular, its cost was about one-fifth to one-third the cost of daisy wheel, Selectric®, or similar printer mechanisms. The original dot matrix printers (mid-1970s) used a 5 × 7 matrix of dots [Fig. 10–18(a)] to form alphanumeric characters.

The dot matrix machine used a print head [Fig. 10–18(b)] to cause the correct dot elements to be energized to make a mark on the paper. Two different methods were once popular, although one has since faded almost to obscurity. Some of the earliest machines were thermally based. The dots were thermally connected to heating coils and could be heated when needed. Special temperature-sensitive paper was used to receive the text.

**FIGURE 10–18**

(a) 5 × 7 dot matrix pattern, (b) print head (with pins extended), (c) print head positioned on paper, the concept depicted schematically.

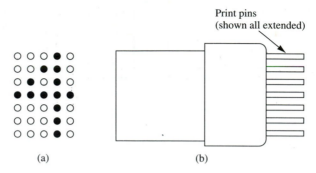

(a)

(b)

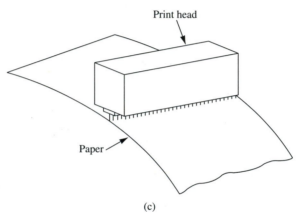

(c)

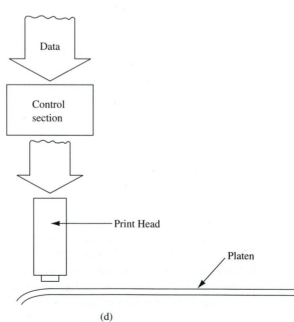

(d)

This method is no longer widely used. The second method used an array of 7 print hammers (actually pins). The pins would either extend or retract depending upon whether that particular dot was active for the character being printed. An advantage of the pin method is that ordinary paper can be used. The pins impact an inked ribbon to leave the impression. Although the original low-resolution printers were 7-pin models (as shown in Figs. 10–18(a) and 10–18(b), higher-resolution models are now available with 9-, 18-, and 24-pin print heads.

Clever programmers rapidly discovered that a dot matrix print head could do graphics as well as alphanumerics. The spate of newsletters, church and school bulletins, and other low-cost (but often creatively done) publications are a testimony to the graphics capability of modern dot matrix technology.

Dot matrix printing can be used to make analog recorders that outperform most older mechanical analog recorders. Figures 10–18(c) and 10–18(d) show the concept schematically. A dot matrix print head with a very large number of pins is arrayed over a platen. The action depends upon whether thermal, electro-arc, or plain strip chart paper is used. Regardless of the particulars, however, the result is a strip chart recording that mixes analog and digital data on the same chart.

Figure 10–19 shows some actual dot matrix analog recordings. The digital computer backing up this system can print the appropriate grid. [Notice the difference between Figs. 10–19(a) and 10–19(b); these recordings were made sequentially on the same machine without changing paper.] Printed as alphanumeric characters along the top of the strip are data including ICU bed number, time, date, ECG lead number, heart rate, and blood pressure values. Figure 10–19(b) is similar to Fig. 10–19(a), but it includes an arterial blood pressure waveform along with the ECG waveform.

## 10–9  LASER AND DOT MATRIX COMPUTER PRINTERS AS RECORDERS

The laser printer is not a classical chart recorder, but when integrated with a graphics-capable computer it can act as such. The dot matrix printer will also work, but not as well as the laser printer (although 24-pin models provide reasonable performance). A computer equipped with an analog-to-digital (A/D) converter at a parallel or serial input will convert analog signals to binary form that can be input to the computer. These signals can then be processed, if needed, and then output to a graph made on a laser printer. There are many cases where the mechanical chart recorder is preferred, and it is usually lower in cost. However, if a computer and laser printer are already part of the equipment in a data acquisition or electronic measurement system, then the use of the laser printer should be considered.

## SUMMARY

1. The PMMC galvanometer is analogous to a d'Arsonval meter movement in which the pointer is replaced with a pen or other writing instrument.
2. The recording potentiometer uses a null-seeking servo system to drive the pen assembly.

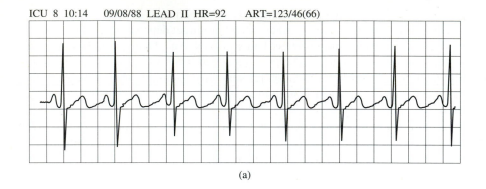

ICU 8  10:14     09/08/88  LEAD  II  HR=92      ART=123/46(66)

(a)

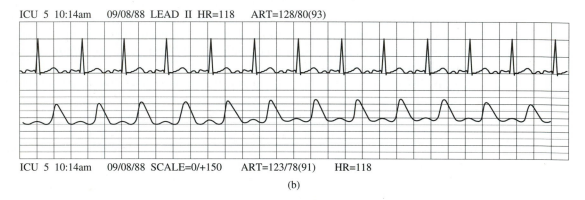

ICU 5  10:14am    09/08/88  LEAD  II  HR=118     ART=128/80(93)

ICU 5  10:14am    09/08/88  SCALE=0/+150     ART=123/78(91)     HR=118

(b)

**FIGURE 10–19**
Dot matrix analog recordings.

3. Several writing systems are used: **direct pressure, thermal, pen and ink, ink jet, optical,** and **CRT camera.**
4. Laser and dot matrix mechanisms can be used as analog chart recorders.

## RECAPITULATION

Now go back and try to answer the questions at the beginning of this chapter. When you are finished, answer the questions and work the problems below. Place a mark beside each problem or question that you cannot answer, and then go back and reread the appropriate sections of the text.

## QUESTIONS

1. Recording VOMs often use the _____ writing system.
2. Describe the operation of a PMMC galvanometer. Draw a comparison between the PMMC galvanometer and the d'Arsonval or taut-band meter movements.

3. What is meant by "curvilinear recording"? Give the advantages and disadvantages of this writing system.

4. Describe a method by which a PMMC galvanometer can be constructed to render a rectilinear tracing.

5. Describe a "pseudorectilinear" PMMC recorder mechanism. Use a picture if you prefer or if your instructor asks for it.

6. How does a writing edge (i.e., a knife edge) render a rectilinear trace from the curvilinear motion of the PMMC pen? What types of writing systems can use this approach?

7. Describe in your own words the **direct pressure** and **thermal** writing systems. Describe any special paper that is required in either of these systems.

8. Describe the **pen and ink** and **ink jet** writing systems, and compare their properties.

9. Describe the optical PMMC galvanometer and CRT camera types of writing systems. Include in your discussion the wet and dry developing methods.

10. Use a block diagram to describe the self-balancing potentiometer recorder.

11. (a) What is a deadband? (b) What problems does it cause? (c) How are these problems overcome?

12. Describe the **overshoot** and **undershoot** phenomena. How are they related to **damping?**

13. What is a **slide wire potentiometer?** How is it constructed? (Use a diagram if desired.)

14. X-Y recorders typically use two _____ recorder mechanisms positioned at right angles to each other.

15. How can X-Y recorders be made into Y-T (i.e., Y-time) recorders?

16. On quality recorders the deadband zone is typically no more than _____% to _____% of full scale.

17. How can a position signal be used to adjust the damping of a PMMC circuit?

## BIBLIOGRAPHY

Brown, John M., and Carr, Joseph J. *Introduction to Biomedical Equipment and Technology,* 2nd ed. Englewood Cliffs, NJ: Prentice-Hall, 1993.

Carr, Joseph J. *Biomedical Equipment: Use, Maintenance, and Management.* Englewood Cliffs, NJ: Prentice-Hall, 1992.

Diefenderfer, James. *Principles of Electronic Instrumentation.* Philadelphia: W. B. Saunders, 1972.

Hewlett-Packard. *X-Y Recorder Applications Notes 1, 2, and 3.* Palo Alto, CA: Hewlett-Packard Corp., n.d.

Strong, Peter. *Biophysical Measurements.* Measurements Concepts Series. Beaverton, OR: Tektronix, Inc., 1973.

# 11

## Special-Purpose Laboratory Amplifiers

### 11-1 OBJECTIVES

1. To learn the different forms of commercial amplifiers.
2. To learn how to specify and apply laboratory amplifiers.

### 11-2 SELF-EVALUATION

Before studying the material in this chapter, try to answer the questions given below. These questions test your knowledge of the subject. If you cannot answer a particular question, look for the answer as you read the text.

1. An isolation amplifier typically has an input impedance of _____ ohms.
2. Chopper amplifiers are used where _____ and noise are a problem because of high voltage gains.
3. An amplifier used for exciting a transducer with ac is called a _____ amplifier.
4. The lock-in amplifier is particularly useful when the signal is in the presence of a large amount of _____ .
5. A differential voltage amplifier with a gain of 100 would be considered a _____ (low-, medium-, high-) gain amplifier.

### 11-3 BASIC AMPLIFIERS

There are many applications for special-purpose amplifiers. Such instruments are typically used in laboratories, industry, and field locations for data collection and distribution and other chores. Although it is quite possible for a skilled engineer to design amplifiers at substantially lower cost than the commercial equivalents, it is rarely cost-effective to do so given certain realities (time, compatibility, workload, etc.). In this chapter we will examine some of the basic forms of amplifiers that can be purchased from instrument manufacturers.

275

Figure 11–1 shows the three basic symbols used to represent amplifiers. These symbols are typically used both for commercial amplifiers that are whole and complete in themselves and for integrated circuit operational amplifiers. The context in which the symbol is used will help you determine which is the case.

The standard **single-ended amplifier** symbol is shown in Fig. 11–1(a). This form of amplifier requires an input signal that is referenced to ground or a common bus; such an input is said to be **unbalanced.** Likewise, the output signal is also unbalanced. In some cases the gain of the amplifier will be written inside the symbol using a format such as "×100" (i.e., gain is times 100) or "$A_v = 100$" (i.e., voltage gain = 100). Generally, we cannot tell whether an amplifier such as that shown in Fig. 11–1(a) is inverting or noninverting unless additional information is provided. For example, an indication of "−100" or "$A_v = -100$" would denote an inverting amplifier. An inverting amplifier provides an output that is 180° out of phase with the input signal, while a noninverting amplifier produces a signal that is in phase with its input signal.

The transfer function of an inverting amplifier is written

$$-\frac{V_o}{V_{in}} = A_v \tag{11–1}$$

where
$V_o$ = the output signal voltage
$V_{in}$ = the input signal voltage
$A_v$ = the voltage gain factor of the amplifier

The transfer function of a noninverting amplifier is the same, but without the minus sign.

A **differential amplifier** symbol is shown in Fig. 11–1(b). This type of amplifier uses two inputs, inverting and noninverting. The inverting input produces an output that is 180° out of phase with the input signal. The noninverting input, on the other hand, produces an output that is in phase with the input signal. Since both inputs see the same gain from the amplifier, an implication of applying the same signal to both inputs simultane-

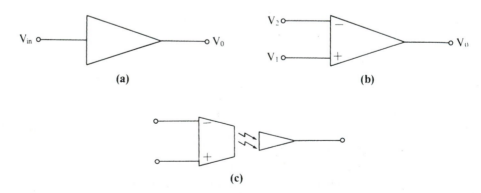

(a)        (b)

(c)

**FIGURE 11–1**

Amplifier symbols. (a) Standard single-ended amplifier, (b) differential amplifier, (c) isolation amplifier.

ously is that the output is zero. This fact can be seen from the transfer equation for a differential amplifier:

$$V_o = (V1 - V2) \times A_{vd} \qquad (11-2)$$

where
$V_o$ = the output signal voltage
$V1$ = the signal voltage applied to the noninverting input
$V2$ = the signal voltage applied to the inverting input
$A_{vd}$ = the differential voltage gain

The last generic type of amplifier symbol we will consider is the **isolation amplifier,** shown in Fig. 11–1(c). This type of amplifier isolates the input circuitry from the output circuitry. A typical hybrid isolation amplifier will have an input impedance (as measured to the ac power supply serving the nonisolated side) of more than $10^{12}$ Ω.

The principal applications for isolation involve safety of either people or instruments. For example, electrical shock considerations in biomedical amplifiers require the use of isolation amplifiers for signal acquisition. An electrocardiograph amplifier, for example, connects a patient to an ac-powered recording or display device. Since as little as 20 µA is considered dangerous to certain electrically compromised patients, leakage current between the chassis and the ac line could place the patient at extreme risk. Thus, an isolation amplifier provides safety. We also see isolation amplifiers used when the environment, often a high voltage, would damage the recording circuitry. In biochemistry, for example, high voltages (1000 V dc) are sometimes used for electrophoresis experiments. In this case a constant current at high voltage will cause different proteins to migrate to different levels in a glass cylinder placed in the electrical field. If we needed to record temperature inside the electrophoretic solution, we would need to connect the thermocouple to the external recorders or data acquisition system via an isolation amplifier. Otherwise, the 1000-V dc field would damage the electronic circuits.

## 11–4    AMPLIFIER CLASSIFICATIONS

Laboratory amplifiers are often classified according to gain. Typical categories are low-gain, medium-gain, and high-gain. Although there is no official system of classification, and definitions tend to vary from one manufacturer to another, the following guidelines generally apply:

**Low-gain.** Voltage gain of 1 to 50.

**Medium-gain.** Voltage gain of 50 to 500.

**High-gain.** Voltage gain over 500.

We also categorize amplifiers as either ac or dc, or both. An ac amplifier will have a lower frequency response limit and will not respond to dc signals. In the case of certain biophysical amplifiers, the lower −3 dB point in the frequency response characteristic might be as low as 0.05 Hz but is not dc. The reason is that there are very low frequency components in these signals, but electrode drift problems argue against dc for the lower end frequency.

## 11–5        LABORATORY AMPLIFIER SYSTEMS

Laboratory amplifiers are often integrated into systems such as shown in Figs. 11–2 and 11–3. The Hewlett-Packard mainframe shown in Fig. 11–2 is one of several made by that company; it finds applications in a wide range of jobs from medicine to industrial signals acquisition. The amplifiers are in the form of plug-in modules. The mainframe accepts the plug-ins and provides them with mechanical support, dc power, ac carrier excitation (for transducer amplifiers), output signal routing, and (sometimes) input signal routing. In most cases the input signals are applied through the front panel of each amplifier, although that is not universally true. This particular H-P configuration has the mainframe as part of a four-channel analog pen recorder that makes a permanent record of the waveforms and is mounted in a portable roll-around cart. Other configurations can be rigged to customer specifications.

**FIGURE 11–2**
Laboratory instrumentation system by Hewlett-Packard.

(a)

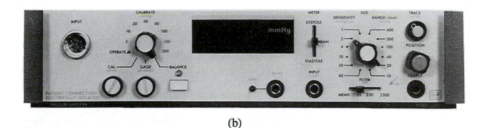

(b)

**FIGURE 11–3**
(a) Laboratory instrumentation system using a recording camera, (b) plug-in pressure amplifier
[(a) courtesy of Electronics for Medicine, Inc.].

Figure 11–3(a) shows another form of instrument package, this one by Electronics
for Medicine. In this case the plug-in amplifiers [example in Fig. 11–3(b)] are mounted in
a housing with a long-persistence analog CRT oscilloscope. The lower unit (horizontal) is
a CRT camera recorder that makes copies of the trace appearing on the upper CRT. The
plug-in shown in Fig. 11–3(b) is a blood pressure amplifier, so the output meter is calibrated in pressure units (millimeters of mercury, mmHg). Note that the gain control on
the right is also calibrated in mmHg.

## 11–6     CHOPPER AMPLIFIERS

Direct current amplifiers have a certain inherent *drift* and tend to be *noisy*. These factors are not too important in low- and medium-gain applications (i.e., gains less than 500), but they loom very large indeed at high gain. For example, a 50-$\mu$V/°C drift figure in an ×100 amplifier produces an output voltage of

$$(50 \ \mu V/°C) \times 100 = 5 \ mV/°C \qquad\qquad \textbf{(11–3)}$$

which is tolerable in most cases. But in an × 100,000 amplifier the output voltage would be

$$(50 \ \mu V/°C) \times 10^5 = 5 \ V/°C$$

and that amount of drift would probably obscure any real signals in a very short time.

Similarly, noise can be a problem in high-gain applications, where it had been negligible in most low- to medium-gain applications. Amplifier noise is usually specified in terms of **nanovolts per square root** hertz [i.e., noise$_{rms}$ = nV(Hz)$^{1/2}$]. A typical low-cost operational amplifier has a noise specification of 100 nV(Hz)$^{1/2}$, so at a bandwidth of 10 kHz the noise amplitude will be

$$\begin{aligned} Noise_{rms} &= 100 \ nV(10^4 \ Hz)^{1/2} \\ &= 100 \ nV(10^2) \\ &= 10^4 \ nV = 10^{-5} \ V \end{aligned}$$

In an ×100 amplifier, without low-pass filtering, the output amplitude will be only 1 mV, but in an ×100,000 amplifier it will be 1 V.

A circuit called a **chopper amplifier** will solve both problems because it makes use of an ac-coupled amplifier.

The drift problem is cured because of two properties of ac amplifiers; one is the inability to pass low-frequency (i.e., near-dc) changes such as those caused by drift, and the other is the ability to regulate the stage through the use of negative feedback.

However, many low analog signals are very low frequency (i.e., in the dc-to-30-Hz range) and thus will not pass through such an amplifier. The answer to this problem is to **chop** the signal at a high frequency so that it passes through the ac amplifier, and then to demodulate the amplifier output signal to recover the original wave shape, but at a higher amplitude.

Figure 11–4(a) shows the basic chopper circuit. The traditional chopper is a vibrator-driven SPDT switch (*S*1) connected so that it alternately grounded first the input and then the output of the ac amplifier. An example of a chopped waveform is shown in Fig. 11–4(c). A low-pass filter following the amplifier filters out any residual chopper hash and any miscellaneous noise signals that may be present.

Most of these mechanical choppers use a chop rate of 400 Hz, although 60-, 100-, 200-, and 500-Hz choppers are also known. The main criterion for the chop rate is that it

**FIGURE 11–4**
(a) Basic chopper amplifier.

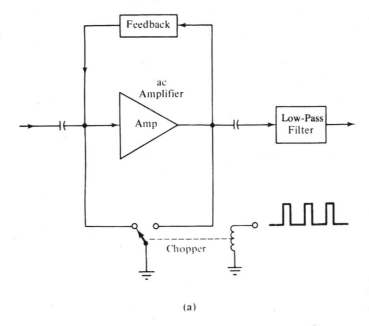

(a)

be twice the highest component frequency that is present in the input waveform. In other words, it must obey Nyquist's criterion.

A differential chopper amplifier is shown in Fig. 11–4(b). In this circuit an input transformer with a center-tapped primary is used. One input terminal is connected to the transformer center tap, while the other input terminal is switched back and forth between the two ends of the primary winding.

A synchronous demodulator following the ac amplifier detects the signal and restores the original, but now amplified, wave shape. Again, a low-pass filter smoothes out the signal.

The modern chopper amplifier may not use mechanical vibrator switches as the chopper. A pair of CMOS or JFET electronic switches driven out of phase with each other will perform the same job. Some monolithic or hybrid function module chopper amplifiers use a varactor switching bridge for the chopper.

The chopper amplifier limits noise because of the low-pass filter and because the amplifier can have a narrow bandpass centered around the chopper frequency.

## 11–7    CARRIER AMPLIFIERS

A **carrier amplifier** is any type of signal-processing amplifier in which the signal carrying the desired information is modulated onto another signal, that is, a "carrier." The chopper amplifier is considered by many to fit this definition, but it is usually regarded as a type in its own right. The two principal carrier amplifiers are the dc-excited and ac-excited varieties.

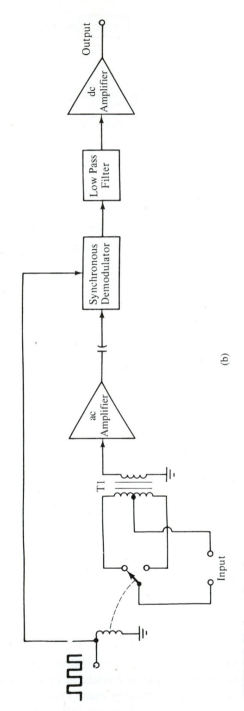

**FIGURE 11–4, *continued***
(b) Differential chopper amplifier.

**FIGURE 11–4,** *continued*
(c) Continuous-vs.-sampled analog
waveforms.

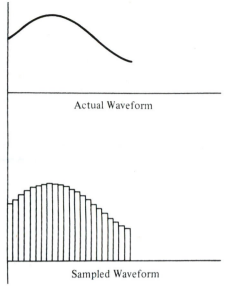

Actual Waveform

Sampled Waveform

(c)

Figure 11–5 shows a dc-excited carrier amplifier. The Wheatstone bridge transducer is excited by dc potential *E*. The output of the transducer, then, is a small dc voltage that varies with the value of the stimulating parameter. The transducer signal is usually of very low amplitude and is noisy. An amplifier builds up the amplitude, and a low-pass filter removes much of the noise. In some models the first stage is actually a composite of these two functions, being a filter with gain.

The signal at the output of the amplifier-filter section is used to amplitude-modulate a carrier signal. Typical carrier frequencies range from 400 Hz to 25 kHz, with 1 kHz and 2.4 kHz being very common. The signal frequency response of a carrier amplifier is a function of the carrier frequency and is usually considered to be one-fourth of the carrier frequency. A carrier of 400 Hz, then, is capable of a signal frequency response to 100 Hz, while the 25-kHz carrier will support a frequency response of 6.25 kHz. Further amplification of the signal is provided by an ac amplifier.

The key to the performance of any carrier amplifier worthy of the name is the phase-sensitive detector (PSD) that demodulates the amplified ac signal. Envelope detectors, while very simple and of low cost, suffer from an inability to discriminate between the real signal and spurious signals.

Figure 11–6 shows a simplified PSD circuit. Transistors *Q*1 and *Q*2 provide a return path to ground for the opposite ends of the secondary winding of input transformer *T*1. These transistors are alternately switched into an out-of-conduction by the reference signal in such a way that *Q*1 is off when *Q*2 is on, and vice versa. The output waveform of the PSD is a full-wave rectified version of the input signal.

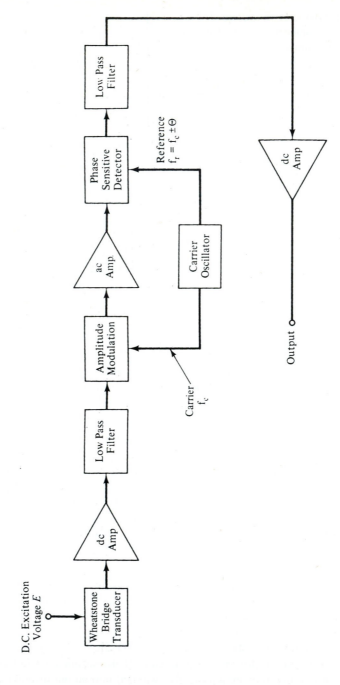

**FIGURE 11-5**
dc-excited carrier amplifier.

284

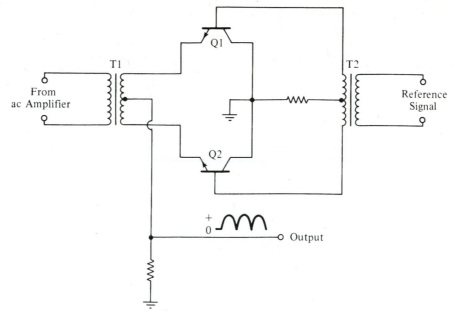

**FIGURE 11–6**
Phase-sensitive detector.

Other electronic switching circuits are also used in PSD design. All systems are designed using the fact that a PSD is essentially an electronic DPDT switch. The digital PSD circuit most often seen uses a CMOS electronic IC switch such as the CD4016/CD4066. These switches are toggled by the reference frequency in such a way that the output is always positive going, regardless of the phase of the input signal.

The advantages of the PSD include the fact that it rejects signals not of the carrier frequency and certain signals that *are* of the carrier frequency. The PSD, for example, will reject even harmonics of the carrier frequency and those components that are out of phase with the reference signal. The PSD will, however, respond to odd harmonics of the carrier frequency. Some carrier amplifiers seem to neglect this problem altogether. But in some cases manufacturers will design the ac amplifier section to be a bandpass amplifier with a response limited to $F_c \pm (F_c/4)$. This response will eliminate any 3rd- or higher-order odd harmonics of the carrier frequency before they reach the PSD. It is then necessary only to assure the purity of the reference signal.

An alternate, but very common, form of carrier amplifier is the ac-excited circuit shown in Fig. 11–7. In this circuit the transducer is ac excited by the carrier signal, eliminating the need for the amplitude modulator. The small ac signal from the transducer is amplified and filtered before being applied to the PSD circuit. Again, some designs use a bandpass ac amplifier to eliminate odd-harmonic response. This circuit allows adjustment of transducer offset errors in the PSD circuit instead of in the transducer, by varying the phase of the reference signal.

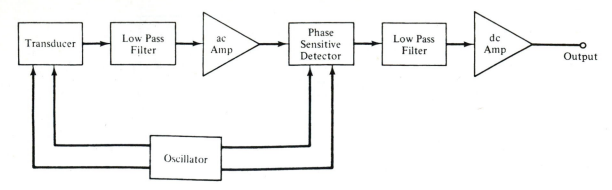

**FIGURE 11–7**
ac-excited carrier amplifier.

## 11–8      LOCK-IN AMPLIFIERS

The amplifiers discussed so far in this chapter produce relatively large amounts of noise and will respond to noise present in the input signal. They suffer from shot noise, thermal noise, H-field noise, E-field noise, ground loop noise, and so forth. The noise voltage or power at the output is directly proportional to the square root of the circuit bandwidth. The **lock-in amplifier** is a special case of the carrier amplifier idea in which the bandwidth is very narrow. Some lock-in amplifiers use the carrier amplifier circuit of Fig. 11–7

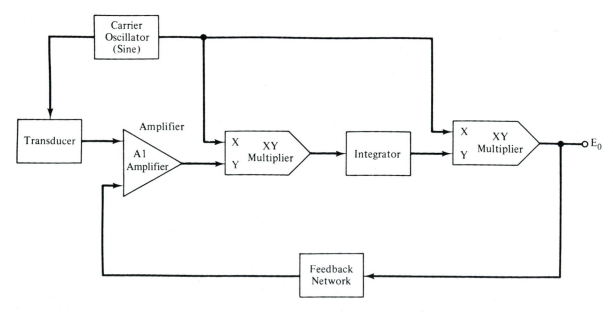

**FIGURE 11–8**
Autocorrelation amplifier.

but use an input amplifier with a very high $Q$ bandpass. The carrier frequency may be anything between 1 and 200 kHz. The lock-in principle works because the information signal is made to contain the carrier frequency in a way that is easy to demodulate and interpret. The ac amplifier accepts only a narrow band of frequencies centered about the carrier frequency. The narrowness of the bandwidth, which makes possible the improved signal-to-noise ratio, also limits the lock-in amplifier to very low frequency input signals. Even then, it is sometimes necessary to time-average the signal for several seconds to obtain the needed data.

Lock-in amplifiers are capable of thinning out the noise and retrieving signals that are otherwise "buried" in the noise level. Improvements of up to 85 dB are relatively easily obtained, and up to 100 dB is possible if cost is no factor.

There are actually several different forms of lock-in amplifiers. The type discussed above is perhaps the simplest type. It is merely a narrow-band version of the ac-excited carrier amplifier. The lock-in amplifier of Fig. 11–8, however, uses a slightly different technique. In this **autocorrelation amplifier,** the carrier is modulated by the input signal and is then integrated (i.e., time averaged). The output of the integrator is demodulated in a product detector circuit. The circuit in Fig. 11–8 produces very low output voltages for input signals that are not in phase with the reference signal, but it produces a relatively high output at the proper frequency.

# 12

# Operational Amplifiers

## 12–1    OBJECTIVES

1. To learn basic operational amplifier theory.
2. To learn the major operational amplifier configurations used in instrumentation circuits.
3. To learn to derive the elementary operational amplifier transfer functions.
4. To learn how to use operational amplifiers in instrumentation.

## 12–2    SELF-EVALUATION

Before studying the material in this chapter, try to answer the questions given below. These questions test your knowledge of the subject matter. If you cannot answer a particular question, then look for the answer as you read the text.

1. Write the transfer equation for (a) an **inverting follower** and (b) a **noninverting follower.**
2. Draw a circuit for an active **integrator** using an operational amplifier.
3. How would you cure **offset voltage** problems?
4. Write the transfer equation for an active operational amplifier **differentiator.**
5. What should be the value of a **compensation resistor, $R_c$?**

## 12–3    OPERATIONAL AMPLIFIERS: AN INTRODUCTION

The operational amplifier has been in existence for at least 5 decades, but only in the last 20 years has it come into its own as an almost universal electronic building block. The term **operational** is derived from the fact that these devices were originally designed for use in analog computers to solve *mathematical operations.* The range of circuit applications today, however, has increased immensely, so the operational amplifier has survived and prospered, even though analog computers, in which they were once a principal constituent, are now almost in eclipse.

Keep in mind, however, that even though the programmable analog computer is no longer used, many analog instruments, and the analog subsystem of digital instruments, are little more than nonprogrammable, dedicated-to-one-chore analog computers.

In this chapter we will examine the properties of the basic operational amplifier and will learn to derive the transfer equations for most common operational amplifier circuits using only Ohm's law, Kirchhoff's law, and the basic properties of all operational amplifiers.

One of the profound beauties of the modern integrated-circuit (IC) operational amplifier is its simplicity when viewed from the outside world. Of course, the inner workings are complex, but they are of little interest in our discussion of the operational amplifier's gross properties. We will limit our discussion somewhat by considering the operational amplifier as a **black box,** and that allows for a very simple analysis in which we relate the performance to the universal transfer function for all electronic circuits, namely, $E_{out}/E_{in}$.

## 12–4        PROPERTIES OF THE IDEAL OP-AMP

An **ideal operational amplifier** is a **gain block,** or black box if you prefer, that has the following general properties:

1. Infinite **open-loop** (i.e., no feedback) gain ($A_{vol} = \infty$)
2. Infinite input impedance ($Z_{in} = \infty$)
3. Zero output impedance ($Z_o = 0$)
4. Infinite bandwidth ($f_0 = \infty$)
5. Zero noise generation

Of course, it is not possible to obtain a real IC operational amplifier that meets these properties—they are *ideal*—but if we read "infinite" as "very, very high," and "zero" as "very, very low," then the approximations of the ideal situation are very accurate. Real IC operational amplifiers, for example, can have an open-loop voltage gain from 20,000 to over 2,000,000. Thus, they can be classed as *relatively infinite,* so the equations work nearly ideally in most cases.

## 12–5        DIFFERENTIAL INPUTS

Figure 12–1 shows the basic symbol for the common operational amplifier, including power terminals. In many schematics of operational amplifier circuits, the $V_{CC}$ and $V_{EE}$ power terminals are deleted, so the drawing will be less "busy."

Note that there are two input terminals, labeled (−) and (+). The terminal labeled (−) is the **inverting** input. The output signal will be out of phase with signals applied to this input terminal; that is, there will be a 180° phase shift. The terminal labeled (+) is the **noninverting** input, so output signals will be in phase with signals applied to this input. It is important to remember that these inputs look into *equal* open-loop gains, so they will have equal but opposite effects on the output voltage.

**FIGURE 12–1**
Symbol for an operational amplifier.

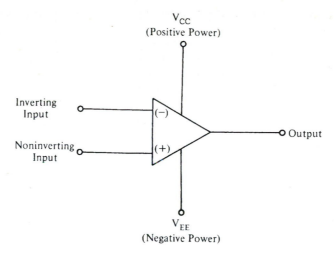

At this point let us add one further property to our list of ideal properties:

**6.** Differential inputs follow each other.

This property implies that the two inputs will behave as if they were at the same potential, especially under static conditions. In Fig. 12–2 we see an inverting follower circuit in which the noninverting (+) input is grounded. The sixth property allows us, in fact requires us, to treat the inverting (−) input as if it were *also* grounded. Many textbooks and magazine articles like to call this phenomenon a "virtual" ground. It is a basic axiom of operational amplifier circuitry that, for purposes of calculation and voltage measurement, the (−) input will act as if it were grounded if the (+) input is actually grounded.

**FIGURE 12–2**
Inverting follower.

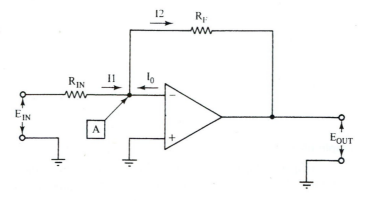

## 12–6        ANALYSIS USING KIRCHHOFF'S AND OHM'S LAWS

We know from Kirchhoff's current law that the algebraic sum of all currents entering and leaving a point in a circuit must be zero. The total current flow into and out of point A in Fig. 12–2, then, must be *zero*. Three possible currents exist at this point: input current $I1$, feedback current $I2$, and any currents flowing into or out of the (−) input terminal of the operational amplifier, $I_0$. But according to ideal property 2, the input impedance of this type of device is infinite. Ohm's law tells us that by

$$I_0 = \frac{E}{Z_{in}} \tag{12–1}$$

current $I_0$ is zero, because $E/Z_{in}$ is zero. So, if current $I_0$ is equal to zero, we conclude that $I1 + I2 = 0$ (Kirchhoff's law). Since this is true, then

$$I2 = -I1 \tag{12–2}$$

We also know that

$$I1 = \frac{E_{in}}{R_{in}} \tag{12–3}$$

and

$$I2 = \frac{E_{out}}{R_f} \tag{12–4}$$

By substituting Equations (12–3) and (12–4) into Equation (12–2), we obtain the result

$$\frac{E_{out}}{R_f} = \frac{-E_{in}}{R_{in}} \tag{12–5}$$

Solving for $E_{out}$ gives us the transfer function normally given in operational amplifier literature for an inverting amplifier:

$$E_{out} = -E_{in} \times \frac{R_f}{R_{in}} \tag{12–6}$$

### ■ Example 12–1

Calculate the output voltage from an inverting operational amplifier circuit if the input signal is 100 mV, the feedback resistor is 100 kΩ, and the input resistor is 10 kΩ.

*Solution*

$$E_{\text{out}} = -E_{\text{in}} \times (R_f / R_{\text{in}}) \tag{12-6}$$
$$= -(0.1 \text{ V})(10^5 \text{ }\Omega/10^4 \text{ }\Omega)$$
$$= -(0.1 \text{ V})(10) = \mathbf{-1 \text{ V}}$$

The term $R_f / R_{\text{in}}$ in Equation (12–6) is the voltage gain factor and is usually designated by the symbol $A_V$, which is written as

$$A_v = -\frac{R_f}{R_{\text{in}}} \tag{12-7}$$

We sometimes encounter Equation (12–6) written using the left-hand side of Equation (12–7):

$$E_{\text{out}} = -A_V E_{\text{in}} \tag{12-8}$$

When designing simple inverting followers using operational amplifiers, use Equations (12–7) and (12–8). Let's look at a specific example. Suppose that we have a requirement for an amplifier with a gain of 50. We want to drive this amplifier from a source that has an output impedance of 1000 $\Omega$. A rule of thumb for designers to follow is to make the input impedance not less than 10 times the source impedance, so in this case the amplifier must have an input source impedance that is equal to or greater than 10,000 $\Omega$ (10 k$\Omega$). This requirement sets the value of the input resistor at 10 k$\Omega$ or higher, but in this example we select a 10-k$\Omega$ value for $R_{\text{in}}$:

$$A_V = R_f / R_{\text{in}} \tag{12-9}$$

$$50 = R_f / 10,000 \text{ }\Omega \tag{12-10}$$
$$R_f = 500,000 \text{ }\Omega$$

Our gain-of-50 amplifier will look like Fig. 12–3.

**FIGURE 12–3**
Gain-of-50 inverting follower.

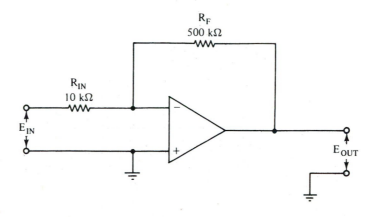

**12–7**  ## NONINVERTING FOLLOWERS

The inverting follower circuits of Figs. 12–2 and 12–3 suffer badly from low input impedance, especially at higher gains, because the input impedance is the value of $R_{in}$. This problem becomes especially acute when we attempt to obtain even moderately high gain figures from low-cost devices. Although some types of operational amplifiers allow the use of 500-kΩ to 2-MΩ input resistors, they are costly and often uneconomical. The **noninverting follower** of Fig. 12–4 solves this problem by using the input impedance problem very nicely, because the input impedance of the op-amp is typically very, very high (ideal property 2).

We may once again use Kirchhoff's law to derive the transfer equation from our basic ideal properties. By property 6 we know that the inputs tend to follow each other, so the inverting input can be treated as if it were at the same potential as the noninverting input, which is $E_{in}$, the input signal voltage. We know that

$$I1 = I2 \tag{12–11}$$

$$I1 = E_{in}/R_{in} \tag{12–12}$$

$$I2 = (E_{out} - E_{in})/R_f \tag{12–13}$$

By substituting Equations (12–12) and (12–13) into Equation (12–11), we obtain

$$I1 = I2 \tag{12–14}$$

$$\frac{E_{in}}{R_{in}} = \frac{E_{out} - E_{in}}{R_f} \tag{12–15}$$

Solving Equation (12–15) for $E_{out}$ results in the transfer equation for the noninverting follower amplifier circuit.

**FIGURE 12–4**
Noninverting follower with gain.

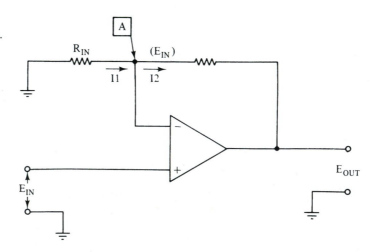

**1.** Multiply both sides by $R_f$:

$$\frac{R_f E_{in}}{R_{in}} = E_{out} - E_{in} \tag{12–16}$$

**2.** Add $E_{in}$ to both sides:

$$\frac{R_f E_{in}}{R_{in}} + E_{in} = E_{out} \tag{12–17}$$

**3.** Factor out $E_{in}$:

$$E_{in} \times \left[\frac{R_f}{R_{in}} + 1\right] = E_{out} \tag{12–18}$$

■ **Example 12–2**

Calculate the output voltage for a 100-mV (0.1-V) input in a noninverting follower amplifier if $R_f = 100$ k$\Omega$ and $R_{in} = 10$ k$\Omega$.

*Solution*

$$
\begin{aligned}
E_{out} &= E_{in}[(R_f/R_{in}) + 1] \tag{12–18}\\
&= (0.1 \text{ V})[(10^5 \ \Omega/10^4 \ \Omega) + 1]\\
&= (0.1 \text{ V})(10 + 1)\\
&= (0.1 \text{ V})(11) = \textbf{1.1 V}
\end{aligned}
$$

In this discussion we have arrived at both of the transfer functions commonly used in operational amplifier design, using only the basic properties, Ohm's law, and Kirchhoff's current law. We may safely assume that the operational amplifier is merely a feedback device that generates a current that exactly cancels the input current. Figure 12–5 gives a synopsis of the characteristics of the most popular operational amplifier configurations. The unity-gain noninverting follower of Fig. 12–5(c) is a special case of the circuit in Fig. 12–5(b), in which $R_f/R_{in} = 0$. In this case the transfer equation becomes

$$E_{out} = E_{in}(0 + 1) \tag{12–19}$$

$$E_{out} = E_{in}(1) \tag{12–20}$$

$$E_{out} = E_{in} \tag{12–21}$$

**12–8        OPERATIONAL AMPLIFIER POWER SUPPLIES**

Although almost every circuit using operational amplifiers uses a dual polarity power supply, it is possible to operate the device with a single polarity supply. An example of

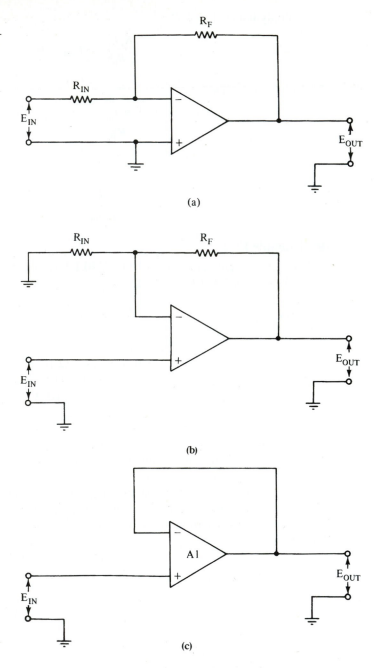

(a)

(b)

(c)

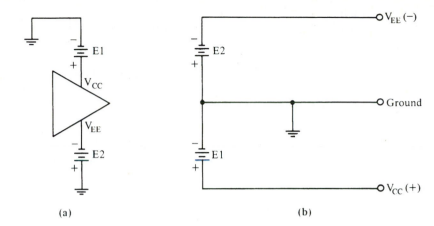

**FIGURE 12–6**

(a) Power supply requirements of an operational amplifier, (b) typical operational amplifier configuration.

single supply operation might be in equipment designed for mobile operation, or in circuits where the other circuitry requires only a single polarity supply and an op-amp or two are but minority features in the design. It is, however, generally better to use the bipolar supplies as intended by the manufacturer.

There are two separate power terminals on the typical operational amplifier device, and these are marked $V_{CC}$ and $V_{EE}$. The $V_{CC}$ supply is connected to a power supply that is *positive* to ground, while the $V_{EE}$ supply is *negative* with respect to ground. These supplies are shown in Fig. 12–6. Keep in mind that although batteries are shown in the example, regular power supplies may be used instead. Typical values for $V_{CC}$ and $V_{EE}$ range from ±1.5 to ±44 V dc. In many cases, perhaps most, the value selected for these potentials will be between ±9 and ±15 V dc.

One further constraint is placed on the operational amplifier power supply: $V_{CC} - V_{EE}$ must be less than some specified voltage, usually 30 V. So, if $V_{CC}$ is +18 V dc, then $V_{EE}$ must be not greater than 30 − 18, or 12 V dc.

## 12–9      PRACTICAL DEVICES: SOME PROBLEMS

Before we can properly apply operational amplifiers in real equipment, we must learn some of the limitations of real-world devices. The devices that we have considered up until now have been *ideal,* so they do not exist. Real IC operational amplifiers carry price tags of less than half a dollar up to several dozen dollars each. The lower the cost, generally, the less ideal the device.

Three main problems exist in real operational amplifiers: **offset current, offset voltage,** and **frequency response.** Of less importance in many cases is noise generation.

In real operational amplifier devices the input impedance is less than infinite, and this implies that a small input bias current exists. The input current may flow into or out of

**FIGURE 12–7**
Use of a compensation resistor.

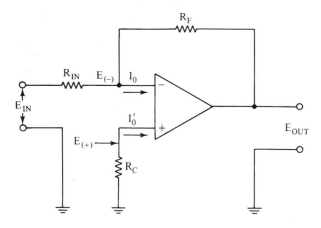

the input terminals of the operational amplifier. In other words, current $I_0$ of Fig. 12–2 is *not really* zero, so it will produce an output voltage equal to $-I_0 \times R_f$. The cure for this problem, shown in Fig. 12–7, involves placing a **compensation resistor** between the non-inverting input terminal and ground. This tactic works because the currents in the respective inputs are approximately equal. Since resistor $R_c$ is equal to the parallel combination of $R_f$ and $R_{in}$, it will generate the same *voltage drop* that appears at the inverting input. The resultant output voltage, then, is zero, because the two inputs have equal but opposite polarity effect on the output.

Output offset voltage is the value of $E_{out}$ that will exist if the input end of $R_{in}$ is grounded (i.e., $E_{in} = 0$). In the ideal device, $E_{out}$ would be zero under this condition, but in real devices there may be some offset potential present. This output potential can be forced to zero by any of the circuits in Fig. 12–8.

The circuit in Fig. 12–8(a) uses a pair of **offset null terminals** found on many, but not all, operational amplifiers. Although many IC operational amplifiers use this technique, some do not. Alternatively, the offset range may be insufficient in some cases. In either event, we may use the circuit of Fig. 12–8(b) to solve the problem.

The offset null circuit of Fig. 12–8(b) creates a current flowing in resistor $R1$ to the summing junction of the operational amplifier. Since the offset current may flow either *into* or *out of* the input terminal, the null control circuit must be able to supply currents of both polarities. Because of this requirement, the ends of the potentiometer ($R1$) are connected to $V_{CC}$ and $V_{EE}$.

In many cases the offset is small compared with normally expected values of input signal voltage. This is especially true in low-gain applications, in which case the nominal offset current will create such a low output error that no action need be taken. In still other cases the offset of each stage in a cascade chain of amplifiers may be small, but their cumulative effect may be a large offset error. In this type of situation it is usually sufficient to null only one of the stages late in the chain (i.e., close to the output stage).

In circuits where the offset is small but critical, it may be useful to replace $R1$ and $R2$ of Fig. 12–8(b) with one of the resistor networks of Figs. 12–8(c) through 12–8(e). These perform essentially the same function but have superior resolution. That is, there is

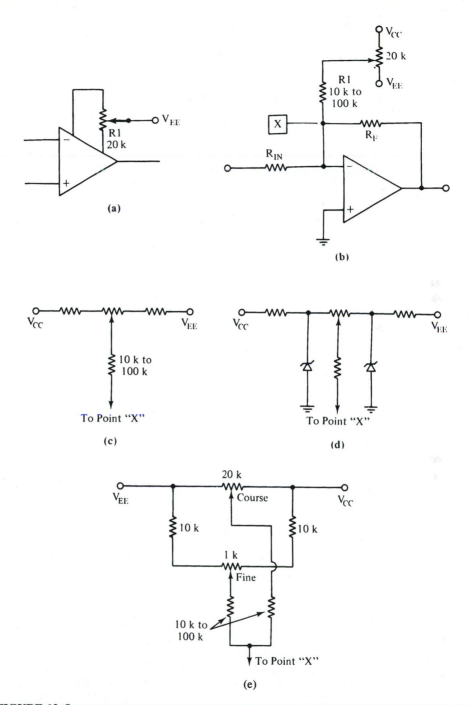

**FIGURE 12–8**

(a) Use of offset terminals to null output, (b) use of summing current to null output. (c)–(e) High-resolution offset null circuits.

a smaller change in output voltage for a single turn of the potentiometer. This type of circuit will have a superior resolution in any event, but even further improvement is possible if a 10-turn (or more) potentiometer is used.

## 12–10    DC DIFFERENTIAL AMPLIFIERS

The fact that an IC operational amplifier has two complementary inputs, inverting and noninverting, makes it a natural for application as a **differential amplifier.** These circuits produce an output voltage that is proportional to the *difference* between two ground-referenced input voltages. Recall from our previous discussion that the two inputs of an operational amplifier have equal but opposite effect on the output voltage. If the same voltage or two equal voltages are applied to the two inputs (i.e., a **common mode** voltage, $E3$ in Fig. 12–9), then the output voltage will be zero. The transfer equation for a differential amplifier is

$$E_{\text{out}} = A_{\text{v}}(E1 - E2) \qquad (12\text{–}22)$$

So, if $E1 = E2$, then $E_{\text{out}} = 0$.

The circuit of Fig. 12–9 shows a simple differential amplifier using a single IC operational amplifier. The voltage gain of this circuit is given by

$$A_{\text{V}} = \frac{R3}{R1} \qquad (12\text{–}23)$$

provided that $R1 = R2$ and $R3 = R4$.

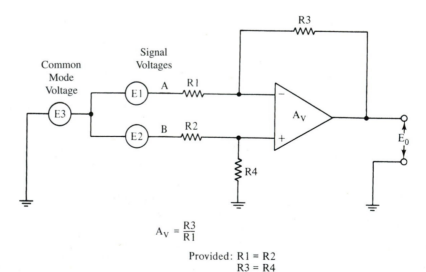

$$A_{\text{V}} = \frac{R3}{R1}$$

Provided: R1 = R2
R3 = R4

**FIGURE 12–9**
Differential amplifier with input voltages.

The main appeal of this circuit is that it is economical, requiring but one IC operational amplifier. It will reject common mode voltages reasonably well if the equal resistors are well matched. A glaring problem exists, however, and that is a low input impedance. Additionally, with the problems existing in real operational amplifiers, this circuit may be a little difficult to tame in high-gain applications. As a result, designers frequently use an alternate circuit in these cases.

In recent years the instrumentation amplifier (IA) of Fig. 12–10 has become popular because it alleviates most of the problems associated with the circuit of Fig. 12–9. The input stages are noninverting followers, so they will have a characteristically high input impedance. Typical values run to as much as 1000 MΩ.

The instrumentation amplifier is relatively tolerant of different resistor ratios used to create voltage gain. In the simplest case the differential voltage gain is given by

$$A_V = \frac{2R3}{R1} + 1 \qquad\qquad (12\text{--}24)$$

provided that $R3 = R2$ and $R4 = R5 = R6 = R7$.

It is interesting to note that the common mode rejection ratio (CMMR) is not seriously degraded by mismatch of resistors $R2$ and $R3$; only the gain is affected. If these resistors are mismatched, then a differential voltage gain error will be introduced.

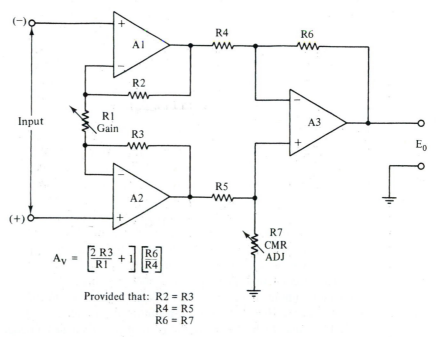

$$A_V = \left[\frac{2\,R3}{R1} + 1\right]\left[\frac{R6}{R4}\right]$$

Provided that:  R2 = R3
R4 = R5
R6 = R7

**FIGURE 12–10**
Instrumentation amplifier.

The situation created by Equation (12–24) results in having the gain of $A3$ equal to unity (i.e., 1), and that is a waste. If gain in $A3$ is desired, then Equation (12–24) must be rewritten into the form

$$A_v = \left[ \frac{2R3}{R1} + 1 \right]\left[ \frac{R7}{R5} \right] \qquad (12\text{–}25)$$

■ **Example 12–3**

Calculate the differential voltage gain of an instrumentation amplifier that uses the following resistor values: $R3 = 33$ k$\Omega$, $R1 = 2.2$ k$\Omega$, $R5 = 3.3$ k$\Omega$, and $R7 = 15$ k$\Omega$.

*Solution*

$$A_V = [(2R3/R1) + 1](R7/R5) \qquad (12\text{–}25)$$

$$= \left[ \frac{(2)(33 \text{ k}\Omega)}{(2.2 \text{ k}\Omega)} + 1 \right]\left[ \frac{(15 \text{ k}\Omega)}{(3.3 \text{ k}\Omega)} \right]$$

$$= \mathbf{141}$$

---

One further equation that may be of interest is the general expression from which the other instrumentation amplifier transfer equations are derived:

$$A_V = \frac{R7(R1 + R2 + R3)}{R1R6} \qquad (12\text{–}26)$$

which remains valid provided that the ratio $R7/R6 = R5/R4$.

Equation (12–26) is especially useful because you need not be concerned with whether the precision resistors are matched pairs, but only that their ratios are equal.

**12–11          PRACTICAL CIRCUIT**

In this section we will consider a practical design example using the instrumentation amplifier circuit. The particular problem required that the frequency response be up to 100 kHz and that the input lines be shielded. But the latter requirement would also deteriorate the signal at high frequencies because of the shunt capacitance of the input cables. To overcome this problem, a **high-frequency compensation** control is built into the amplifier. Voltage gain is approximately 10.

The circuit to the preamplifier is shown in Fig. 12–11. It is, of course, the instrumentation amplifier of Fig. 12–10 with some modifications. When the frequency response is less than 8 kHz or so, we may use any of the 741-family devices (i.e., 741, 747, 1456, and 1458), but premium performance demands a better operational amplifier. In this case one of the most economical is the CA3140, although an L156 would also suffice.

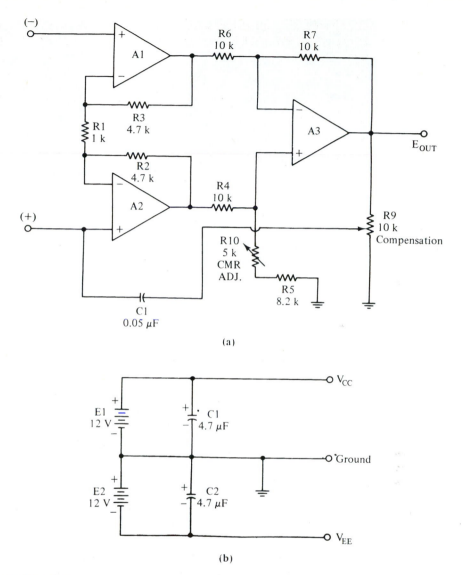

**FIGURE 12–11**

(a) Instrumentation amplifier with capacitance null, (b) power supply for (a).

Common mode rejection can be adjusted to compensate for any mismatch in the resistors or IC devices by adjusting $R10$. This potentiometer is adjusted for zero output when the same signal is applied simultaneously to both inputs.

The frequency response characteristics of this preamplifier are shown in Figs. 12–12 through 12–16. The input in each case was a 1000-Hz square wave from a function generator. The waveform in Fig. 12–12 shows the output signal when resistor $R9$ is set with its wiper closest to ground. Notice that it is essentially square, showing only a small

**FIGURE 12–12**
Square wave.

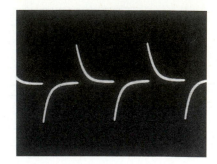

**FIGURE 12–13**
Differentiated square wave.

**FIGURE 12–14**
Voltage gain vs. frequency.

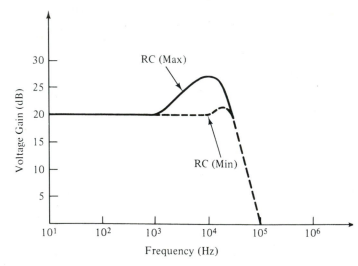

amount of roll-off of high frequencies. The waveform in Fig. 12–13 is the same signal when R9 is at maximum resistance. This creates a small amount of regenerative (i.e., positive) feedback, although it is not sufficient to start oscillation but will enhance amplification of high frequencies.

The problem of oscillation can be quite serious (Fig. 12–15), however, if certain precautions are not taken, most of which involve limiting the amplitude of the feedback signal. This goal is realized by using a 2200-$\Omega$ resistor in series with the potentiometer.

Another source of oscillation is the value of C1. When a 0.001-$\mu$F capacitor is used at C1, an 80-kHz oscillation is created (see Fig. 12–16). The frequency response is shown in Fig. 12–14. To obtain any particular response curve, modify the values of C1 and R9.

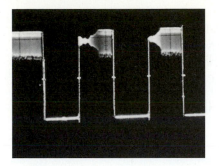

**FIGURE 12–15**
Ringing on square wave.

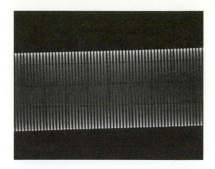

**FIGURE 12–16**
80-kHz oscillation.

## 12–12    IC INSTRUMENTATION AMPLIFIERS

The operational amplifier truly revolutionized analog circuit design. For a long time, however, the only additional advances were that op-amps became better and better (they became nearer the ideal op-amp of textbooks). While those developments were exciting, there were no really new devices. The next big breakthrough came when the analog device designers made an IC version of the instrumentation amplifier of Fig. 12–10, called the **integrated circuit instrumentation amplifier (ICIA).** Today, manufacturers are offering better and better ICIA devices; we can truly say as in an early op-amp textbook, "The contriving of contrivances is a game for all."

The Burr-Brown INA-101 is an ICIA device. This amplifier is also simple to connect. There are only dc power connections, differential input connections, offset adjust connections, ground, and an output. The gain of the circuit is set by

$$A_{vd} = \frac{40 \text{ k}\Omega}{R_g} + 1 \qquad (12\text{–}27)$$

The INA-101 is basically a low-noise, low-input bias current integrated circuit version of the IA of Fig. 12–11(a). The resistors labeled $R2$ and $R3$ in Fig. 12–11(a) are 20 k$\Omega$, hence the "40 k" term in Fig. 12–17.

Potentiometer $R1$ in Fig. 12–17 is used to null the offset voltages appearing at the output. An **offset voltage** is a voltage that exists on the output at a time when it should be zero (i.e., when $V1 = V2$, so that $V1 - V2 = 0$). The offset voltage might be internal to the amplifier or a component of the input signal. Direct current offsets in signals are common, especially in biopotential amplifiers such as the ECG and EEG.

Still another ICIA is the LM-363 device shown in Fig. 12–18. The miniDIP version is shown in Fig. 12–18(a) (an 8-pin metal can is also available), while a typical circuit is

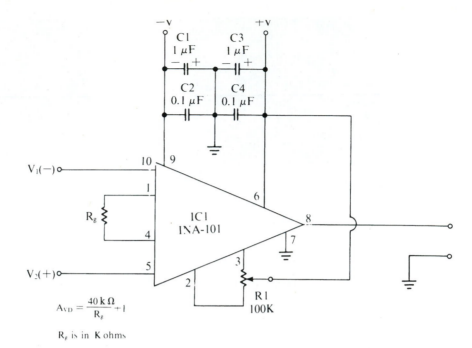

**FIGURE 12–17**
Burr-Brown INA-101 ICIA.

shown in Fig. 12–18(b). The LM-363 device is a fixed-gain ICIA. There are three versions:

| Designation | Gain |
|-------------|------|
| LM-363-10   | 10   |
| LM-363-100  | 100  |
| LM-363-500  | 500  |

The LM-363-x is useful in places where one of the standard gains is required and there is minimum of space available. Two examples spring to mind. We can use the LM-363-x as a transducer preamplifier, especially in noisy signal areas; the LM-363-x can be built onto (or into) the transducer to build up its signal before sending it to the main instrument or signal acquisition computer. The other example is in bioamplifiers. The biopotentials are typically very small, especially in lab animals. The LM-363-x can be mounted on the subject, and a higher-level signal can be sent to the main instrument to a little exotic, but useful nonetheless.

A selectable-gain version of the LM-363 device is shown in Fig. 12–19. The 16-pin DIP package is shown in Fig. 12–19(a), while a typical circuit is shown in Fig. 12–19(b). The type number of this device is LM-363-AD, which distinguishes it from the

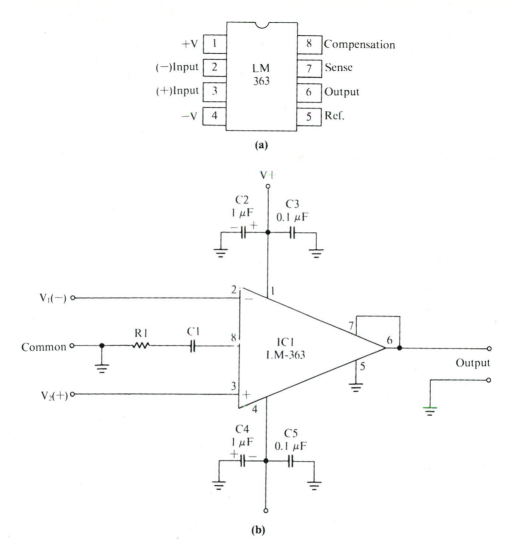

**FIGURE 12–18**

National Semiconductor LM-363 ICIA. (a) Pinouts, (b) circuit.

LM-363-x devices. The gain can be ×10, ×100, or ×1000, depending upon the programming of the gain setting pins (2, 3, and 4). The programming protocol is as follows:

| Gain Desired | Jumper Pins |
|:---:|:---:|
| ×10 | (all open) |
| ×100 | 3 and 4 |
| ×1000 | 2 and 4 |

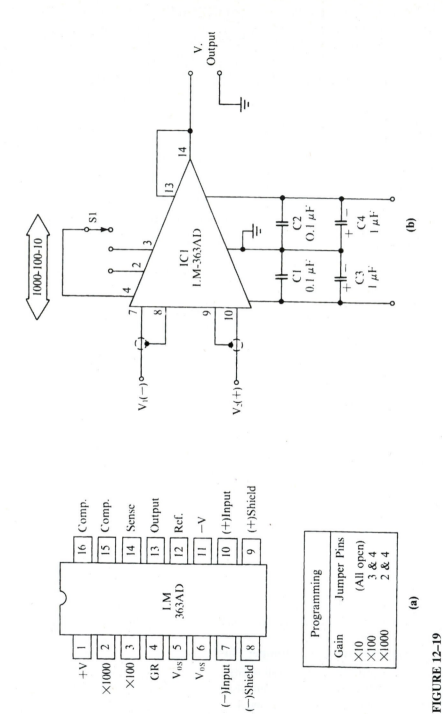

**FIGURE 12–19**

National Semiconductor LM-363AD ICIA. (a) Pinouts, (b) circuit.

Switch $S1$ in Fig. 12–19(b) is the **gain select** switch. This switch should be mounted close to the IC device, but it is quite flexible in mechanical form. The switch could also be made from a combination of CMOS electronic switches (e.g., 4066).

The dc power supply terminals are treated in a manner similar to the other amplifiers. Again, the 0.1-μF capacitors need to be mounted as closely as possible to the body of the LM-363-AD.

Pins 8 and 9 are guard shield outputs. These pins make the LM-363-AD more useful for many instrumentation problems than other models. By outputting a signal sample back to the shield of the input lines, we can increase the common mode rejection ratio. This feature is frequently used in bipotential amplifiers and in other applications where a low-level signal must pass through a strong-interference (high-noise) environment.

The LM-363 devices will operate with dc supply voltages of ±5 to ±18 V dc, with a common mode rejection ratio (CMRR) of 130 dB. The 7 nV/[SQR(Hz)] noise figure makes the device useful for low-noise applications (a 0.5-nV model is available at premium cost).

## 12–13    ISOLATION AMPLIFIERS

Many applications for instrumentation amplifiers are dangerous for either the circuit or the user. In biomedical applications the issue is patient safety. There are numerous analog signal acquisition needs in biomedical instrumentation where the patient is at risk. Even the simple ECG machine, which measures and records the heart's electrical activity, was once implicated in patient safety problems. Another problem area in biomedical applications is in catheterization instruments. There are several tests where doctors insert an electrode or transducer into the body and then measure the resulting signal. For example, the intracardiac ECG places an electrode inside the heart by way of a blood vein; the cardiac output computer uses a signal from a thermistor inside a catheter placed in the heart (also through a vein); and simple electronic blood pressure monitors use a transducer that connects to an artery. In all of these cases, we do not want the patient exposed to small differences of potential due to current leakage from the 60-Hz ac power lines. The solution is use of an **isolation amplifier.**

Another application is signals acquisition in high-voltage circuits. We do not want to mix high-voltage sources with low-voltage electronics because we don't want the low-voltage circuits to blow out. Again, the solution is the isolation amplifier.

Figure 12–20 shows the basic symbol for the isolation amplifier. The break in the triangle used to represent any amplifier denotes that there is an extremely high impedance (typically $10^{12}$ Ω) between the inputs and the output terminal of the isolation amplifier.

Notice that there are two sets of dc power supply terminals. The $V-$ and $V+$ terminals are the same as found on all ICIA or op-amp devices. These dc power supply terminals are connected to the regular dc supply of the equipment where the device is used. Such a power supply derives its dc potential from the ac power mains by way of a 60-Hz transformer. The isolated dc power supply inputs ($VI-$ and $VI+$) are used to power the input amplifier stages and must be isolated from the main dc power supply of the equipment. The $VI-$ and $VI+$ terminals are usually either battery powered or powered from a dc-to-dc converter that produces a dc output from the main power supply by using a high-

**FIGURE 12–20**
Isolation amplifier symbol.

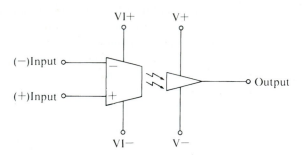

frequency (50- to 500-kHz) oscillator. The high-frequency "power supply" transformer does not pass 60-Hz signals well, so the isolation is maintained.

Figure 12–21 shows the circuit of an isolation amplifier based on the Burr-Brown 3652 device. The dc power for both the isolated and the nonisolated sections of the 3652 is provided by the 722 dual dc-to-dc converter. This device produces two independent ±15 V dc supplies that are each isolated from the 60-Hz ac power mains and from each other. The 722 device is powered from a +12 V dc source that is derived from the ac power mains. In some cases the nonisolated section (which is connected to the output terminal) is powered from a bipolar dc power supply that is derived from the 60-Hz ac

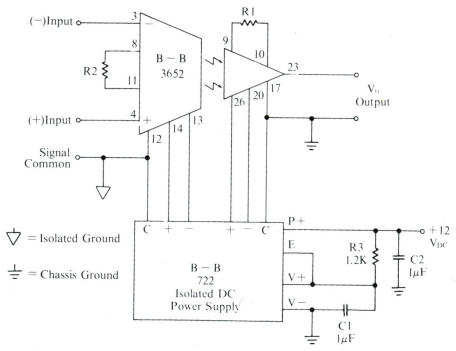

**FIGURE 12–21**
Isolation amplifier circuit.

mains, such as a ±12 V dc or ±15 V dc supply. In no instance, however, should the isolated dc power supplies be derived from the ac power mains.

There are two separate ground systems in this circuit, symbolized by the small triangle and the regular three-bar "chassis" ground symbol. The isolated ground is not connected to either the dc power supply ground/common or the chassis ground. It is kept floating at all times and becomes the signal common for the input signal source.

The gain of the circuit is approximately

$$\text{Gain} = \frac{1,000,000}{R1 + R2 + 115} \tag{12–28}$$

In most design cases the issue is the unknown values of the gain-setting resistors. We can rearrange Equation (12–28) to solve for $(R1 + R2)$:

$$(R1 + R2) = \frac{1,000,000 - (115 \times \text{Gain})}{\text{Gain}} \tag{12–29}$$

where   $R1$ and $R2$ are in ohms ($\Omega$)
            Gain = the voltage gain desired

Let's work an example. Suppose we need a differential voltage gain of 1000. What combination of $R1$ and $R2$ will provide that gain figure? If Gain = 1000, then

$$(R1 - R2) = \frac{1,000,000 - (115 \times 1000)}{1000} \tag{12–30}$$

$$(R1 + R2) = \frac{1,000,000 - (115,000)}{1000} \tag{12–31}$$

$$(R1 + R2) = \frac{885,000}{1000} \tag{12–32}$$

$$(R1 + R2) = 885 \ \Omega \tag{12–33}$$

In this case we need some combination of $R1$ and $R2$ that adds to 885 $\Omega$. The value 440 $\Omega$ is "standard" and will result in only a tiny gain error if used.

The IC instrumentation amplifier and the isolation amplifier open new applications that the simple op-amp cannot match.

## 12–14   DIFFERENTIAL AMPLIFIER APPLICATIONS

Differential amplifiers find application in many different instrumentation situations. Of course, you should realize that they are required wherever a differential signal voltage is found. Less obvious, perhaps, is that they are used to acquire signals or to operate in control systems, in the presence of large noise signals. Many medical applications, for example, use the differential amplifier because they look for minute biopotentials in the presence of strong 60-Hz fields from the ac power mains.

**312**    Chapter 12

Another class of applications is the amplification of the output signal from a Wheatstone bridge, as shown in Fig. 12–22. If one side of the bridge's excitation potential is grounded, then the output voltage is a differential signal voltage. This signal can be applied to the inputs of a differential amplifier or instrumentation amplifier to create an amplified, single-ended, output voltage.

A "rear end" stage suitable for many operational amplifier instrumentation projects is shown in Fig. 12–23. This circuit consists of three low-cost operational amplifier ICs. Since they follow most of the circuit gain, we may use low-cost devices such as the 741 in this circuit. The gain of this circuit is given by $R2/10^4$.

**FIGURE 12–22**
Differential amplifier used to amplify output of Wheatstone bridge.

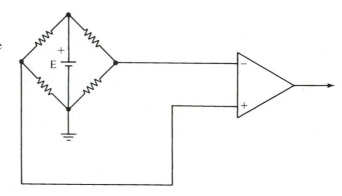

**FIGURE 12–23**
Universal "rear end" for instrumentation amplifiers and other purposes.

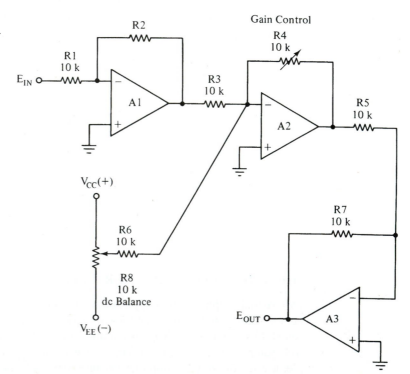

## 12–15 INTEGRATORS

Figure 12–24 shows the basic operational amplifier **integrator** circuit. The transfer equation for this circuit may be derived in the same manner as before, with due consideration for $C1$:

$$I2 = -I1 \tag{12–34}$$

but

$$I1 = \frac{E_{in}}{R1} \tag{12–35}$$

and

$$I2 = C1\left(\frac{dE_0}{dt}\right) \tag{12–36}$$

Substituting Equations (12–35) and (12–36) into (12–34) results in

$$\frac{C1\ dE_0}{dt} = \frac{-E_{in}}{R1} \tag{12–37}$$

We may now solve Equation (12–37) for $E_0$ by integrating both sides:

$$\int \frac{C1\ dE_0}{dt}\ dt = -\int \frac{E_{in}}{R1}\ dt \tag{12–38}$$

$$C1E_0 = \frac{-1}{R1} \int_0^t E_{in}\ dt \tag{12–39}$$

$$E_0 = \frac{-1}{R1C1} \int_0^t E_{in}\ dt \tag{12–40}$$

**FIGURE 12–24**
Integrator circuit.

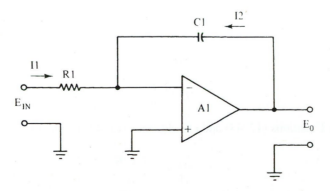

Equation (12–40), then, is the transfer equation for the operational amplifier integrator circuit.

### ■ Example 12–4

A constant potential of 2 V is applied to the input of the integrator in Fig. 12–24 for 3 s. Find the output potential if $R1 = 1$ M$\Omega$ and $C1 = 0.5$ μF$(5 \times 10^{-7}$ F).

*Solution*

$$E_0 = \frac{-1}{R1C1} \int_0^t E_{in}\ dt \qquad\qquad (12\text{–}41)$$

$$= \frac{-E_{in}}{R1C1} \int_0^3 dt$$

$$= \frac{(-2\ \text{V})(t\ )}{(10^6\ \Omega)(5 \times 10^{-7}\ \text{F})}\Bigg|_0^3$$

$$= \frac{(-2\ \text{V})(3\ \text{s})}{0.5\ \text{s}} - 0 = \mathbf{-12\ V}$$

Note that the *gain* of the integrator is given by the term $1/R1C1$. If small values of $R1$ and $C1$ are used, then the gain can be very large. For example, if $R1 = 100$ k$\Omega$ and $C1 = 0.001$ μF, then the gain is 10,000. A very small input voltage in that case will saturate the output very quickly. In general, the **time constant** $R1C1$ should be longer than the period of the input waveform by a factor of 5.

## 12–16    DIFFERENTIATORS

An operational amplifier **differentiator** is formed by reversing the roles of $R1$ and $C1$ in the integrator, as shown in Fig. 12–25. We know that

$$I2 = -I1 \qquad\qquad (12\text{–}42)$$

$$I1 = \frac{C1\ dE_{in}}{dt} \qquad\qquad (12\text{–}43)$$

and

$$I2 = \frac{E_0}{R1} \qquad\qquad (12\text{–}44)$$

Substituting Equations (12–43) and (12–44) into (12–42) results in

$$\frac{E_0}{R1} = \frac{-C1\ dE_{in}}{dt} \qquad\qquad (12\text{–}45)$$

**FIGURE 12–25**
Differentiator circuit.

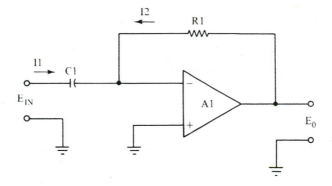

Solving Equation (12–45) for $E_0$ gives us the transfer equation for an operational amplifier differentiator circuit:

$$E_0 = -R1C1 \, \frac{dE_{in}}{dt} \tag{12–46}$$

■ **Example 12–5**

A 12-V/s ramp function voltage is applied to the input of an operational amplifier differentiator, in which $R1 = 1 \text{ M}\Omega$ and $C1 = 0.2 \text{ }\mu\text{F}$. What is the output voltage?

***Solution***

$$
\begin{aligned}
E_0 &= -R1C1(dE_{in}/dt) \tag{12–47}\\
&= -(10^6 \text{ }\Omega)(2 \times 10^{-7} \text{ F})(12 \text{ V/s})\\
&= -(2 \times 10^{-1} \text{ s})(12 \text{ V/s}) = \textbf{-2.4 V}
\end{aligned}
$$

The differentiator time constant $R1C1$ should be set very short relative to the period of the waveform being differentiated, or in the case of square waves, triangle waves, and certain other signals, the time constant should be short compared with the *rise time* of the leading edge.

12–17          **LOGARITHMIC AND ANTILOG AMPLIFIERS**

Figure 12–26(a) shows an elementary **logarithmic amplifier** circuit using a bipolar transistor in the feedback loop. We know that the collector current bears a logarithmic relationship to the base-emitter potential, $V_{be}$:

$$V_{be} = \frac{KT}{q} \ln\left[\frac{I_c}{I_s}\right] \tag{12–48}$$

where   $V_{be}$ = the base-emitter potential in volts (V)
            $K$ = Boltzmann's constant, $1.38 \times 10^{-23}$ joules/kelvin (J/K)

**FIGURE 12–26**
(a) Logarithmic amplifier, (b) improved logarithmic amplifier.

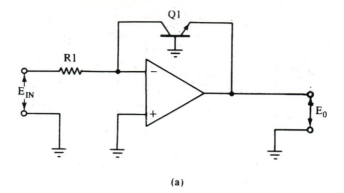

(a)

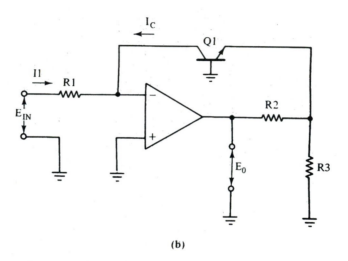

(b)

$T$ = the temperature in kelvin (K)
$q$ = the electronic charge ($1.6 \times 10^{-19}$ C)
$I_c$ = the collector current in amperes (A)
$I_s$ = the reverse saturation current for the transistor in amperes (A)

At 27°C (300 K, i.e., room temperature), the term $KT/q$ evaluates to approximately 26 mV (0.026 V), so Equation (12–48) becomes

$$V_{be} = 26 \text{ mV } \ln\left(\frac{I_c}{I_s}\right) \tag{12–49}$$

But $V_{be} = E_0$, and $I_c = E_{in}/R1$, so we may safely say that

$$E_0 = 2\text{x6 mV } \ln\left[\frac{E_{in}}{I_s R1}\right] \tag{12–50}$$

But $I_s$ is a constant if the temperature is also constant, and $R1$ is constant under all conditions. Thus, by the rule that the logarithm of a constant is also a constant, we may state that Equation (12–50) is the transfer function of the natural logarithmic amplifier. For base-10 logarithms,

$$E_0 = 60 \text{ mV } \log_{10}\left[\frac{E_{in}}{I_s R1}\right] \tag{12–51}$$

The relationship of Equations (12–50) and (12–51) allows us to construct amplifiers with logarithmic properties. If $Q1$ is in the feedback loop of an operational amplifier, then the output voltage $E_0$ will be proportional to the logarithm of input voltage $E_{in}$. If, on the other hand, the transistor is connected in series with the input of the operational amplifier (see Fig. 12–27), then the circuit becomes an antilog amplifier.

Both of these circuits exhibit a strong dependence on temperature, as evidenced by the $T$ term in Equation (12–48). In actual practice, then, some form of temperature correction must be used. Two forms of temperature correction are commonly used: **compensation** and **stabilization.**

The compensation method uses temperature-dependent resistors (i.e., **thermistors**) to regulate the gain of the circuit with changes in temperature. For example, it is common practice to make $R3$ in Fig. 12–26(b) a thermistor.

The stabilization method requires that the temperature of $Q1$, and preferably the op-amp also, be held constant. In the past this has meant that the components must be kept inside an electrically heated oven, but today other techniques are used. One manufacturer builds a temperature-controlled hybrid logarithmic amplifier by nesting the op-amp and transistor on the same substrate as a class A amplifier. Such an amplifier, under zero-signal conditions, dissipates very nearly constant heat. After the chip comes to equilibrium, the temperature will remain constant.

The antilog amplifier is the inverse of the log amplifier, so the principal components are reversed (Fig. 12–27). A similar analysis can be used. In the case of the antilog amplifier

$$I_c = \frac{E_0}{R1} \tag{12–52}$$

**FIGURE 12–27**
Antilog amplifier.

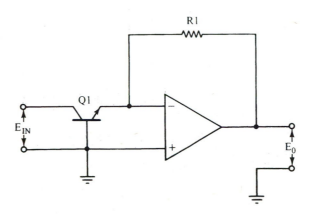

and

$$E_{in} = V_{be} \tag{12–53}$$

$$E_{in} = 26 \text{ mV } \ln\left(\frac{E_0}{R1I_s}\right) \tag{12–54}$$

## 12–18        CURRENT-TO-VOLTAGE CONVERTERS

Most analog recording devices, such as oscilloscopes and graphic recorders, are voltage-input devices. That is to say, they require a *voltage* for an input signal. When measuring or recording a current, however, they require some sort of **current-to-voltage converter** circuit.

Two examples of operational amplifier versions are shown in Fig. 12–28. In the first example, Fig. 12–28(a), a small-value resistor $R$ is placed in series with the current $I$, which produces a voltage equal to $IR$. This potential is seen by the operational amplifier as a valid input voltage. The output voltage is

$$E_0 = \frac{-IRR_f}{R_{in}} \tag{12–55}$$

**FIGURE 12–28**
(a) Current-to-voltage converter, (b) current-to-voltage converter for small currents.

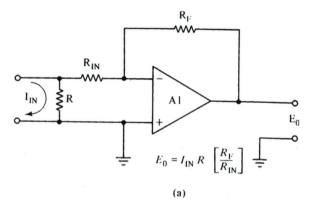

(a)

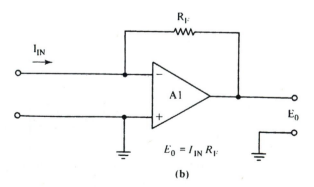

(b)

provided that

$$R << R_{in}$$
$$R << R_f$$

The circuit shown in Fig. 12–28(b) is used for small currents. The output voltage in that circuit is given by

$$E_0 = -I_{in}R_f \qquad\qquad\qquad \textbf{(12–56)}$$

# SUMMARY

1.  The transfer function of an operational amplifier circuit can be determined by the basic properties, Ohm's law, and Kirchhoff's law.
2.  Operational amplifiers can be used as amplifiers, integrators, differentiators, and so forth, to provide numerous functions, both linear and nonlinear.
3.  Operational amplifiers can be configured to provide differential inputs.
4.  The gain of an operational amplifier circuit can be set by the ratio of two resistors.

# RECAPITULATION

Now go back and try to answer the questions at the beginning of the chapter. When you are finished, answer the questions and work the problems given below. Place a mark beside any problem or question that you cannot answer, and then go back to the text and reread appropriate sections.

# QUESTIONS

1.  Write the transfer function for an inverting follower.
2.  Write the transfer function for a noninverting follower.
3.  Write the transfer function for a differential amplifier.
4.  Write the transfer function for an instrumentation amplifier.
5.  Write the transfer function for an operational amplifier integrator.
6.  Write the transfer function for an operational amplifier differentiator.
7.  State four of the basic properties of an ideal operational amplifier.
8.  Describe **virtual ground** in your own words. State the property of ideal operational amplifiers on which this term is based.
9.  An operational amplifier is designed to operate from two power supplies. $V_{CC}$ is ————————— with respect to ground, while $V_{EE}$ is ————————— with respect to ground.
10.  Name three mechanisms by which an output offset voltage is created.
11.  List two different methods for eliminating the output offset voltage.
12.  The common mode voltage gain for a perfect operational amplifier is —————————.

13. The gain of an operational amplifier integrator is given by the following expression: _____.

14. Write the transfer equations for (a) base-*e* and (b) base-10 logarithmic amplifiers.

15. What steps may be taken to reduce the effect of temperature on the logarithmic amplifier?

## PROBLEMS

1. Calculate the voltage gain of an inverting follower if the feedback resistor is 560 kΩ and the input resistor is 82 kΩ.

2. Calculate the voltage gain of an inverting follower if the feedback resistor is 100 kΩ and the input resistor is 10 kΩ.

3. Calculate the voltage gain of an inverting follower if the feedback resistor is 5.6 kΩ and the input resistor is 560 Ω.

4. Calculate the voltage gain of an inverting follower if the feedback resistor is 5 kΩ and the input resistor is 10 kΩ.

5. Calculate the voltage gain of a noninverting follower if the feedback resistor is 120 kΩ and the input resistor is 56 kΩ.

6. Calculate the voltage gain of a noninverting follower if the feedback resistor is 10 kΩ and the input resistor is 10 kΩ.

7. Calculate the voltage gain of a noninverting follower if the feedback resistor is 0 Ω and the input resistor is 100 kΩ.

8. Calculate the voltage gain of a noninverting follower if the feedback resistor is 2.2 kΩ and the input resistor is 4.7 kΩ.

In Problems 9 through 11 refer to Fig. 12–29.

9. In Fig. 12–29 point A is at a potential of +1 V, point B is at +6 V, and point D is at 0 V. Find the output voltage if $R2 = 10$ kΩ and $R1 = 9.1$ kΩ.

10. In Fig. 12–29 point A is grounded, point B is at a dc potential of −3 V, and point D is at a potential of +1.5 V dc. Find the output potential if $R1 = R2 = 100$ kΩ and $R3 = (R1)/2$.

11. In Fig. 12–29 points A and B are grounded, and point D is at a potential of −1 V dc. Find $R2$ if $R3 = 2.2$ kΩ and $E_0 = +10$ V.

12. Find the gain of an instrumentation amplifier such as in Fig. 12–10 if $R1 = 500$ Ω, $R2 = 16$ kΩ, $R4 = 3.9$ kΩ, and $R7 = 27$ kΩ.

13. An operational amplifier integrator uses a 0.69-μF capacitor and an 820-kΩ resistor. What will the output voltage be if a 100-mV dc level is applied to the input for 4 s?

14. What is the gain of an integrator that uses a 0.1-μF capacitor and a 47-kΩ resistor?

15. An operational amplifier differentiator uses a 1-MΩ resistor and a 0.68-μF capacitor. What output voltage exists when an 8-V/s ramp function is applied to the input?

16. Find $V_{be}$ for a transistor operated at 40°C if the reverse saturation current at that temperature is $10^{-12}$ A and the collector current is 1.2 mA.

**FIGURE 12–29**
Circuit for Problems 9–11.

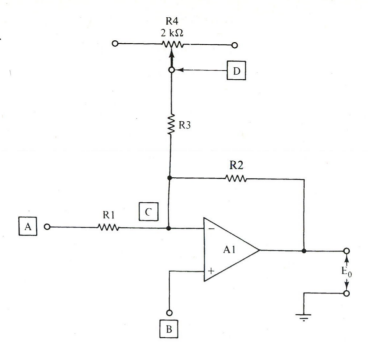

**17.** A logarithmic amplifier is adjusted for the natural log operation mode. Find the output voltage if the reverse saturation current is $10^{-13}$ A, the input resistor is 100 k$\Omega$, and $E_{in}$ = 100 mV.

**18.** What is the **slope factor** of a natural logarithmic amplifier at 37°C?

**19.** A current-to-voltage converter consists of an operational amplifier and a 100-k$\Omega$ resistor in the negative feedback loop. Find $E_0$ if the input current is 20 μA.

**20.** Derive the transfer function for an antilog amplifier without referring to the text.

## BIBLIOGRAPHY

Bannon, Edward. *Operational Amplifiers: Theory and Servicing.* Reston, VA: Reston Publishing Co., 1975.

Barna, Arpad. *Operational Amplifiers.* New York: Wiley-Interscience, 1971.

Carr, Joseph J. *Op-Amp Circuit Design & Applications.* Blue Ridge Summit, PA: TAB Books, 1976.

Graeme, Jerald G. *Applications of Operational Amplifiers—Third Generation Techniques.* New York: McGraw-Hill Book Company, 1973.

Graeme, Jerald G., Tobey, Gene, and Huelsman, L. P., eds. *Operational Amplifiers—Design and Applications.* New York: McGraw-Hill Book Company, 1973.

Jung, Walter. *IC Op-Amp Cookbook.* Indianapolis: Howard W. Sams & Co., 1974.

Stout, David F., and Kaufman, Milton, eds. *Handbook of Operational Circuit Design.* New York: McGraw-Hill Book Company, 1976.

# 13

# Sensors, Electrodes, and Transducers

## 13–1    OBJECTIVES

1. To learn the principles behind the operation of common types of transducers.
2. To learn how to specify and apply transducers.
3. To learn the limitations of certain transducers.
4. To learn about certain application problems and how they are solved.

## 13–2    SELF-EVALUATION

Before studying the material in this chapter, try to answer the questions given below. These questions test your knowledge of the subject matter. If you cannot answer a particular question, then look for the answer as you read the text.

1. Describe the differences between **bonded** and **unbonded** piezoresistive strain gages.
2. List three different types of temperature transducers.
3. List two types of gas flow transducers.
4. Define **transducer** in your own words.
5. How may a force be measured if only displacement transducers are available?

## 13–3    TRANSDUCERS AND TRANSDUCTION

Not all of the physical variables that must be measured lend themselves to direct input into electronic instruments and circuits. Unfortunately, electronic circuits operate only with inputs that are currents and voltages. So, when one is measuring nonelectrical physical quantities, it becomes necessary to provide a device that converts physical parameters such as *force, displacement, temperature,* and so forth, into proportional voltages or currents. The **transducer** is such a device.

> **Definition:** A **transducer** is a device or apparatus that converts nonelectrical physical parameters into electrical signals (i.e., currents or voltages) that are proportional to the value of the physical parameter being measured.

Transducers take many forms and may be based on a wide variety of physical phenomena. Even when one is measuring the *same* parameter, different instruments may use different types of transducers.

This chapter will not be an exhaustive catalogue treatment covering all transducers—manufacturers' data sheets may be used for that purpose—but we will discuss some of the more common *types* of transducers used in scientific, industrial, medical, and engineering applications.

## 13–4   THE WHEATSTONE BRIDGE

Many forms of transducers create a variation in an electrical resistance, inductance, or capacitance in response to some physical parameter. These transducers are often in the form of a Wheatstone bridge or one of the related ac bridge circuits. In many cases where the transducer itself is not in the form of a bridge, it is used in a bridge circuit with other components forming the other arms of the bridge. At this point, then, it may be advisable for the reader to quickly review Chapter 4, "dc and ac Bridge Circuits."

## 13–5   STRAIN GAGES

All electrical conductors possess some amount of electrical resistance. A bar or wire made of such a conductor will have an electrical resistance that is given by

$$R = \rho\left(\frac{L}{A}\right) \tag{13–1}$$

where   $R$ = the resistance in ohms ($\Omega$)
   $\rho$ = the **resistivity constant,** a property specific to the conductor material, given in units of ohm-centimeters ($\Omega$-cm)
   $L$ = the length in centimeters (cm)
   $A$ = the cross-sectional area in square centimeters (cm$^2$)

■ **Example 13–1**

A constantan (i.e., 55% copper, 45% nickel) round wire is 10 cm long and has a radius of 0.01 mm. Find the electrical resistance in ohms. (*Hint:* The resistivity of constantan is $44.2 \times 10^{-6}$ $\Omega$-cm).

*Solution*

$$R = \rho(L/A) \tag{13–1}$$

$$= \frac{(44.2 \times 10^{-6} \text{ } \Omega\text{-cm})(10 \text{ cm})}{\pi\left[0.01 \text{ mm} \times \dfrac{1 \text{ cm}}{10 \text{ mm}}\right]^2}$$

$$= \frac{(4.42 \times 10^{-4} \ \Omega\text{-cm}^2)}{\pi(0.001 \ \text{cm})^2}$$

$$= \frac{(4.42 \times 10^{-4} \ \Omega\text{-cm}^2)}{\pi 10^{-6} \ \text{cm}^2} = \mathbf{141 \ \Omega}$$

Note that the resistivity factor $\rho$ in Equation (13–1) is a constant, so if length $L$ or area $A$ can be made to vary under the influence of an outside parameter then the electrical resistance of the wire will change. This phenomenon, called **piezoresistivity,** is an example of a transducible property of a material.

**Definition: Piezoresistivity** is the change in the electrical resistance of a conductor due to changes in length and cross-sectional area. In piezoresistive materials, mechanical deformation of the material produces changes in electrical resistance.

Figure 13–1 shows how an electrical conductor can use the piezoresistivity property to measure **strain,** that is, *forces* applied to it in *compression* or *tension*. Figure 13–1(a) shows a conductor at rest, in which no forces are acting. The length is given as $L_0$ and the cross-sectional area as $A_0$. The resistance of this conductor, from Equation (13–1), is

$$R_0 = \rho\left(\frac{L_0}{A_0}\right) \tag{13–2}$$

where  $\rho$ = the resistivity as defined previously
$R_0$ = the resistance in ohms ($\Omega$) when no forces are applied
$L_0$ = the resting (i.e., no force) length in cm
$A_0$ = the resting cross-sectional area in cm$^2$

But Figure 13–1(b) shows the situation where a compression force of magnitude $F$ is applied along the axis in the inward direction. The conductor will deform, causing the length $L_1$ to decrease to $L_0 - \Delta L$, and the cross-sectional area to increase to $A_0 + \Delta A$. The electrical resistance decreases to $R_0 - \Delta R$:

$$R_1 = (R_0 - \Delta R) \propto \frac{(L_0 - \Delta L)}{(A_0 + \Delta A)} \tag{13–3}$$

Similarly, when a tension force of the same magnitude ($F$) is applied (i.e., a force that is directed outward along the axis), the length increases to $L_0 + \Delta L$, and the cross-sectional area decreases to $A_0 - \Delta A$. The resistance will increase to

$$R_2 = (R_0 - \Delta R) \propto \frac{(L_0 + \Delta L)}{(A_0 + \Delta A)} \tag{13–4}$$

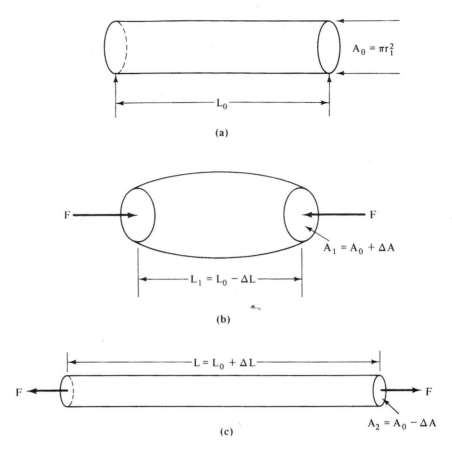

**FIGURE 13–1**
(a) Unstrained metal bar, (b) metal bar in compression, (c) metal bar in tension.

The **sensitivity** of the strain gage is expressed in terms of unit change of electrical resistance for a unit change in length and is given in the form of a **gage factor** $S$:

$$S = \frac{(\Delta R/R)}{(\Delta L/L)} \qquad \text{(13–5)}$$

where    $S$ = the gage factor (dimensionless)
   $R$ = the unstrained resistance of the conductor
   $\Delta R$ = the change in resistance due to strain
   $L$ = the unstrained length of the conductor
   $\Delta L$ = the change in length due to strain

■ **Example 13–2**
Find the gage factor of a 128-$\Omega$ conductor that is 24 mm long if the resistance changes 13.3 $\Omega$ and the length changes 1.6 mm under a tension force.

*Solution*

$$S = (\Delta R/R)/(\Delta L/L) \tag{13–5}$$

$$= \frac{(13.3 \ \Omega/128 \ \Omega)}{(1.6 \ \text{mm}/24 \ \text{mm})}$$

$$= \frac{1.04 \times 10^{-1}}{6.67 \times 10^{-2}} = \mathbf{1.56}$$

We may also express the gage factor in terms of the length and diameter of the conductor. Recall that the diameter is related to the cross-sectional area ($A = \pi d^2/4 = \pi r^2$), so the relationship between the gage factor $S$ and these other factors is given by

$$S = 1 + 2\left[\frac{(\Delta d/d)}{(\Delta L/L)}\right] \tag{13–6}$$

■ **Example 13–3**

Calculate the gage factor $S$ if a 1.5-mm-diameter conductor that is 24 mm long changes length by 1 mm and diameter by 0.02 mm under a compression force.

*Solution*

$$S = 1 + 2\left[\frac{(\Delta d/d)}{(\Delta L/L)}\right] \tag{13–6}$$

$$= 1 + \left[\frac{2[(0.02 \ \text{mm})/(1.5 \ \text{mm})]}{(1 \ \text{mm})/(24 \ \text{mm})}\right]$$

$$= 1 + \left[\frac{(2)(1.3 \times 10^{-2})}{(4.2 \times 10^{-2})}\right]$$

$$= 1 + (2)(0.31) = \mathbf{1.62}$$

Note that the expression $\Delta L/L$ is sometimes denoted by the Greek letter $\epsilon$, so Equations (13–5) and (13–6) become

$$S = 1 + \frac{2(\Delta d/d)}{\epsilon}$$

$$S = \frac{(\Delta R/R)}{\epsilon}$$

Gage factors for various metals vary considerably. Constantan, for example, has a gage factor of approximately 2, while certain other common alloys have gage factors between 1 and 2. At least one alloy (92% platinum, 8% tungsten) has a gage factor of 4.

Semiconductor materials such as germanium and silicon can be doped with impurities to provide custom gage factors between 50 and 250. The problem with semiconductor strain gages, however, is that they exhibit a marked sensitivity to temperature changes. Where semiconductor strain gages are used, either a thermally controlled environment or temperature compensating circuitry must be provided.

## 13–6    BONDED AND UNBONDED STRAIN GAGES

Strain gages can be classified as **unbonded** or **bonded.** These categories refer to the method of construction used. Figure 13–2 shows both methods of construction.

The unbonded type of strain gage, shown in Fig. 13–2(a), consists of a wire resistance element stretched taut between two flexible supports. These supports are configured in such a way as to place tension or compression forces on the taut wire when external forces are applied. In the particular example shown, the supports are mounted on a thin metal diaphragm that flexes when a force is applied. Force $F1$ will cause the flexible supports to spread apart, placing a tension force on the wire and increasing its resistance. Alternatively, when force $F2$ is applied, the ends of the flexible supports tend to move closer together, effectively placing a compression force on the wire element, thereby reducing its resistance. In actuality, the wire's resting condition is **tautness,** which

**FIGURE 13–2**
(a) Unbonded strain gage,
(b) bonded strain gage.

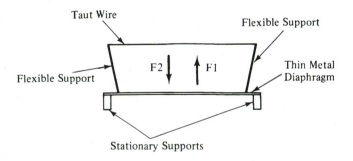

(a)

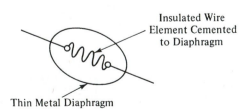

(b)

implies a tension force. Thus, *F*1 increases the tension force from normal, and *F*2 decreases the normal tension.

The bonded strain gage is shown in Fig. 13–2(b). In this type of device a wire or semiconductor element is cemented to a thin metal diaphragm. When the diaphragm is flexed, the element deforms to produce a resistance change.

The linearity of both types can be quite good, provided that the elastic limits of the diaphragm and the element are not exceeded. It is also necessary to ensure that the $\Delta L$ term is only a very small percentage of *L*.

In the past it has been "standard wisdom"—the opinions of those who make purchase decisions—that bonded strain gages are more rugged, but less linear, than unbonded models. Although this may have been true at one time, recent experience has shown that modern manufacturing techniques produce reliable linear instruments of both types.

**13–7**        **STRAIN GAGE CIRCUITRY**

Before a strain gage can be useful, it must be connected into a circuit that will convert its resistance changes to a current or voltage output. Most applications are voltage output circuits.

Figure 13–3(a) shows the **half-bridge** (so called because it is actually half of a Wheatstone bridge circuit) or **voltage divider** circuit. The strain gage element of resistance *R* is placed in series with a fixed resistance *R*1 across a stable and well-regulated voltage source *E*. The output voltage $E_0$ is found from the voltage divider equation:

$$E_0 = \frac{ER}{R + R1} \tag{13–7}$$

Equation (13–7) describes the output voltage $E_0$ when the transducer is at rest, that is, when nothing is stimulating the strain gage element. When the element is stimulated, however, its resistance changes a small amount $\Delta R$. To simplify our discussion we will adopt the standard convention used in many texts of letting $h = \Delta R$. Thus,

$$E_0 = \frac{E(R + h)}{(R \pm h) + R1} \tag{13–8}$$

Another half-bridge is shown in Fig. 13–3(b), but in this case the strain gage is in series with a **constant current source (CCS),** which will maintain current *I* at a constant level regardless of changes in strain gage resistance. The normal output voltage $E_0$ is

$$E_0 = IR \tag{13–9}$$

for nonstimulated conditions, and

$$E_0 = I(R \pm h) \tag{13–10}$$

under stimulated conditions.

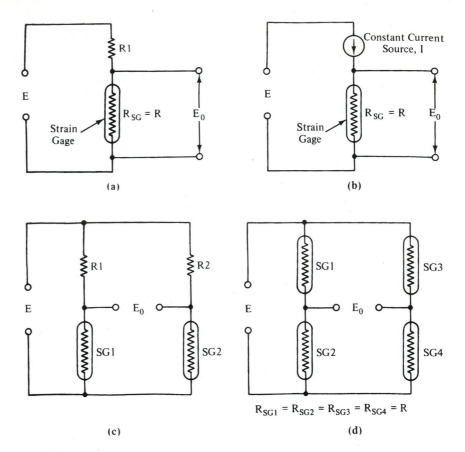

**FIGURE 13–3**
(a) Constant voltage strain gage circuit (half-bridge type); (b) constant current strain gage circuit (half-bridge type); (c) two strain-gage elements in Wheatstone bridge; (d) four-active-element strain gage Wheatstone bridge.

The half-bridge circuits suffer from one major defect: output voltage $E_0$ will always be present regardless of the stimulus. Ideally, in any transducer system, we want $E_0$ to be zero when the stimulus is also zero, and take a value proportional to the stimulus when the stimulus value is nonzero. A Wheatstone bridge circuit in which one or more strain gage elements form the bridge arms has this property.

Figure 13–3(c) shows a circuit in which strain gage elements $SG1$ and $SG2$ form two bridge arms and fixed resistors $R1$ and $R2$ form the other two arms. Usually $SG1$ and $SG2$ will be configured so that their actions oppose each other; that is, under stimulus, $SG1$ will have a resistance $R + h$, and $SG2$ will have a resistance $R - h$, or vice versa.

One of the most linear forms of transducer bridges is the circuit of Fig. 13–3(d) in which all four bridge arms contain strain gage elements. In most such transducers all four strain gage elements have the same resistance, that is, *R,* which has a value between 100 and 1000 Ω in most cases.

Recall from Chapter 4 that the output voltage from a Wheatstone bridge is the difference between the voltages across the two half-bridge dividers. The following equations hold true for bridges in which one, two, or four equal active elements are used.

*One active element:*

$$E_0 = \frac{E}{4}\left[\frac{h}{R}\right] \tag{13-11}$$

(accurate to ±5%, provided that $h \le 0.1$)

*Two active elements:*

$$E_0 = \frac{E}{2}\left[\frac{h}{R}\right] \tag{13-12}$$

*Four active elements:*

$$E_0 = \frac{Eh}{R} \tag{13-13}$$

where, for all three equations,  $E_0$ = the output potential in volts (V)
$E$ = the excitation potential in volts (V)
$R$ = the resistance of all bridge arms
$h$ = the quantity $\Delta R$, the change in resistance of a bridge arm under stimulus

(These equations apply only for the case where all the bridge arms have equal resistances under zero stimulus conditions.)

■ **Example 13–4**

A transducer that measures force has a nominal resting resistance of 300 Ω and is excited by +7.5 V dc. When a 980-dyne force is applied, all four equal-resistance bridge elements change resistance by 5.2 Ω. Find the output voltage $E_0$.

*Solution*

$$E_0 = E(h/R) \tag{13-13}$$
$$= (7.5 \text{ V})(5.2 \text{ }\Omega/300 \text{ }\Omega)$$
$$= (7.5 \text{ V})(5.2)/(300) = \mathbf{0.13 \text{ V}}$$

**13–8**     ## TRANSDUCER SENSITIVITY ($\psi$)

When designing electronic instrumentation systems involving strain gage transducers, it is convenient to use the **sensitivity factor** (denoted by $\psi$, the Greek letter "psi"), which relates the output voltage in terms of the excitation voltage and the applied stimulus. In most cases we see a specification giving the number of microvolts or millivolts output per volt of excitation potential per unit of applied stimulus (i.e., $\mu V/V/Q_0$ or $mV/V/Q_0$). Thus,

$$\psi = E'_0/V/Q_0 \qquad \text{(13–14a)}$$

and

$$\psi = \frac{E'_0}{V \times Q_0} \qquad \text{(13–14b)}$$

where  $E'_0$ = the output potential
  $V$ = one unit of potential, that is, 1 V
  $Q_0$ = one unit of stimulus

The sensitivity is often given as a specification by the transducer manufacturer. From it we can predict output voltage for any level of stimulus and excitation potential. The output voltage, then, is found from

$$E_0 = \psi EQ \qquad \text{(13–15)}$$

where  $E_0$ = the output potential in volts (V)
  $\psi$ = the sensitivity in $\mu V/V/Q_0$
  $E$ = the excitation potential in volts (V)
  $Q$ = the stimulus parameter

■  **Example 13–5**

A well-known medical arterial blood pressure transducer uses a four-element piezoresistive Wheatstone bridge with a sensitivity of 5 $\mu V$ per volt of excitation per torr of pressure, that is, 5 $\mu V/V/T$ (*Note:* 1 torr = 1 mmHg). Find the output voltage if the bridge is excited by 5 V dc and 120 T of pressure is applied.

*Solution*

$$E_0 = \psi EQ \qquad \text{(13–15)}$$

$$= \frac{5\ \mu V}{V - T} \times (5\ V) \times (120\ T)$$

$$= (5 \times 5 \times 120)\ \mu V = \mathbf{3000\ \mu V}$$

**13–9**        ## BALANCING AND CALIBRATING THE BRIDGE

Few, if any, Wheatstone bridge strain gages meet the ideal condition in which all four arms have exactly equal resistances. In fact, the bridge resistance specified by the manufacturer is a *nominal* value only. There will inevitably be an **offset voltage,** that is, $E_0 \neq 0$ when $Q = 0$. Figure 13–4 shows a circuit that will balance the bridge when the stimulus is zero. Potentiometer $R1$, usually a type with 10 or more turns of operation, is used to inject a balancing current $I$ into the bridge circuit at one of the nodes. $R1$ is adjusted, with the stimulus at zero, for zero output voltage.

The best calibration method is to apply a precisely known value of stimulus to the transducer and adjust the amplifier following the transducer for the output proper for that level of stimulus. But that may prove unreasonably difficult in some cases, so an *artificial* calibrator is needed to simulate the stimulus. This function is provided by $R3$ and $S1$ in Fig. 13–4. When $S1$ is open, the transducer is able to operate normally, but when $S1$ is closed, it *unbalances* the bridge and produces an output voltage $E_0$ that simulates some standard value of the stimulus. The value of $R3$ is given by

$$R3 = \left[ \frac{R}{4Q\psi} - \frac{R}{2} \right] \qquad \textbf{(13–16)}$$

where   $R3$ = the resistance of R3 in ohms ($\Omega$)
        $R$ = the nominal resistance of the bridge arms in ohms ($\Omega$)
        $Q$ = the calibrated stimulus parameter
        $\psi$ = the sensitivity factor in V/V/$Q$ (*Note:* Different units of $\psi$ are used: V instead of $\mu$V.)

■   **Example 13–6**

An arterial blood pressure transducer has a sensitivity of 10 $\mu$V/V/T and a nominal bridge arm resistance of 200 $\Omega$. Find a value for $R3$ in Fig. 13–4 if we want the calibration to simulate an arterial pressure of 200 mmHg (i.e., 200 T).

*Solution*

$$R3 = \left[ \frac{R}{4Q\psi} - \frac{R}{2} \right] \qquad \textbf{(13–16)}$$

$$= \frac{200\ \Omega}{4 \times \left( \dfrac{10^{-5}\ \text{V}}{\text{V} - \text{T}} \right) \times 200} - \frac{200\ \Omega}{2}$$

$$= \frac{200\ \Omega}{(4)(10^{-5})(200)} - 100\ \Omega = \textbf{24,900}\ \Omega$$

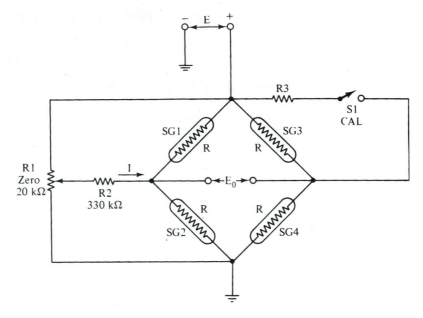

**FIGURE 13–4**
Circuit for using Wheatstone bridge transducer.

**13–10      TEMPERATURE TRANSDUCERS**

A large number of physical phenomena are temperature dependent, so we find quite a variety of electrical temperature transducers on the market. In this discussion, however, we will discuss only three basic types: **thermistor, thermocouple,** and **semiconductor pn junctions.**

**13–11      THERMISTORS**

Metals and most other conductors are temperature sensitive and will change electrical resistance with changes in temperature, namely,

$$R_t = R_0[1 + \alpha(T - T_0)] \qquad (13\text{–}17)$$

where   $R_t$ = the resistance in ohms ($\Omega$) at temperature $T$
$R_0$ = the resistance in ohms ($\Omega$) at temperature $T_0$ (often a standard reference temperature)
$T$ = the temperature of the conductor
$T_0$ = a previous temperature of the conductor at which $R_0$ was determined
$\alpha$ = the **temperature coefficient** of the material, a property of the conductor, in °C$^{-1}$

The temperature coefficients of most metals are positive, as are the coefficients for most semiconductors; for example, gold has a value of +0.004/°C. Ceramic semiconductors used to make **thermistors** (i.e., thermal resistors) can have either negative or positive temperature coefficients depending upon their composition.

The resistance of a thermistor is given by

$$R_t = R_0 e^{\beta[(1/T) - 1/T_0)]}$$

where   $R_t$ = the resistance of the thermistor at temperature $T$
$R_0$ = the resistance of the thermistor at a reference temperature (usually the icc point, 0°C, or room temperature, 25°C)
$e$ = the base of the natural logarithms
$T$ = the thermistor temperature in kelvin (K)
$T_0$ = the reference temperature in kelvin (K)
$\beta$ = a property of the material used to make the thermistor (*Note:* $\beta$ will usually have a value between 1500 K and 7000 K.)

■  **Example 13–7**

Calculate the resistance of a thermistor at 100°C if the resistance at 0°C is 18 kΩ. The material of the thermistor has a value of 2200 K. (*Note:* 0°C = 273 K, so 100°C = 373 K.)

*Solution*

$$R_t = R_0 \exp \beta \left( \frac{1}{T} - \frac{1}{T_0} \right) \tag{13–18}$$

$$= (1.8 \times 10^4 \ \Omega) \exp\left[ (2200 \ \text{K}) \left( \frac{1}{373 \ \text{K}} - \frac{1}{273 \ \text{K}} \right) \right]$$

$$= 1.8 \times 10^4 \ \Omega \ e^{-2.15} = \textbf{2084} \ \Omega$$

Equation (13–18) demonstrates that the response of a thermistor is exponential, as shown in Fig. 13–5. Note that both curves are nearly linear over a portion of their ranges, but then become decidedly nonlinear in the remainder of the region. If a wide measurement range is needed, then a linearization network will be required.

Thermistor transducers will be used in any of the circuits in Fig. 13–3. They will also be found using many packaging arrangements. Figure 13–6 shows a bead thermistor used in medical instruments to continuously monitor a patient's rectal temperature.

The equations governing thermistors usually apply if there is little **self-heating** of the thermistor, although there are applications where self-heating is used. But in straight temperature measurements it is to be avoided. To minimize self-heating it is necessary to control the power dissipation of the thermistor.

Also of concern in some applications is the **time constant** of the thermistor. The resistance does not jump immediately to the new value when the temperature changes but requires a small amount of time to stabilize at the new resistance value. This is expressed

**FIGURE 13–5**
Thermistor temperature-vs.-resistance curves.

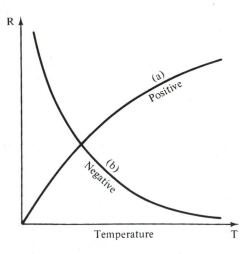

**FIGURE 13–6**
Thermistor in a medical rectal probe (courtesy of Electronics-for-Medicine).

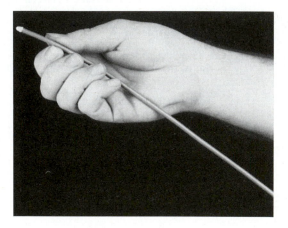

in terms of the time constant of the thermistor in a manner that is reminiscent of capacitors charging in RC circuits.

## 13–12    THERMOCOUPLES

When two dissimilar metals are joined together to form a "vee" [as in Fig. 13–7(a)], it is possible to generate an electrical potential merely by heating the junction. This phenomenon, first noted by Seebeck in 1823, is due to different *work functions* for the two metals. Such a junction is called a **thermocouple,** The Seebeck EMF generated by the junction is proportional to the junction temperature and is reasonably linear over wide temperature ranges.

A simple thermocouple, shown in Fig. 13–7(b), uses two junctions. One junction is the measurement junction and is used as the thermometry probe. The other junction is a reference and is kept at a reference temperature such as the ice point (0°C) or room temperature.

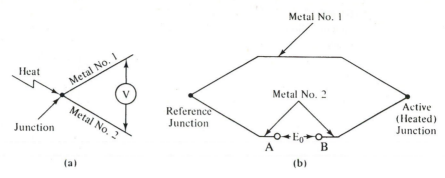

**FIGURE 13–7**
(a) Thermocouple junction, (b) two-thermocouple temperature transducer.

Interestingly enough, there is an inverse thermocouple phenomenon, called the **Peltier effect,** in which an electrical potential applied across A–B in Fig. 13–7(b) will cause one junction to absorb heat (i.e., get hot) and the other to lose heat (i.e., get cold). Semiconductor thermocouples have been used in small-scale environmental temperature chambers, (e.g., drink coolers), and it is reported that one company researched the possibility of using Peltier devices to cool submarine equipment. Ordinary air conditioning equipment proves too noisy in submarines desirous of "silent running."

## 13–13    SEMICONDUCTOR TEMPERATURE TRANSDUCERS

Ordinary pn junction diodes exhibit a strong dependence upon temperature. This effect can be easily demonstrated by using an ohmmeter and an ordinary rectifier diode such as the 1N4000-series devices. Connect the ohmmeter so that it forward-biases the diode, and note the resistance at room temperature. Next hold a soldering iron or other heat source close to the diode's body, and watch the electrical resistance change. In a circuit such as Fig. 13–8 the current is held constant, so output voltage $E_0$ will change with temperature-caused changes in diode resistance.

Another solid-state temperature transducer is shown in Fig. 13–9. In this version the temperature-sensor device is a pair of diode-connected transistors. In any transistor the base-emitter voltage $V_{be}$ is

$$V_{be} = \frac{kT}{q} \ln\left[\frac{I_c}{I_s}\right] \tag{13–19}$$

where  $V_{be}$ = the base-emitter potential in volts (V)
  $k$ = Boltzmann's constant ($1.38 \times 10^{-23}$ J/K)
  $T$ = the temperature in kelvin (K)
  $Q$ = the electronic charge ($1.6 \times 10^{-19}$ C)
  ln denotes the natural logarithms
  $I_c$ = the collector current in amperes (A)
  $I_s$ = the reverse saturation current in amperes (A)

**FIGURE 13–8**
pn junction diode as a temperature transducer.

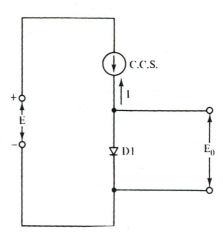

**FIGURE 13–9**
Two transistors connected as diodes from a temperature transducer if $I1 = I2/2$.

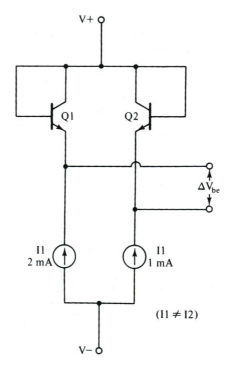

Note that the $k$ and $Q$ terms in Equation (13–19) are constants, and both currents can be made constant. The only variable, then, is *temperature*.

The circuit of Fig. 13–9 uses two transistors connected to provide a differential output voltage $\Delta V_{be}$ that is the difference between $V_{be(Q1)}$ and $V_{be(Q2)}$. Combining the expressions for $V_{be}$ for both transistors yields the expression

$$\Delta V_{be} = \frac{kT}{Q} \ln\left[\frac{I1}{I2}\right] \tag{13-20}$$

Note that since ln 1 = 0, currents $I1$ and $I2$ *must not be equal.* In general, designers set a ratio of 2 : 1; that is, $I1 = 2$ mA and $I2 = 1$ mA. Since currents $I1$ and $I2$ are supplied from constant current sources, the ratio $I1/I2$ is a constant. Also, the logarithm of a constant is a constant. Therefore, all terms in Equation (13–20) are constants except temperature $T$. Equation (13–20), therefore, may be written in the form

$$\Delta V_{be} = KT \tag{13-21}$$

where  $K = (k/Q) \ln(I1/I2)$
$K = (1.38 \times 10^{-23})/(1.6 \times 10^{-19}) \ln(2/1)$
$K = 5.98 \times 10^{-5}$ V/K = 59.8 µV/K

We may now rewrite Equation (13–21) in the following form:

$$\Delta V_{be} = 59.8 \text{ µV/K} \tag{13-22}$$

■ **Example 13–8**

Calculate the output voltage from a circuit such as Fig. 13–9 if the temperature is 35°C. (*Hint:* K = °C + 273.)

*Solution*

$$\Delta V_{be} = KT \tag{13-21}$$

$$= \frac{59.8 \text{ µV}}{\text{K}} \times (35 + 273) \text{ K}$$

$$= (59.8)(308) \text{ µV} = 18{,}418 \text{ µV} = \mathbf{0.0184 \ V}$$

In most thermometers using the circuit of Fig. 13–9, an amplifier increases the output voltage to a level that is numerically the same as a unit of temperature, so that the temperature may be easily read from a digital voltmeter. The most common scale factor is 10 mV/K, so for our transducer the postamplifier requires a gain of

$$A_v = \frac{10 \text{ mV/K}}{59.8 \text{ µV} \times \dfrac{1 \text{ mV}}{10^3 \text{ µV}}} = 167$$

**13–14**     **INDUCTIVE TRANSDUCERS**

Inductance $L$ and inductive reactance $X_L$ are transducible properties because they can be *varied* by certain mechanical methods.

Figure 13–10(a) shows an example of an inductive Wheatstone bridge. Resistors $R1$ and $R2$ form two fixed arms of the bridge, while coils $L1$ and $L2$ form variable arms. Since inductors are used, the excitation voltage must be ac. In most cases the ac excitation source will have a frequency between 400 and 5000 Hz and an rms amplitude of 5 to 10 V.

The inductors are constructed coaxially, as shown in Fig. 13–10(b), with a common core. It is a fundamental property of any inductor that a ferrous core increases its inductance. In the rest condition (i.e., zero-stimulus) the core will be positioned equally inside both coils. If the stimulus moves the core in the direction shown in Fig. 13–10(b), the core tends to move out of $L1$ and further into $L2$. This action reduces the inductive reactance of $L1$ and increases that of $L2$, unbalancing the bridge.

**13–15**      ## LINEAR VARIABLE DIFFERENTIAL TRANSFORMERS (LVDTs)

Another form of inductive transformer is the **linear variable differential transformer (LVDT)** shown in Fig. 13–11. The construction of the LVDT is similar to that of the inductive bridge, except that it also contains a primary winding.

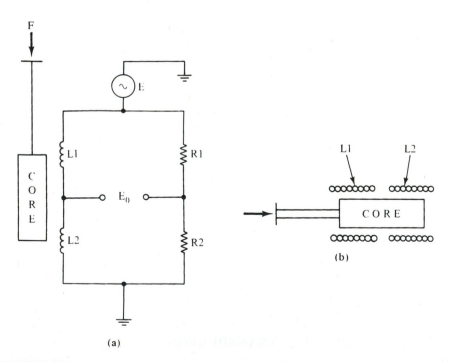

**FIGURE 13–10**
(a) Inductive Wheatstone bridge transducer, (b) mechanical form.

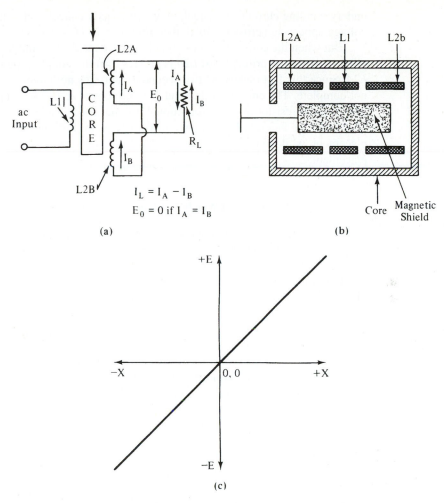

**FIGURE 13–11**
(a) LVDT, (b) LVDT construction, (c) output transfer function.

One advantage of the LVDT over the bridge-type transducer is that it provides higher output voltages for small changes in core position. Several commercial models are available that produce 50 to 300 mV/mm. In the latter case a 1-mm displacement of the core produces a voltage output of 300 mV.

In normal operation the core is equally inside both secondary coils, *L2a* and *L2b*, and an ac carrier is applied to the primary winding. This carrier typically has a frequency between 40 Hz and 20 kHz and an amplitude in the range 1 to 10 V rms.

Under rest conditions the coupling between the primary and each secondary is equal. The currents flowing in each secondary, then, are equal to each other. Note in Fig. 13–11(a) that the secondary windings are connected in series opposing, so if the secondary winding currents are equal, they will exactly cancel each other in the load. The ac voltage appearing across the load, therefore, is *zero* ($I_a = I_b$).

But when the core is moved so that it is more inside *L2b* and less inside *L2a*, the coupling between the primary and *L2b* is greater than the coupling between the primary and *L2a*. Since this fact makes the two secondary currents no longer equal, the cancellation is not complete. The current in the load $I_L$ is no longer zero. The output voltage appearing across load resistor $R_L$ is proportional to the core displacement, as shown in Fig. 13–11(c). The *magnitude* of the output voltage is proportional to the *amount* of core displacement, while the *phase* of the output voltage is determined by the *direction* of the displacement.

## 13–16     POSITION-DISPLACEMENT TRANSDUCERS

A position transducer will create an output signal that is proportional to the position of some object along a given axis. For very small position ranges we could use a strain gage (i.e., Fig. 13–12), but note that the range of such transducers is necessarily very small. Most strain gages either are nonlinear for large displacements or are damaged by large displacements.

The LVDT can be used as a position transducer. Recall that the output polarity indicates the direction of movement from a zero-reference position, and the amplitude indicates the magnitude of the displacement. Although the LVDT will accommodate larger displacements than the strain gage, it is still limited in range.

The most common form of position transducer is the potentiometer. For applications that are not too critical, ordinary linear taper potentiometers are often sufficient. Rotary models are used for curvilinear motion, and slide models for rectilinear motion.

We have already seen an example of the potentiometer as position transducers in our discussion (Chapter 10) on servomechanism strip chart recorders.

In precision applications designers use either regular precision potentiometers or special potentiometers designed specifically as position transducers.

Figure 13–13 shows two possible circuits using potentiometers as position transducers. In Fig. 13–13(a) we see a single-quadrant circuit for use where the zero point (i.e., the starting reference) is at one end of the scale. The pointer will always be at some point such that $0 \leq x \leq X_m$. The potentiometer is connected so that one end is grounded and the other is connected to a precision, regulated voltage source $V+$. The value of $V_x$ represents $X$ and will be $0 \leq V_x \leq V+$, such that $V_x = 0$ when $X = 0$, and $V_x = V+$ when $X = X_m$.

**FIGURE 13–12**
Beam transducer using a strain gage element.

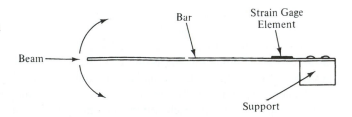

**FIGURE 13–13**
(a) Position transducer using a
potentiometer (one quadrant),
(b) position transducer using a
potentiometer (two quadrants).

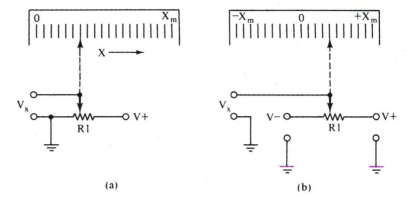

A two-quadrant system, shown in Fig. 13–13(b), is similar to the previous circuit except that instead of grounding one end of the potentiometer, it is connected to a precision, regulated, *negative-to-ground* power source, *V–*. Figure 13–14 shows the output functions of these two transducers. Figure 13–14(a) represents the circuit of Fig. 13–13(a), while Figure 13–14(b) represents the circuit of Fig. 13–13(b).

A four-quadrant transducer can be made by placing two circuits such as Fig. 13–13(b) at right angles to each other and arranging linkage so that the output signal varies appropriately.

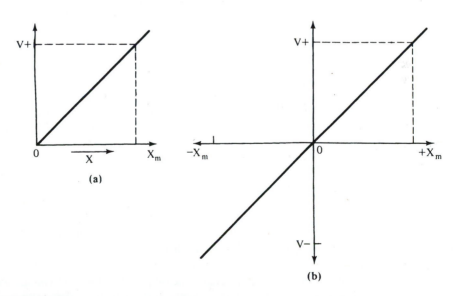

**FIGURE 13–14**
(a) *V+* vs. *X* for Fig. 13–13(a); (b) *V* vs. *X* for Fig. 13–13(b).

**FIGURE 13–15**
Example of derived signals using integration or differentiation.

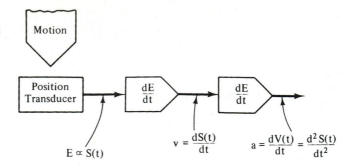

## 13–17    VELOCITY AND ACCELERATION TRANSDUCERS

**Velocity** can be defined as displacement per unit of time, and **acceleration** is the time rate of change of velocity. Since both velocity $v$ and acceleration $a$ can be related to position $s$, position transducers are often used to *derive* velocity and acceleration signals. The relationships are as follows:

$$v = \frac{ds}{dt} \qquad \textbf{(13–23)}$$

$$a = \frac{dv}{dt} \qquad \textbf{(13–24)}$$

$$a = \frac{d^2 s}{dt^2} \qquad \textbf{(13–25)}$$

Velocity and acceleration are the first and second time derivatives of displacement (i.e., change of position), respectively. We may derive electrical signals proportional to $v$ and $a$ by using an operational amplifier differentiator circuit (see Fig. 13–15). The output of the transducer is a time-dependent function of position (i.e., displacement). This signal is differentiated by the stages following to produce the velocity and acceleration signals.

## 13–18    TACHOMETERS

Alternating current and direct current generators are also used as velocity transducers. In their basic form they will transduce rotary motion—that is, produce an angular velocity signal—but with appropriate mechanical linkage they will also indicate rectilinear motion.

In the case of a dc generator, the output signal is a dc voltage with a magnitude that is proportional to the angular velocity of the armature shaft.

The ac generator, or **alternator**, maintains a relatively constant output voltage, but its ac *frequency* is proportional to the angular velocity of the armature shaft.

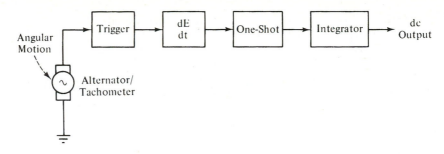

**FIGURE 13–16**
Alternator tachometer.

If a dc output is desired instead of an ac signal, then a circuit similar to Fig. 13–16 is used. The ac output of the tachometer is fed to a trigger circuit (either a comparator or a Schmitt trigger) so that squared-off pulses are created. These pulses are then differentiated to produce spike-like pulses to trigger the monostable multivibrator (one-shot). The output of the one-shot is integrated to produce a dc level proportional to the tachometer frequency.

The reason for using the one-shot stage is to produce output pulses that have a *constant amplitude* and *duration.* Only the pulse repetition rate (i.e., the number of pulses per unit of time) varies with the input frequency. This fact allows us to integrate the one-shot output to obtain our needed dc signal. If either duration or amplitude varied, then the integrator output would be meaningless. This technique, incidentally, is widespread in electronic instruments, so you should understand it well.

## 13–19    FORCE AND PRESSURE TRANSDUCERS

Force transducers can be made by using strain gages or either LVDT or potentiometer displacement transducers. The displacement transducer [Fig. 13–17(a)] becomes a force transducer by causing a power spring to either compress or stretch. Recall Hooke's law, which tells us that the force required to compress or stretch a spring is proportional to a *constant* and the *displacement* caused by the compression or tension force applied to the spring. So by using a displacement transducer and a calibrated spring, we are able to measure force.

Strain gages connected to flexible metal bars (refer to Fig. 13–12) are also used to measure force, because a certain amount of force is required to deflect the bar any given amount. Several transducers on the market use this technique, and they are advertised as "force-displacement" transducers. Such transducers form the basis of the digital bathroom scales now on the market.

Do not be surprised to see such transducers, especially the smaller types, calibrated in *grams.* We all know that the gram is a unit of *mass,* not force, so what this usage refers to is the gravitational force on one gram (1 g) at the earth's surface, roughly 980 dynes. A 1-g weight suspended from the end of the bar in Fig. 13–12 would represent a force of 980 dynes.

**FIGURE 13–17**
(a) Force from a displacement transducer, (b) force/pressure transducer.

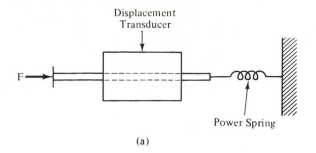

(a)

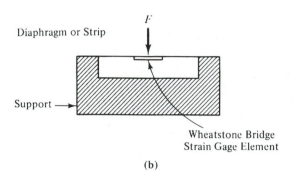

(b)

A side view of a cantilever force transducer is shown in Fig. 13–17(b). In this device a flexible strip is supported by mounts at either end, and a piezoresistive strain gage is mounted to the underside of the strip. Flexing the strip unbalances the gage's Wheatstone bridge, producing an output voltage.

A related device uses a cup- or barrel-shaped support and a circular diaphragm instead of the strip. Such a device will measure force or pressure, that is, *force per unit of area.*

## 13–20     FLUID PRESSURE TRANSDUCERS

Fluid pressures are measured in a variety of ways, but the most common involve a transducer such as those shown in Figs. 13–18 and 13–19.

In the example of Fig. 13–18(a), a strain gage or LVDT is mounted inside a housing that has a bellows or an aneroid assembly exposed to the fluid. More force is applied to the LVDT or gage assembly as the bellows compresses. The compression of the bellows is proportional to the fluid pressure.

An example of the **Bourdon tube** pressure transducer is shown in Fig. 13–18(b). Such a tube is hollow and curved, but flexible. When a pressure is applied through the inlet port, the tube tends to straighten out. If the end tip is connected to a position/displacement transducer, then the transducer output will be proportional to the applied pressure.

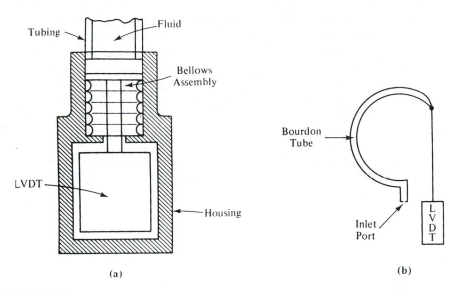

**FIGURE 13–18**
(a) Fluid pressure transducer, (b) Bourdon tube fluid pressure transducer.

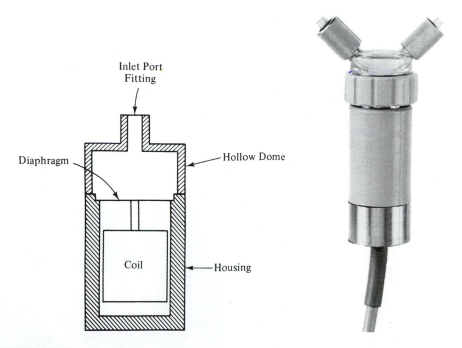

**FIGURE 13–19**
(a) Dome-type fluid pressure transducer, (b) commercial dome-type transducer [(b) courtesy of Hewlett-Packard].

Figure 13–19(a) shows another popular form of fluid transducer. In this version a diaphragm is mounted on a cylindrical support similar to that in Fig. 13–18. In some cases a bonded strain gage is attached to the underside of the diaphragm, or flexible supports to an unbonded type are used. In the example shown, the diaphragm is connected to the core drive bar of an inductive transducer or LVDT.

Figure 13–19(b) shows the Hewlett-Packard type-1280 transducer used in medical electronics to measure human blood pressure. In this device the hollow fluid-filled dome is fitted with **Luer-lock** fittings, standards in medical apparatus.

The fluid transducers shown so far will measure gage pressure (i.e., pressure above atmospheric pressure) because one side of the diaphragm is open to air. A **differential** pressure transducer will measure the difference between pressures applied to the two sides of the diaphragm. Such devices will have two ports marked "$P1$" and "$P2$" or something similar.

## 13–21    LIGHT TRANSDUCERS

There are several different phenomena for measuring light, and they create different types of transducers. For this chapter we will limit the discussion to **photoresistors, photovoltaic cells, photodiodes,** and **phototransistors.**

A photoresistor can be made because certain semiconductor elements show a marked decrease in electrical resistance when exposed to light. Most materials do not change linearly with increased light intensity, but certain combinations such as cadmium sulfide (CdS) and cadmium selenide (CdSe) are effective. These cells operate over a spectrum from "near-infrared" through most of the visible light range, and they can be made to operate at light levels of $10^{-3}$ to $10^{+3}$ footcandles (i.e., $10^{-3}$ to 70 mW/cm$^2$). Figure 13–20(a) shows the photoresistor circuit symbol, and Figure 13–20(b) shows an example of a photoresistor.

A photovoltaic cell, or "solar cell," as it is sometimes called, will produce an electrical current when connected to a load. Both silicon (Si) and selenium (Se) types are known. The Si type covers the visible and near-infrared spectrum, at intensities between $10^{-3}$ and $10^{+3}$ mW/cm$^2$. The selenium cell, on the other hand, operates at intensities of $10^{-1}$ to $10^2$ mW/cm$^2$, but it accepts a spectrum of near-infrared to the ultraviolet.

**FIGURE 13–20**
(a) Symbol for photoresistor cell,
(b) actual photoresistor cell.

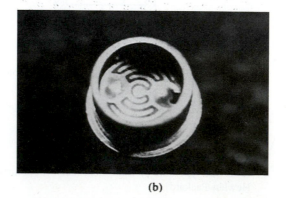

(a)                 (b)

Semiconductor pn junctions under sufficient illumination will respond to light. Interestingly enough, they tend to be photoconductive when heavily reverse biased, and photovoltaic when forward biased. These phenomena have led to a whole family of photodiodes and phototransistors.

**13–22**  ## CAPACITIVE TRANSDUCERS

A parallel plate capacitor can be made by positioning two conductive planes parallel to each other. The capacitance is given by

$$C = \frac{kKA}{d} \tag{13–26}$$

where  $C$ = the capacitance in *farads* (F) or a subunit ($\mu$F, pF, etc.)
$k$ = a units constant
$K$ = the dielectric constant of the material used in the space between the plates ($K$ for air is 1)
$A$ = the area of the plates "shading" each other
$d$ = the distance between the plates

Figure 13–21 shows several forms of capacitance transducers. In Fig. 13–21(a) we see a rotary plate capacitor that is not unlike the variable capacitors used to tune radio transmitters and receivers. The capacitance of this unit is proportional to the amount of area on the fixed plate that is covered (i.e., "shaded") by the moving plate. This type of transducer will give signals proportional to curvilinear displacement or angular velocity.

A rectilinear capacitance transducer, shown in Fig. 13–21(b), consists of a fixed cylinder and a moving cylinder. These pieces are configured so that the moving piece fits inside the fixed piece but is insulated from it.

Two types of capacitive transducers discussed so far vary capacitance by changing the shaded area of two conductive surfaces. Figure 13–21(c), on the other hand, shows a transducer that varies the *spacing* between surfaces, that is, the $d$ term in Equation (13–26). In this device the metal surfaces are a fixed plate and a thin diaphragm. The dielectric is either air or a vacuum. Such devices are often used as capacitance microphones.

Capacitance transducers can be used in several ways. One method is to use the varying capacitance to frequency-modulate an RF oscillator. This method is employed with capacitance microphones [those built as in Fig. 13–21(c), not the electrotet type]. Another method is to use the capacitance transducer in an ac bridge circuit.

Figure 13–22 shows an application in which a differential capacitance transducer is used to provide a position signal in a PMMC galvanometer chart recorder (Chapter 10). This type of transducer is superior to others in this application because it can be constructed with very low mass, so it will not significantly dampen the pen motion. The position signal is used in a feedback control circuit that gives the pen controlled damping. In one model the differentiated position signal warns the instrument that the pen will slam into the mechanical limit stop, and it puts the brakes on to prevent pen damage.

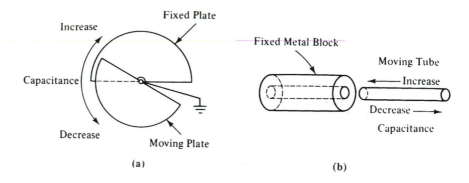

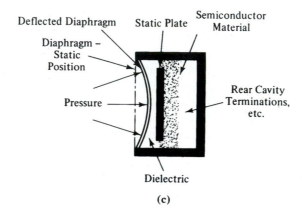

**FIGURE 13–21**
Capacitance transducers (from Harry Thomas, *Handbook of Biomedical Instrumentation and Measurement,* Figure 1–6, p. 13, Reston Publishing Co.).

## SUMMARY

1. A **transducer** is a device that converts a physical stimulus such as displacement, force, temperature, and so on, to an electrical voltage or current proportional to the magnitude of the stimulus.
2. A **strain gage** depends upon the **piezoresistance**—that is, resistance change due to mechanical deformation—phenomenon.
3. There are two basic forms of inductive transducers: $X_L$ bridge and LVDT.
4. Capacitive transducers depend upon varying the space between, or the shaded area of, two conductive surfaces.
5. Several types of temperature transducers are common: thermistor, thermocouple, and semiconductor junctions.

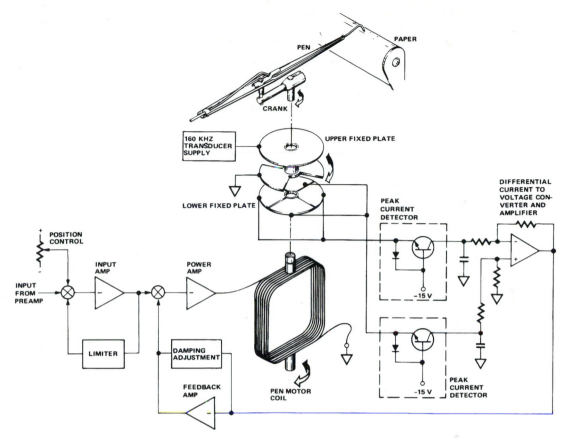

**FIGURE 13–22**
Capacitance transducers used as a position/velocity sensor (courtesy of Hewlett-Packard).


## RECAPITULATION

Now go back and try to answer the questions at the beginning of the chapter. When you are finished, answer the questions and work the problems given below. Place a mark beside each problem or question that you cannot answer, and then go back to the text and reread appropriate sections.


## QUESTIONS

1. Define **transducer** in your own words.
2. List several electrical parameters that can be used to transduce physical parameters.
3. Many transducers are often used in _____ bridge circuits.
4. Define **piezoresistivity** in your own words.

5. **Tension** forces in a cylindrical conductor _____ the electrical resistance.
6. **Compression** forces in a cylindrical conductor _____ the electrical resistance.
7. Changes in electrical resistance due to deformation are called _____.
8. Define **gage factor** in your own words.
9. Write the two equations for gage factor.
10. Discuss the differences between **bonded** and **unbonded** strain gages.
11. Write the equation for $E_0$ in a Wheatstone bridge transducer that uses *one* active element.
12. What are the *units* for the sensitivity of the fluid pressure transducer?
13. Draw a circuit showing a **balance control** in a Wheatstone bridge strain gage.
14. List three types of temperature transducers.
15. Define **Seebeck effect** and **Peltier effect** in your own words. How are these effects related?
16. The temperature coefficients for most elemental metals are _____.
17. Specially "doped" semiconductor materials can have both _____ and _____ temperature coefficients.
18. A thermistor used as an electronic thermometer must be operated to minimize _____.
19. A thermocouple is formed of two _____ metals and works because the metals have different _____ _____.
20. List three types of semiconductor temperature transducers.
21. List two types of inductive transducers. Draw appropriate circuit diagrams.
22. The secondary windings of an LVDT are connected in _____.
23. List two types of position/displacement transducers.
24. Draw circuits for one- and two-quadrant position transducers.
25. How can a displacement transducer derive velocity and acceleration signals?
26. What types of devices are used to measure angular velocity?
27. List three types of phototransducers.
28. List two types of photoconductive semiconductor elements.

## PROBLEMS

1. A constantan cylinder with a diameter of 0.05 mm is 14.6 cm long. Find its electrical resistance.
2. Predict the **gage factor** if a 12-mm-long strain gage element changes length 0.7 mm and its resistance changes from 78 to 82 $\Omega$. What type of force is applied, **tension** or **compression?**
3. What is the gage factor of a cylindrical wire 0.05 mm in diameter and 18 mm long if the diameter changes 0.03 mm and the length changes 2.9 mm when a tension force is applied?
4. A Wheatstone bridge strain gage is excited by a 6-V dc source. When it is stimulated by an outside parameter, each element changes resistance by a factor of 0.042$R$, and all four arms have equal resistance under zero-stimulus conditions. Calculate the output voltage $E_0$ for (a) one, (b) two, and (c) three active elements.

5. A 5-g mass hangs from the end of a bar-type strain gage (Fig. 13–12). Express this mass as a force in *dynes*.

6. A fluid pressure transducer has a sensitivity of 50 μV/V/cmHg and is excited by a +7.5-V dc source. Find the output voltage if a 150-mmHg pressure is applied.

7. The transducer in Problem 6 is a Wheatstone bridge strain gage, with four active elements that each have a zero-pressure resistance of 500 Ω. Calculate the value of a calibration resistance (to shunt one arm) that will give an artificial pressure of 120 mmHg.

8. A thermistor is known to have a resistance of 100 kΩ at 25°C and a temperature coefficient of $0.02°C^{-1}$. Calculate the resistance at (a) 30°C and (b) 56°C.

9. Find the resistance of a thermistor at 100°C if the ice-point resistance is 500 kΩ and the value of β is 2900 K.

10. Find the base-emitter voltage of a transistor at 33°C if the collector current is 10 mA and the reverse saturation current is $10^{-13}$ A.

11. Find the slope factor of a dual-transistor temperature transducer such as Fig. 13–9 if $I1 = 10$ mA and $I2 = 3$ mA.

12. What amplification following the transducer in Problem 11 results in a scale factor of 10 mV/K?

## BIBLIOGRAPHY

Cobbold, Richard S. C. *Transducers for Biomedical Measurements.* New York: Wiley-Interscience, 1974.

Cooper, W. D. *Electronic Instrumentation and Measurement Techniques.* Englewood Cliffs, NJ: Prentice-Hall, Inc., 1970.

Oliver, Frank J. *Practical Instrumentation Transducers.* New York: Hayden Book Co., 1971.

Prensky, Sol. *Electronic Instrumentation,* 2nd ed. Englewood Cliffs, NJ: Prentice-Hall, Inc., 1971.

*Reference Data for Radio Engineers.* Indianapolis, IN.: Howard W. Sams & Co., Inc., 1968.

Sheingold, D. H., ed. *Nonlinear Circuits Handbook.* Norwood, MA.: Analog Devices, Inc., 1974.

Strong, Peter. *Biophysical Measurements.* Beaverton, OR.: Tektronix, Inc., Measurement Concepts Series, 1973.

Thomas, Harry E. *Handbook of Biomedical Instrumentation and Measurements.* Reston, VA: Reston Publishing Co., 1974.

Tusinski, Joseph. "Strain Gages Come of Age." *Electronics World* (March 1969), pp. 35–37.

# 14

# Probes and Connectors

## 14–1    OBJECTIVES

1. To learn about the types of connectors and probes used with instrumentation.
2. To learn some of the problems associated with probes.
3. To learn which probe or connector is appropriate in certain measurements.
4. To learn the correct *use* of probes.

## 14–2    SELF-EVALUATION

Before studying the material in this chapter, try to answer the questions given below. These questions test your knowledge of the subject matter. If you cannot answer a particular question, then look for the answer as you read the text.

1. Draw the circuit for a low-capacitance probe.
2. What is the *purpose* of a low-capacitance probe?
3. Why does a "twisted pair" sometimes perform better than test leads?
4. Why is it improper to use the power line grounding instead of direct probe grounding?

## 14–3    WHAT ARE "PROBES"?

A **probe,** as used in the context of electronic instrumentation, is a device used to connect the input of a measurement instrument such as an oscilloscope or electronic voltmeter to a point in the circuit where the measurement is to be made.

A probe may consist of alligator clip leads, or it may be a complex circuit including amplifiers and other active components. Along with probes, we will consider certain other hardware that is frequently used to interconnect instruments. We will also discuss when and where certain types of interconnecting devices are suitable.

## 14–4    TEST LEADS

A **test lead** may be a length of hook-up wire with alligator clips at either end, or it may have a special probe-type end (Fig. 14–1). The probe usually makes it easier to access points in the circuit and is generally safer to use in live circuits than are alligator clips. Besides the electrical shock hazard, there is always the possibility of accidentally slipping off the test point and damaging the circuit. Solid-state components are notoriously unforgiving of such "goofs."

Test lead wire is a special type of hook-up wire that is well insulated (500–600 V or more) but is flexible enough to allow easy use in making measurements.

The simple test lead is found most frequently in ac and dc voltmeter/multimeter applications and in certain other applications where the ac frequency is low. At higher frequencies the distributed capacitance and noise pickup are too great for accurate measurement.

Figure 14–2(a) shows a mechanism by which interference can occur when open test leads are used. Note that one of the test lead wires is grounded, while the other carries a signal from a generator. A nearby conductor carrying an electric current generates a magnetic field, and flux from this field cuts the signal wire, inducing a spurious signal current. The "nearby" wire need not actually be in extremely close proximity to the signal wire, but it is often the building power wiring located a dozen or so feet away from the test site. Indeed, the high current loads usually encountered in building power distribution systems render them quite capable of generating interference over a distance.

Figures 14–2(b) and 14–2(c) show the effect of building-wiring-originated, 60-Hz interference on a 100-kHz signal circuit such as Fig. 14–2(a). The trace on the oscilloscope in Fig. 14–2(b) shows the 100-kHz, 100-mV signal *as it should be*. But in Fig. 14–2(c) we have the same trace modulated by the 60-Hz signal. Note well that this effect is very often far more severe. It is possible to obtain up to several volts of 60-Hz signal in high-impedance (>1MΩ) circuits, as demonstrated by touching the tip of an unterminated oscilloscope probe.

Besides frequency, two other factors must be considered when one is using test leads: **signal amplitude** and **circuit impedance.** If the amplitude is high enough, then the arti-

**FIGURE 14–1**
Test probes.

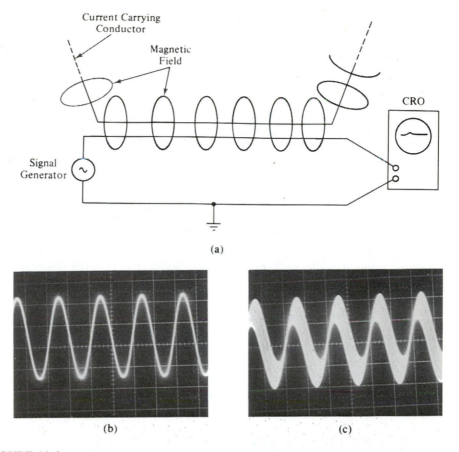

**FIGURE 14–2**

(a) Pickup of interference on unshielded test leads, (b) proper reproduction of 100-kHz signal, (c) reproduction with 60-Hz interference.

fact caused by an interfering signal may be negligibly small. But the source and load impedances in the circuit affect the amplitude of the interfering signal. We observe several volts displayed on the screen of an oscilloscope, because the impedance of the connection between the probe tip and dry skin is in the range of 10 kΩ to 1 MΩ and the oscilloscope input impedance is typically 1 MΩ. In the example of Fig. 14–2 the signal generator supplying the 100-kHz signal has an output impedance of 600 Ω, so the circuit impedance is much lower. A much smaller amplitude signal, therefore, is displayed on the oscilloscope screen.

Note that a *current* is induced by the magnetic field, so by Ohm's law, if the impedance is high, then so is the voltage developed (and vice versa). A popular rule of thumb regarding 60-Hz pickup claims 10 m V/in, but this figure is not dependable because of the dependence of the voltage on circuit impedances.

**FIGURE 14–3**
Twisted-pair unshielded test leads.

A possible solution that often proves effective is to use a **twisted pair** of test leads, as shown in Fig. 14–3. Such test leads can be purchased, or they can be made by inserting one end of each of the two wires into the chuck of an electric drill and then anchoring the other ends. Some practice, however, is required before this job can be performed safely.

The reasons for the effectiveness of twisted pairs are that (1) they are self-shielding and (2) the magnetic field affects both conductors equally. The self-shielding effect is particularly noted in unbalanced circuits where one side of the circuit is grounded, while the second reason is more prevalent where the two conductors are balanced with respect to ground.

The criteria for using simple test leads, then, are low-frequency or dc signals, high signal amplitude, and low source and load impedances.

## 14–5        SHIELDED CABLES

One other possible solution is to use **shielded wire,** as shown in Fig. 14–4. A shielded wire is a conductor that is surrounded by and insulated from another conductor; in most cases the outer conductor is a braided cylinder. Coaxial cable, often used as a transmission line in RF circuits (Chapter 21), meets this description and is often used as shielded wire. Note, however, that shielded cables are not necessarily transmission lines unless certain other criteria are met, and that includes coaxial cables.

The coupling between close-proximity conductors is both inductive and capacitive. Capacitor $C1$ in Fig. 14–4 is the capacitance between the shield and a nearby current-carrying conductor, and it represents a path for interfering currents. Capacitor $C2$ in Fig. 14–4 represents the capacitance between the inner conductor and the shield.

There are two main reasons why shielding can reduce coupling. In some systems it is due to the fact that $C1$ and $C2$ are *in series,* so the total capacitance is less than either capacitance alone. In unbalanced systems such as Fig. 14–4, one end of each capacitor is grounded, that is, at zero potential [see Fig. 14–4(b)]. The interfering potential *usually* does not get into the signal circuit. For exceptions see Chapter 15.

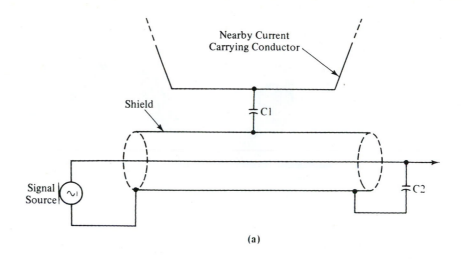

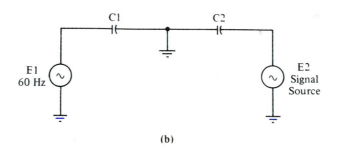

**FIGURE 14–4**
(a) Shielded test lead, (b) equivalent circuit.

## 14–6        CONNECTORS

Quite a few different connectors are used to interconnect electronic instrumentation and circuits, and (it seems to beleaguered engineers and technicians) the probability that two instruments will have different connectors is directly proportional to the urgency of using them together. All electronic facilities seem to keep a rather large number of adapters that are used to allow interfacing instruments that have different input or output connectors.

Figure 14–5 shows several different connectors. Figure 14–5(a) shows the banana plug and jack. These connectors are used on power supplies, audio generators, and other low-frequency instruments including multimeters and low-frequency oscilloscopes.

The PL-259 "UHF" connector is shown in Fig. 14–5(b). This connector mates with the SO-239 connector that is used as the antenna connector in CB and most other two-way radio transmitters. These connectors were once used extensively on signal generators

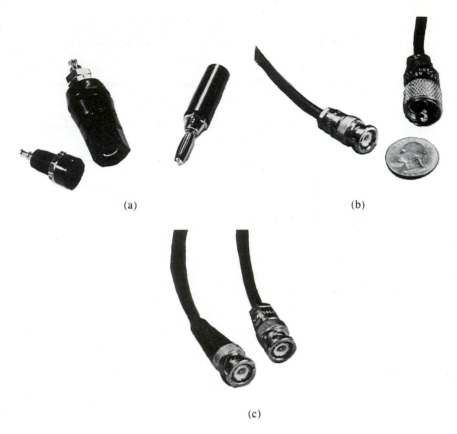

(a)

(b)

(c)

**FIGURE 14–5**
(a) Banana plug and two types of banana jack, (b) BNC (left) and UHF (right) connectors, (c) two types of BNC connectors.

and oscilloscopes built a decade or more in the past. Modern equipment, however, tends to use the BNC connector of Fig. 14–5(c). In current practice the use of the UHF connector is limited to RF power applications below 450 mHz. Most owners of older instruments that were fitted with the SO-239 mate to the PL-259 have long since obtained a BNC-to-UHF adapter so that modern cables and probes will fit. A collection of adapters is shown in Fig. 14–6, although there are literally dozens on the market.

14–7     **PROBLEMS WITH SHIELDED CABLES**

Besides the ground-loop problems to be discussed in Chapter 15, we must also be on guard against the capacitance of shielded cables. The very property of the shielded cable that confers protection against one type of measurement disaster will provide equal

**FIGURE 14–6**
Connector adapters.

opportunity for another. In some cases the capacitance of the cable will severely attenuate high-frequency signals.

Although the figures vary from one type of cable to another, most small-diameter coaxial cables have a capacitance of approximately 30 pF/ft. A 3-ft length of cable, then, has a capacitance of approximately 90 pF. Moreover, the input impedance of instruments such as oscilloscopes is usually specified in the parallel-equivalent form, that is, a resistance shunted by a capacitance. Typical values are 1 MΩ shunted by 30 pF. The source, then, sees a total impedance equivalent to 1 MΩ shunted by 120 pF.

The source is not perfect either; it typically has a certain output impedance and so may be represented as in Fig. 14–7(a) by a resistor $R_s$ in series with an ideal voltage source $E$ (this $E$ is very nearly the open-circuited output potential of the instrument). Figure 14–7(a) is an equivalent circuit that takes into account the various imperfections. Capacitors $C1$ and $C2$ represent the cable and instrument input capacitances, respectively; together they total 120 pF. Resistor $R1$ represents the oscilloscope input resistance. This circuit is essentially a voltage divider, and $E_0$ is the effective potential applied to the instrument input.

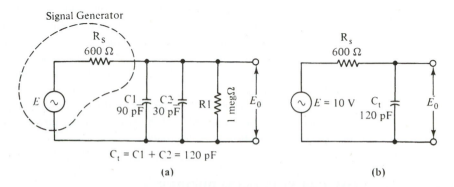

**FIGURE 14–7**
(a) Equivalent instrument input circuit with shielded leads, (b) composite circuit.

To make the circuit arithmetic simpler, set $E$ at 10 V. At dc, the error that is due to voltage division is negligible:

$$E_0 = \frac{ER1}{R1 + R} \tag{14–1}$$

$$= \frac{(10 \text{ V})(1 \times 10^6 \ \Omega)}{[(1 \times 10^6) + (6 \times 10^2)] \ \Omega}$$

$$= \frac{1 \times 10^6 \text{ V}}{1.0006 \times 10^6} = \mathbf{9.99 \text{ V}}$$

The error is approximately 0.1%. But consider what happens at higher frequencies, where the reactance of the capacitors can no longer be simply ignored. The reactance of a 120-pF capacitor at 100 kHz is approximately 13 k$\Omega$, and at 1 mHz it drops to 1.3 k$\Omega$. In both cases the reactance of the capacitor will predominate, and the voltage output is reduced. We may ignore the 1-M$\Omega$ input resistance, so our equivalent circuit is the RC voltage divider of Fig. 14–7(b). At 1 mHz the reactance of the capacitance is 1.3 k$\Omega$, so $E_0$ will be

$$E_0 = \frac{(10 \text{ V})(1300 \ \Omega)}{(1300 + 600) \ \Omega} \tag{14–2}$$

$$= \frac{(10 \text{ V})(1300)}{(1900)} = \mathbf{6.84 \text{ V}}$$

In addition to the voltage drop, there is a *phase shift* of the signal:

$$\theta = \arctan(R_s/X_c) \tag{14–3}$$
$$= \arctan(600/1300)$$
$$= \arctan(0.46) \approx \mathbf{24.7°}$$

These errors may not be terribly important on *some* sine wave measurements because a correction factor based on frequency can be applied, and phase shift may not be deemed important. But in making measurements to almost any nonsinusoidal waveform, however, these problems become acute. In fact, the faster the rise time of the waveform, the more acute they become. The solution to these problems is the use of a low-capacitance probe, described in Sec. 14.8.

Ordinary shielded cables should be used only for the same type of measurements as the test (Sec. 14.5). The shielded lead, however, usually confers the advantage over test leads of superior rejection of interference.

## 14–8    LOW-CAPACITANCE PROBES

The basic low-capacitance probe (Fig. 14–8) contains a parallel RC network consisting of a high-value resistor and a small-value capacitor. This type of probe is called a **passive**

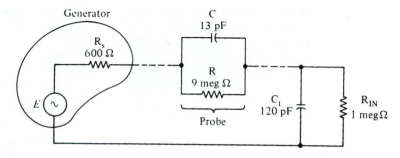

**FIGURE 14–8**
Equivalent instrument input circuit when a low-capacitance probe is used.

probe because it contains no amplifying devices. The probe in Fig. 14–8 presents 1/10 the capacitive load on the signal source as did the shielded cable, and it provides a constant 10:1 voltage division ratio over a wide range of frequencies. Figure 14–9 shows a commercial low-capacitance oscilloscope probe.

The usual designation for a 10:1 **division** probe is "10×," a practice that may tend to confuse. When a 10× probe is used, the displayed voltage is 1/10 the actual voltage, so multiply the indicated voltage by a factor of 10.

The RC combinations in the circuit can still cause a phase shift, so some manufacturers place a variable trimmer capacitor in the circuit. In low-cost instruments the capacitor inside the probe itself is made variable, whereas in most professional-grade instruments the variable capacitor is located inside the molded BNC connector that attaches to the instrument. This capacitor will eliminate most phase shift problems. Figure 14–10(a) shows the normal waveform, while Figure 14–10(b) shows the result of a misadjustment—loss of high frequencies. The opposite problem, enhanced high frequencies (i.e., peaking of the waveform), also occurs. Figure 14–10(c) shows the access hole for the adjustment capacitor in a Tektronix probe.

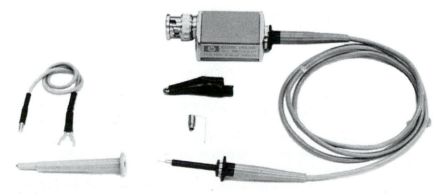

**FIGURE 14–9**
Typical low-capacitance oscilloscope probe (courtesy of Hewlett-Packard).

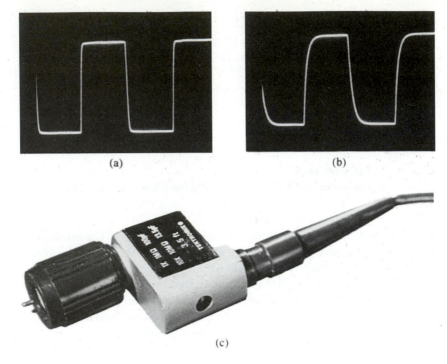

(a)     (b)

(c)

**FIGURE 14–10**
(a) Normal trace, (b) "C" misadjusted, (c) connector end of low-capacitance probe showing access hole to compensating capacitor.

**14–9     PROPER PROBE USE**

The low-capacitance probe is easy to use and for the most part produces good results even in inexperienced hands, but it has limitations and pitfalls. Be sure, for example, to heed the manufacturer's frequency response specification and to set the frequency compensation properly. The fact that a probe was compensated on another instrument is irrelevant.

Some probes have a switch to allow selection of 1×–10× positions. In one position the probe is direct, while in the other it is a divider RC probe. Be sure of which position that switch is set for, or there will be an order of magnitude error in the measurement. This seems like a trivial admonition, but it happens so often that it bears emphasis. In fact, one oscilloscope manufacturer (Tektronix) uses a special BNC connector on its 'scopes to sense which position the switch is set to and to indicate the proper V/div factor either on the CRT screen or on the vertical attenuator switch.

Another pitfall involves the ground wire on the probe: *use it.* Since most modern instruments use a three-wire power cord from the ac mains, their cabinets are at, or *near,* ground potential. Some people tend to remove ground wires from their probes and then depend upon the ac power mains ground between the instruments to make the connection.

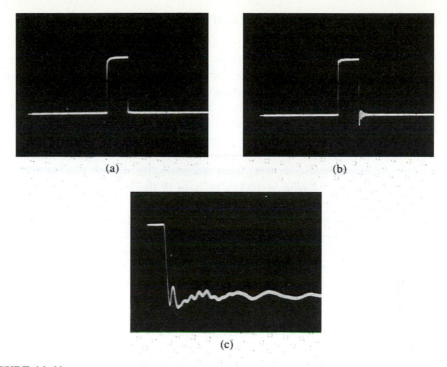

**FIGURE 14–11**
(a) Normal, (b) ringing due to lost ground, (c) ringing portion of (b) expanded.

Although the ground wire on the probe is sometimes a nuisance, there are at least two faults with this system: **ground loops** and **ringing.**

A ground loop occurs because of voltage drops in the power line ground wiring. The alternating currents passing through the ground system cause these voltage drops because of the ohmic resistance of the wires and resistance in connectors. The electrical potential at each chassis, then, is a little different from the potential on nearby chassis. The oscilloscope or voltmeter will see this difference of potential as a valid signal, so an artifact in the measurement is created.

Ringing is shown in Fig. 14–11. The normal pulse is shown in Fig. 14–11(a); the waveform photograph was taken with the probe's ground wire connected properly to the circuit under test. Figure 14–11(b) shows the result of *disconnecting* the probe ground and depending upon the power mains ground to make the return connection. Ringing oscillations follow the leading and trailing edges of the pulse, although it is most pronounced following the trailing edge. This phenomenon is due to the stray capacitances and inductances in the ground circuit, which form a distributed tank circuit. Like any tank circuit energized by a pulse, it will ring, that is, oscillate with an exponentially decaying amplitude. The features of this ringing behavior are shown more clearly in Fig. 14–11(c). This type of artifact is eliminated by the probe ground connection.

## 14–10    HIGH-VOLTAGE PROBES

Very few oscilloscopes or voltmeters have a maximum dc voltage range in excess of 1500 V. But many rather ordinary circuits operate at potentials far in excess of this value. A color television receiver, for example, may use up to approximately 25,000 V on the postdeflection accelerator anode.

A high-voltage probe is a special voltage divider (see Fig. 14–12) that allows these high potentials to be read on an ordinary voltmeter or oscilloscope. Most of these probes use the input resistance of the instrument and a 900-MΩ resistor inside the probe as the divider, although a few include the entire divider inside the probe. For electronic volt-meters with a 10-MΩ input resistance, the divider ratio is 100:1, but for most oscillo-scopes the ratio is 1000:1 because of the lower (i.e., 1-MΩ) input resistance of those instruments.

Note in Fig. 14–12 the construction of the probe; it is built of thick, insulating plas-tic to prevent electrical shock to the user. The finlike structures on the probe effectively lengthen the *surface path* between the tip of the probe and the user's hand. High tension has the nasty habit of creeping along the surfaces of materials that it cannot go through. High-voltage probes are generally considered safe to use but must be regarded with cau-tion under some circumstances. For one thing, always use the ground wire attached to the probe, regardless of whether or not the instrument that it is connected to also provides a ground wire. Second, never use a probe that is dirty or wet or that has grease or other sur-face contaminants. These could form a surface path for high-voltage arcs to follow. Third, never use a probe that is broken or in poor condition. Additionally, high-voltage probes such as that shown in the figure are not recommended for high-altitude use. At high alti-tudes the barometric pressure is lower, so the ionization potential of air is also lower. Equipment used at high altitudes, for example, mountaintop radar installations or airborne

**FIGURE 14–12**
High-voltage divider probe.

equipment, is specially designed for the application, and this type of probe defeats the special precautions taken by the manufacturer of the equipment.

## 14–11     RF DEMODULATOR PROBES

Radio frequency voltages cannot be directly measured on most readily available instruments (i.e., multimeters and oscilloscopes). In both cases a possible solution is the RF demodulator probe shown in Fig. 14–13(a). Diode $D1$ serves to rectify the signal, while the RC network following the diode serves as a filter circuit (as in a power supply). The voltage across capacitor $C2$ will be the peak RF voltage, and the resistor network reduces

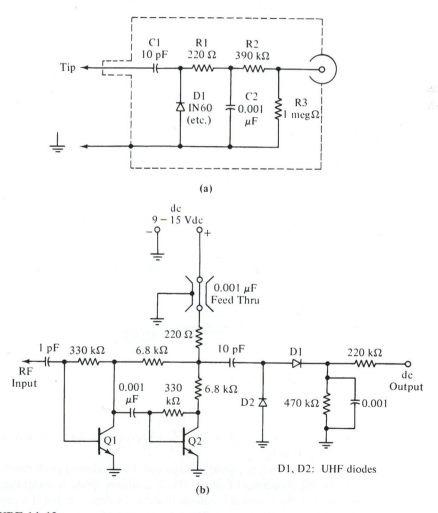

(a)

(b)

**FIGURE 14–13**
(a) RF demodulator probe, (b) amplified RF demodulator probe.

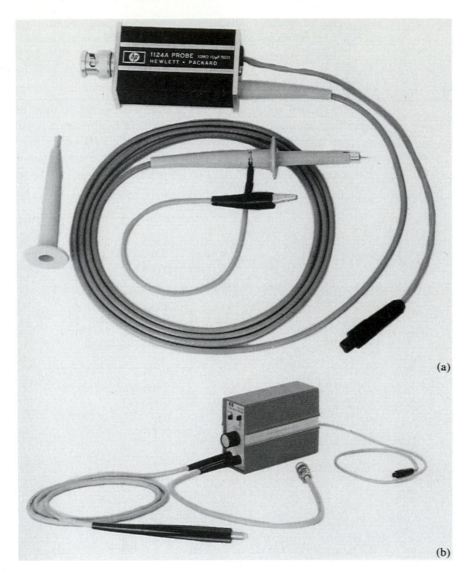

(a)

(b)

**FIGURE 14–14**
Active RF probes.

the peak voltage by 0.707 so that $E_0$ will represent the rms value, *provided that* the waveform is sinusoidal.

The minimum readable voltage and the maximum peak reverse voltage (prv) is limited by the properties of diode $D1$. If a silicon diode is used, then the 600-mV junction potential is the lowest RF voltage that can be detected, but if a germanium diode is used, then RF potentials down to 200 mV are readable.

The prv rating of most VHF and UHF signal diodes that are suitable for use in an RF probe is between 40 and 100 V. This rating can be increased by connecting several identical diodes in parallel, but only at the expense of an increased minimum readable voltage rating.

An **active** RF probe is shown in Fig. 14–13(b); it will amplify weak RF signals before rectification. Transistors $Q1$ and $Q2$ should be selected to have moderate-to-high beta ratings, with an $F_t$ of 700 mHz or higher. If proper attention is paid to "good VHF construction practices," then the probe will be essentially flat out to approximately 100 mHz and will operate at reduced gain to approximately 150 mHz. The circuit of Fig. 14–13(b) should be built inside a shielded probe housing. Examples of commercial active probes are shown in Figs. 14–14(a) and 14–14(b).

**14–12**

## SPECIAL PROBES FOR ICs

Digital and linear integrated circuits (ICs) are not too forgiving of accidental shorts. If a probe should slip off an IC pin, a very frequent occurrence, then a destructive short may be created. In addition to this problem, the extremely close spacing of IC DIP pins makes it difficult to attach normal probes. Most often recommended by equipment makers as a solution to this troublesome problem is the clip-on probe shown in Fig. 14–15. This probe is a clothespin-like assembly that has electrode surfaces on the inner sides of the jaws. The operator "squeezes" the grips to open the jaws and places the electrode surfaces over the IC pins. Multiple clothespin electrodes are often used to troubleshoot or debug digital circuits.

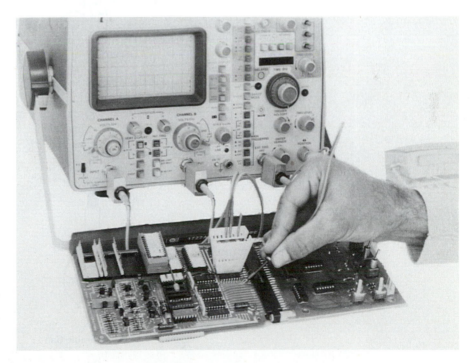

**FIGURE 14–15**
Clip-on probe for integrated circuits.

**14–13** ## CURRENT PROBES

The measurement of a *current* by a current meter requires that the circuit be broken so that the meter can be connected in series with the load. But this is not always either possible or desirable. An active probe, such as one of the current-to-voltage converter circuits of Chapter 12, will allow the measurement of dc and low-frequency ac on a voltage-measuring instrument such as an oscilloscope or electronic voltmeter. But the requirement for breaking the circuit still exists.

A passive current probe uses the magnetic field surrounding a current-carrying conductor to measure the strength of the current in the conductor. Figure 14–16(a) shows the construction of such a probe, using a toroid (i.e., doughnut-shaped) core that is wrapped with several turns of wire. The conductor carrying an alternating current forms the primary of the transformer, while the turns of wire on the core form the secondary. Alternating current in the conductor will induce current in the wire of the probe, so an output voltage $E_0$ will be generated and will be proportional to the current flowing.

**FIGURE 14–16**
(a) Current probe circuit, (b) construction.

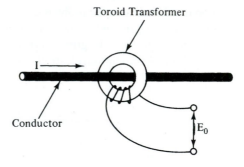

(a)

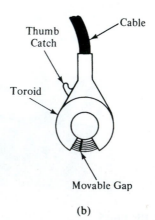

(b)

Toroid cores present a problem as far as wrapping around a conductor is concerned. Figure 14–16(b) shows the constructional details of most typical probes. There is a gap in the toroid. A latch on the side of the core allows the user to spread the gap to admit the conductor. Most of these probes operate over a frequency range of 100 Hz to 3 mHz but require an output amplifier that produces an output scaled in units of mV/mA.

## SUMMARY

1. Test leads are used at dc and low ac frequencies, where signal amplitudes are relatively high and circuit impedances are low.
2. Low-capacitance probes are used for general-purpose measurements with oscilloscopes up to approximately 100 mHz.
3. RF demodulator or detector probes are used to make amplitude measurements of RF signals on dc voltmeters.
4. Active current probes use an *I*-to-*E* converter circuit but require series connection in the circuit.
5. Passive current probes can be constructed to simply wrap around the conductor.

## RECAPITULATION

Now go back and try to answer the questions at the beginning of the chapter. When you are finished, answer the questions and work the problems given below. Place a mark beside each problem or question that you cannot answer, and then go back to the text and reread appropriate sections.

## QUESTIONS

1. What is a **probe?**
2. Under what conditions is it allowable to use test leads?
3. Two mechanisms by which test leads pick up interference are _____ and _____.
4. A _____ -pair is a type of test lead set that reduces some of the problems associated with normal test leads.
5. List two reasons why the leads in Question 4 are effective.
6. What are the advantages and disadvantages of shielded cable test leads?
7. List two types of connectors commonly used with shielded cable.
8. Which of the two basic connectors is currently favored?
9. What is the purpose of the low-capacitance probe?
10. Most low-capacitance probes have a _____ division ratio.
11. Measurements may be made with low-capacitance probes using the power mains ground for the return path. True or false?
12. Describe what is meant by a "ground loop" and how it might affect a measurement.
13. A high-voltage probe is actually a well-insulated _____.

**14.** Draw a circuit for a demodulator probe that produces an output voltage $E_0$ equal to the *peak* RF voltage.

**15.** Describe the differences between a **transmission line** and a **shielded cable.** Assume that both are made from RG-58/U coaxial cable.

**16.** How can VSWR affect the measurement of receiver sensitivity at 150 mHz?

**17.** What effects can SWR have on frequency counters if the input waveform is a fast-rise-time pulse?

**18.** List two types of active current probes.

**19.** Most passive ac current probes depend upon the _____ surrounding the current-carrying conductor.

**20.** If the load impedance of a transmission line does not approximate $Z_0$, then a _____ _____ device should be used.

## PROBLEMS

**1.** A signal generator has a 1000-$\Omega$ output impedance and is connected through 10 ft of coaxial cable to an oscilloscope that has a 1-M$\Omega$/20-pF input impedance. (a) What is the value of the voltage delivered to the oscilloscope input terminal if the signal source's open terminal voltage is 5.8 V ac? (b) What is the *phase shift* at 10 mHz?

**2.** The input probe of most VTVMs contains a 1-M$\Omega$ resistor, while the basic input impedance of the instrument is 10 M$\Omega$. Calculate the *percentage of error* created if a technician calibrates the instrument *without* the probe and fails to realize that the meter markings are valid *with* the probe.

**3.** Calculate the VSWR of a 27-mHz transmission line if the voltage maximum along the line is 2.8 V and the minimum is 0.7 V.

**4.** A signal generator with a 50-$\Omega$ output impedance is connected to a receiver with a 300-$\Omega$ input impedance. Calculate the VSWR.

**5.** A long 50-$\Omega$ transmission line is connected to a frequency counter with a 50-$\Omega$ input impedance. Calculate the SWR.

**6.** A 50-$\Omega$ source is to be connected to a 100-$\Omega$ load through a transmission line. Calculate the surge impedance of coaxial cable that can be used to make a quarter-wavelength impedance transformer. Make a recommendation as to an appropriate type of cable.

**7.** How long should the coaxial cable in Problem 6 be if the frequency is 65 mHz and the coaxial has a foam dielectric?

# 15

# Handling Signals, Sensors, and Instruments

1. To understand the problems of ground loop interference.
2. To understand proper grounding and shielding techniques.
3. To be able to deal effectively with electromagnetic interference (EMI).

15–2      **SELF-EVALUATION**

Before studying the material in this chapter, try to answer the questions given below. These questions test your knowledge of the subject. If you cannot answer a particular question, then look for the answer as you read the text.

1. What are three different types of noise commonly seen in recording low-level analog signals?
2. Calculate the noise voltage generated by a 50-$\Omega$ resistor at room temperature over a bandwidth of 1 MHz.
3. Describe the function of a guard shield, including how a guard shield connection is made.
4. How do ground loop voltages interfere with low-level signal recording?

15–3      **PROBLEMS IN SIGNAL ACQUISITION**

Recording strong signals (i.e., those over about 50 mV) is usually a straightforward job, and few problems will be encountered. But as the full-scale value of the signal amplitude drops, then certain acute problems begin to emerge. Signals in the millivolt and high microvolt range will exhibit some of these problems, while signals in the low microvolt and nanovolt range usually exhibit the problems.

      Noise signals mixed with the desired signal will be recorded and displayed as a valid signal unless steps are taken to eliminate the noise or at least reduce its value to a point of negligible effect. **Noise** is any electrical signal or tracing anomaly that is not part of the desired signal. Several different types of noise are recognized: **white noise, impulse noise,** and **interference noise.**

White noise supposedly contains all frequencies, phases, and amplitudes, so it gets its name from the analogy to white light, which contains all visible colors. Such noise is also called **gaussian noise,** although it is neither truly "white" nor "gaussian" unless there are no bandwidth limits present. True white noise has a bandwidth of dc to daylight and beyond. In most instrumentation systems, however, there are bandwidth limitations to consider, so the noise is actually **pseudogaussian** (also called "pink noise," i.e., bandwidth-limited white noise). The bandwidth limitations are often put in place to limit the effects of noise on the system. Because it integrates to zero given sufficient time, true gaussian noise can be eliminated by low-pass filtering or bandpass filtering. Bandwidth-limited noise, however, does not usually integrate to zero but rather to a very low value. The effect of low-pass filtering, then, is not total on bandwidth-limited noise.

An example of pseudogaussian bandwidth-limited noise is the hiss heard between stations on an FM broadcast band receiver (such noise is present between stations on the AM band as well but is obscured by other forms of noise that are also present). Most such noise in instrumentation systems is due to *thermal* sources, and it has an rms value of

$$E_n = \sqrt{4KTBR} \qquad \qquad \textbf{(15–1)}$$

where   $E_n$ = the rms value of the noise potential in volts (V)
  $K$ = Boltzmann's constant ($1.38 \times 10^{-23}$ J/K)
  $T$ = the temperature in kelvin (K) (use 300 K as standard room temperature)
  $B$ = the bandwidth in hertz (Hz)
  $R$ = the resistance in ohms ($\Omega$)

Note in Equation (15–1) that a simple resistor, with no power applied, will generate a noise signal due to thermal agitation of the atoms within the resistor.

■   **Example 15–1**

Find the amplitude of the noise signal in a circuit at room temperature (300 K) if the bandwidth is 10 kHz and the resistance is 100 k$\Omega$.

*Solution*

$$
\begin{aligned}
E_n &= \sqrt{4KTBR} \\
&= (4)\ (1.38 \times 10^{-23}\ (\text{J/K})\ (300\ \text{K})\ (10^4\ \text{Hz})\ (10^5\ \Omega) \\
&= \sqrt{1.66 \times 10^{-11}}\ \text{V} = 4.1 \times 10^{-6}\ \text{V} = \textbf{4.1 } \mu\textbf{V}
\end{aligned}
$$

In Example 15–1 a signal of 4.1 μV is created by nothing more than thermal agitation of molecules in a circuit resistance. Although this signal may appear to have a low amplitude, keep in mind that many signals recorded on oscilloscopes and graphics recorders are also in the microvolt range. For example, the human electroencephalograph

(EEG) is a recording of brain wave potentials and must be able to deal with signals with peak amplitudes in the 1-to-80-$\mu$V range. In that type of application a 4.1-$\mu$V noise signal represents a sizable artifact.

The answer to this problem is to keep the circuit impedances in the early stages—that is, those stages that most of the gain stages follow—very low so that the $R$ term in Equation (15–1) is low. Additionally, the bandwidth of recording amplifiers should be adjusted to that required to faithfully reproduce the input signal waveform. This tactic will reduce the bandwidth term ($B$) in Equation (15–1).

Several other types of noise are peculiar to solid-state amplifiers: **shot noise, Johnson noise,** and **flicker noise.** In low-cost amplifiers these noise sources can add up to significant amplitude. Although low-pass filtering offers relief, it is better to specify a low-noise preamplifier when one is dealing with low-level signals.

Impulse noise is due to local electrical disturbances such as arcs, lightning bolts, electrical motors, and so forth. Shielding of signal lines will help somewhat, as will low-pass filtering, but the best solution is to eliminate the noise at its source. Filtering, incidentally, is a two-edged sword, and it must be done prudently. Filtering tends to broaden pulse signals, and that can create even more problems than it cures.

Other electrical devices nearby can induce signals into the instrumentation system, the chief among these sources being the 60-Hz ac power lines. It is wise to use only **differential** amplifier inputs, because of their high common mode rejection ratio (CMRR—see Chapters 11 and 12). Signals from the desired source can be connected across the two inputs, and so become a differential signal, while the 60-Hz interference field affects both inputs equally and thus is common (and is thereby suppressed by the amplifier CMRR).

## 15–3–1　Creating Valid Signals from Common Mode Interference

It is sometimes possible to erroneously manufacture a valid differential or single-ended signal from a common mode interference signal. There are two principal ways this is done, and both involve the improper use of shielded wire inputs.

One source of this problem, called a **ground loop,** is shown in Fig. 15–1(a). This problem arises from the use of too many grounds. In this example the shielded source, the shielded input lines, and the dc power supply are all grounded to *different* points on the ground plane. Power supply direct currents $I$ flow from the dc power supply at point A to the amplifier common point E. Since the ground has resistance (as all conductors do), the voltage drops $E1$ through $E4$ are formed when the direct current flows. These voltages are seen by the amplifier as valid signals and can become especially troublesome if $I$ is a varying current.

The solution to this problem is to use **single-point grounding,** also called **star grounding,** as shown in Fig. 15–1(b). Some amplifiers used in sensitive signal acquisition systems keep the power and signal grounds separate except at a single, specified common point. In fact, a few models go further by creating several ground buses, especially where dc power plus analog, digital, video, and RF signals must be mixed together on a printed wiring board.

In some instances the shield on the input lines must *not* be grounded at both ends. In those cases it is usually better to ground the shield only at the amplifier end of the cable.

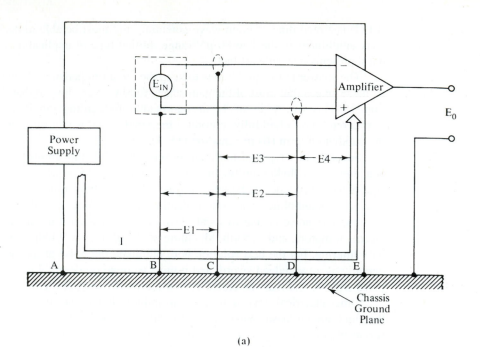

(a)

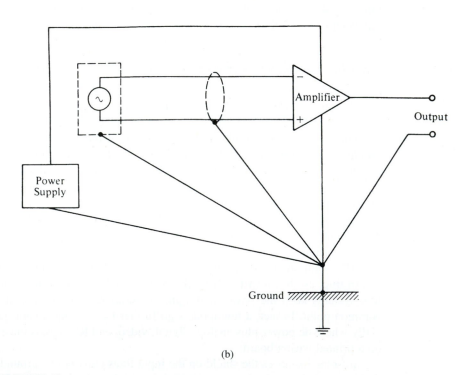

(b)

**FIGURE 15–1**

(a) Improper grounding. (b) Proper, single-point grounding.

Figure 15–2 shows some of the causes of and cures for differential artifact signals manufactured from common mode signal sources. The circuit in Fig. 15–2(a) uses standard single shielding, and the equivalent circuit shown in Fig. 15–2(b) reveals a potential problem. The shield has a capacitance to ground with the input wires ($C1$ and $C2$), and a cable resistance and a source resistance are always present, represented in Fig. 15–2(b) by $R1$ and $R2$. The system works well if $R1 = R2$ and $C1 = C2$, but even small imbalances in the RC networks can allow $E_{CM}$ to manufacture a differential signal. In that case, $E_{C1} \neq E_{C2}$, so the amplifier sees the difference $E_{C1} - E_{C2}$ which it accepts as a valid signal (which may be a dc offset or an ac signal such as 60-Hz power line interference).

A **guard shield** circuit [Fig. 15–2(c)] can be used to overcome this problem. The guard shield is driven by signals from the two input lines through high-value resistors $R_A$

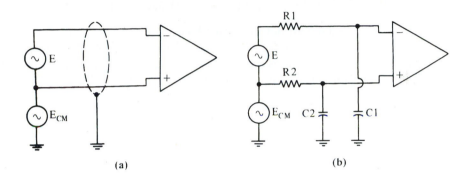

(a)    (b)

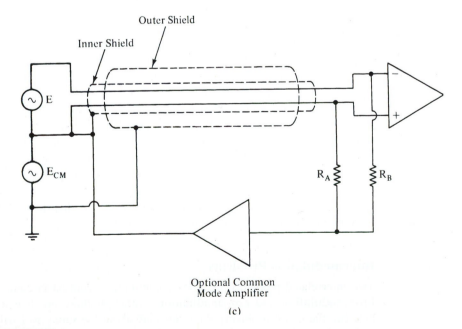

(c)

**FIGURE 15–2**

(a) Differential and common mode signals, (b) equivalent circuit, (c) guard shield connections.

and $R_B$ and, in many cases, through a common mode amplifier. This tactic has the effect of placing both sides of the cable capacitances at the same potential so that $E_{C1} = E_{C2} = 0$. The outer shield is not strictly necessary, but it can serve to reduce coupling from the outside common mode radiation source to the input signal lines, thereby reducing the value of $E_{CM}$.

It is sometimes difficult to predict all possible sources of input interference, so it often pays to try several different input shielding configurations to find the one that works best in a specific application. For other cases, however, the tactics discussed in the next sections will help eliminate interference from outside electromagnetic sources.

## 15–4          ELECTROMAGNETIC INTERFERENCE (EMI) SUPPRESSION

**Electromagnetic interference (EMI)** is the general term covering a wide variety of problems of interoperability of electronic equipment. In one class of EMI problems, a device that produces a large amount of radio frequency (RF) energy (e.g., a radio transmitter) interferes with other equipment (such as the low-level signal acquisition system). In other cases unintentional radiations from one equipment affect another. For example, later in this section the story is recounted of how an FM broadcast radio receiver used by nurses in a post coronary care (radio telemetry) unit grossly interfered with the operation of the telemetry system. In still other cases TV or radio reception is affected by nearby radio transmitters that are operating completely legally. In this section we will deal with these problems.

EMI problems are surfacing in ever-increasing forms. Part of this problem arises from the fact that increasingly large numbers of instruments and other electronic devices are being used. Coupled with the proliferation of electronic devices is a tremendous increase in the use of computers and word processors. If you doubt that theses devices are capable of interference, then try using an AM broadcast band radio receiver in close proximity to a desktop computer. Even further complicating the problem is the fact that many different forms of radio transmitters are on the premises or nearby. For example, in addition to the walkie-talkies used by the security department, there may also be a paging system transmitter on the roof. Nearby buildings and businesses may also have powerful radio transmitters on their own premises, and these transmitters are capable of interfering with electronic instrumentation equipment. And don't forget the potential interference from nearby broadcast stations. Those stations are potentially the most powerful emitters of RF energy around.

The first problems that we will consider are those that derive from the mixture of a large number of radio signals in the local vicinity.

### 15–4–1          Intermodulation Problems

Two interrelated problems often deteriorate radio reception in communications systems: **intermodulation** and **crossmodulation.** Although there are fine technical differences between these two problems, their cures are about the same, so I will take the liberty of calling them all "intermod problems." These problems result in interference on a given frequency due to heterodyning (mixing) between two other unrelated signals.

**Heterodyning** refers to the nonlinear mixing of two signals. Consider the guitar analogy. Pick one string and make it vibrate to produce a tone. That tone has a frequency of $F1$. After that tone dies out, pluck another string and cause a new tone ($F2$). Now pluck both strings simultaneously. What happens? How many tones are present? The initial answer might be "$F1$ and $F2$," but that is wrong. In a nonlinear system (such as hearing) additional frequencies are created. In addition to $F1$ and $F2$ there will be the sum frequency ($F1 + F2$) and the difference frequency ($F1 - F2$). It is the difference frequency, $F1 - F2$, by the way, that some guitar players use to tune the instrument. When two frequencies are close but not exact, the difference frequency is very low, almost subaudible. This low frequency accounts for the slow wavering tone that you hear as the strings are tuned closer to the same pitch. Exact tune is indicated by disappearance of the wavering. This phenomenon is heterodyning.

Other combinations are also produced according to the expression $mF1 \pm nF2$, where $m$ and $n$ are integers. For most purposes these additional frequencies are unimportant. However, a distinctly important problem with these additional frequencies is discussed here; "intermod problems" need not work on only $F1$ and $F2$.

There is a hill near my location that radio technicians and engineers in the area call "Intermod Hill." It happens to be one of the higher locations in the county, so several broadcasters and AT&T have seen fit to build radio towers there. In addition, both of the two main radio towers bristle with landmobile antennas whose owners rent space on the tower in order to get better coverage. In total, there are two 50,000-W FM broadcast stations, a 5000-W AM broadcast station, a many-frequency microwave relay station, and several dozen 30-mHz to 950-mHz VHF/UHF landmobile stations and radio paging stations. Nearby is a major community hospital that operates its own security system radio station and a hospital paging system, in addition to many other instruments. They also have a coronary care unit that uses UHF radio telemetry to keep track of ambulatory heart patients. All of those signals can heterodyne together to produce apparently valid signals on other channels.

The frequencies produced in a signal-rich environment are many and roughly follow the rule given above:

$$F_{\text{unwanted}} = mF1 \pm nF2$$

where     $m$ and $n$ = integers (1,2,3, . . .)
                $F1$, $F2$ = the frequencies present

If the receiver system is linear, then there is little chance for a problem. But nonlinearities do creep in, and when the receiver is nonlinear in a case where extraneous signals are present, then intermods show up. Being in the fields of so many radio transmitter signals almost ensures nonlinearity; hence intermod problems abound on and around Intermod Hill. Imagine the number of possible combinations when there are literally dozens of frequencies floating around the neighborhood!

The ability (or lack of same) to reject these unwanted signals is a good measure of a receiver's performance. A quality, well-designed radio receiver doesn't respond to either out-of-band or in-band off-channel signals, except in the most extreme cases of overload. Vital requirement to specify When requesting quotations and estimates from industry for a new installation, you should specify that the receiver *must* be able to discriminate against these spurious signals.

The linearity or dynamic range of the input RF amplifier of the receiver is the cause of most cases of intermodulation interference. There are other causes, however, and you must consider anything that can cause nonlinearity. For example, a well-known car radio had problems with the AGC rectifier diode. It was leaky, and thus the radio was easily overdriven by external signals. In some sets the manufacturer used a pair of back-to-back, silicon, small-signal diodes across the antenna terminals in order to shunt high-voltage potentials to ground. This method was especially popular a few years ago. A large signal could drive these diodes into conduction and produce nonlinearity.

One of the funniest intermod situations I know of occurred when I worked in a hospital as a bioelectronics engineer. We used a VHF band telemetry unit to monitor patient ECGs in the PCCU, which is the unit that coronary care unit patients go to after they are no longer acute but still bear close watching. The portable ECG transmitters generated 1 to 4 mW of VHF RF energy that was frequency modulated with the patient's electrocardiogram (ECG) signal. The signal level was so low that five 17-in whip antennas sticking down from the ceiling were needed to cover an area that consisted of two corridors each approximately 150 ft in length. Each antenna was connected directly to a 60-dB master TV antenna amplifier. (The ECG transmitter channels were located in the "guard bands" between TV video and audio carriers of commercial VHF TV channels.) One of the whip/amplifier assemblies was right over the receiver console.

One morning about 2:00 A.M., a nurse called me at home complaining that Mr. Jones's ECG was riding in on Mr. Smith's channel. Not quite believing her, I nonetheless went to the hospital and checked the situation out. Swapping receivers, telemetry transmitters, and amplifiers did no good. Finally, after two hours of trying (almost to the point of looking silly to nurses who don't easily tolerate others' failures regarding their equipment), I noticed the FM broadcast receiver sitting on top of the telemetry receiver cabinet less than 18 in from the antenna/amplifier (Fig. 15–3), and it was playing. On a sheer hunch, I turned off the receiver, and Mr. Jones went back to his own channel! Previously, his signal was showing up on both his own channel and Mr. Smith's channel, but now it showed up only where it belonged. Turning the FM receiver back on caused the situation to return. Also, tuning the radio to another channel made the problem go away.

What happened in that situation? The local oscillator in the FM broadcast receiver was heterodyning with Mr. Jones's signal to produce an intermod signal on Mr. Smith's channel. The situation is shown graphically in Fig. 15–3. The six VHF receivers used in the system are installed in a mainframe rack that forms the nurses' station console (along with an oscilloscope and strip chart recorder). The FM receiver was placed on top of the receiver rack such that its telescoping whip antenna was only a short distance from the telemetry receiver antenna.

Because the FM receiver is a superheterodyne, it produces a signal from the internal local oscillator that is 10.7 MHz higher than the received frequency. If the radio is tuned to the station at 99.7 MHz on the FM dial, then the local oscillator operates at 99.7 MHz + 10.7 MHz, or 110.4 MHz. This signal is radiated and is picked up by the telemetry antenna. Because of its close proximity, it is the strongest signal seen by the input of the 60-dB amplifier. The other signals impinging on the antenna are weak signals at 220.8

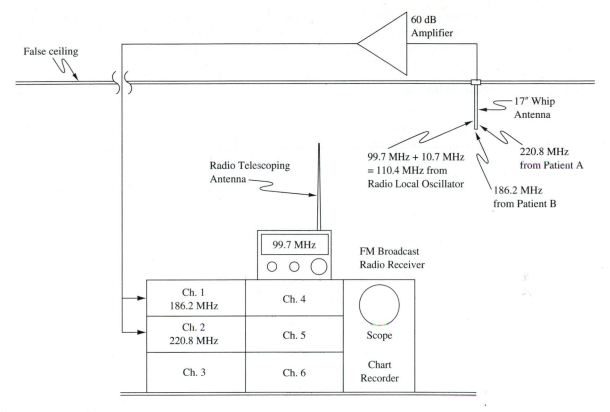

**FIGURE 15–3**
Heterodyning of FM receiver with oscillator.

MHz from Patient A and 186.2 MHz from Patient B. The mixing action at the input of the 60-dB amplifier, then, consists of 110.4 MHz, 186.2 MHz, and 220.8 MHz.

In the case cited it was apparently the 2nd harmonic of the FM radio local oscillator signal (i.e., 110.4 MHz × 2, or 220.8 MHz) that caused the problem. Consider the mathematics:

$$186.2 \text{ MHz} + 220.8 \text{ MHz} = 407 \text{ MHz}$$
$$407 \text{ MHz} - 186.2 \text{ MHz} = 220.8 \text{ MHz}$$

In this scenario the 186.2-MHz transmitter signal will appear on the 220.8-MHz channel!

The rule of thumb for which signal will appear is based on the "capture effect." Telemetry transmitters and receivers are actually frequency modulated (FM). An FM receiver will generally "capture" the stronger of two competing cochannel signals and will exclude the other. As a result, the stronger of the two signals predominates. Thus, in this example 186.2 MHz translated to 220.8 MHz by mixing in the 60-dB amplifier.

After that night the hospital banned FM radio receivers, patient-owned TV receivers, and CB and ham radio sets in the CCU/PCCU area for exactly the same reason they are banned on commercial airliners: interference with the electronic equipment.

**Some Solutions**

There are a number of ways to overcome most intermod problems. Modification of the receiver is possible, especially since poor design is a basic cause of intermods. But that approach is rarely feasible except for the most technically intrepid. There are, however, a few pointers for the rest of the people.

First, make sure that the receiver is well shielded. This problem rarely shows up on costly modern rigs, but it is a strong possibility on lower-cost receivers. Most radio transceivers have adequate shielding because of the requirements of the transmitter in the same cabinet with the receiver. But if there are any holes in the shielding, cover them up with sheet metal or copper foil (available in hobby shops). For others, the best approach is to use one of the methods shown in Figs. 15–4 and 15–5. In all cases you must identify one of the interfering frequencies or (in some cases) at least the band.

**Half-Wave Shorting Stub.**   A nonmatching load impedance attached to the business end of a transmission line repeats itself every half wavelength back down the line toward the input end. For example, if a 250-$\Omega$ antenna is attached to the load end of a half-wavelength piece of 50-$\Omega$ coax, an impedance meter at the input end will measure 250 $\Omega$. Lengths other than integer multiples of half wavelength will see different impedances at the input end. This phenomenon is the basis for transmission line transformers used in antenna matching. Therefore, if the end of a piece of coax is shorted (see Fig. 15–4), then the input end will see a short circuit at the frequency for which the coax is a half wavelength. The interfering frequency will be shorted to ground, while the desired frequency sees a high impedance . . . provided that the two are widely separated. The length $L$ is found from $L_{(ft)} = 492V/F_{(MHz)}$ where $L$ is in ft, $F$ is in MHz, and $V$ is the velocity factor of the coax (usually 0.66 for regular coax and 0.80 for polyfoam coax).

The method of Fig. 15–4 is best suited to cases where the interfering signal is in the VHF region or lower. Because the length of the coax stub is very short relative to the HF wavelength of the desired band, some very untransmission-line-like behavior might take place when transmitting. Therefore, for transceivers, some engineers recommend adding a second antenna jack especially for the stub, but connected to the receiver circuitry (RX CKT).

Another method (Fig. 15–5) is to use a frequency-selective filter. The selection of type of filter, and the cutoff frequency, is determined by the case. In a maze of frequencies like Intermod Hill, it might be wise to use a bandpass filter on the band of choice. Otherwise, use a low-pass filter if at least one of the interfering signals is higher than the desired band, and use a high-pass filter if the interfering signal is lower. For most cases a low-pass TVI filter is desirable on HF, so this solution is automatically taken care of. That is, the 35-to-45-MHz cutoff point of most such filters will attenuate most VHF signals trying to get back in.

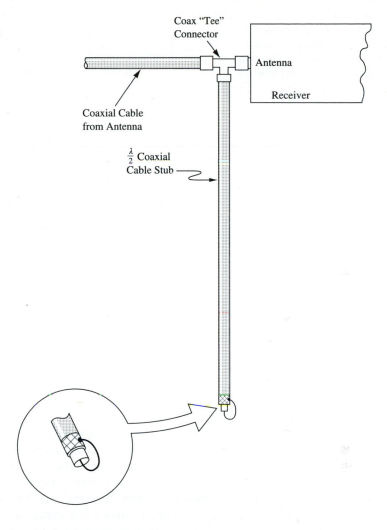

**FIGURE 15–4**
Use of a half-wave shorted stub as a filter.

Coax "Tee"
Connector

Antenna

Receiver

Coaxial Cable
from Antenna

$\frac{\lambda}{2}$ Coaxial
Cable Stub

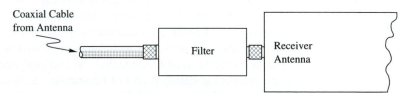

**FIGURE 15–5**
Correct placement of an EMI or TVI filter.

Coaxial Cable
from Antenna

Filter

Receiver
Antenna

**FIGURE 15–6**
Different filter configurations. (a)
Parallel resonant circuit in series
with the signal line, (b) series resonant trap across the signal line, (c)
combination wavetrap.

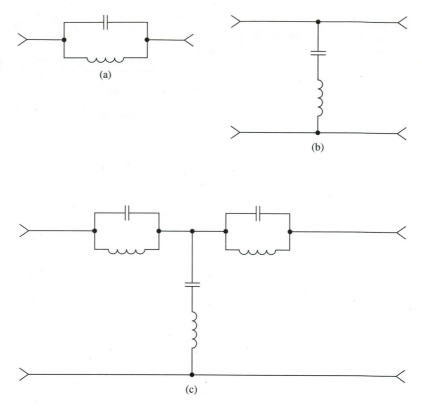

Figure 15–6 shows some of the different types of filter configurations that might be
used. The first wavetrap is a parallel resonant circuit [Fig. 15–6(a)] in series with the signal line. A parallel resonant trap has a high impedance at the resonant frequency but a low
impedance on other frequencies, so it blocks the undesired frequency. Figure 15–6(b)
shows a series resonant trap across the signal line. The series resonant circuit is just the
opposite of the parallel case: it offers a low impedance at its resonant frequency and a
high impedance elsewhere. Thus, it serves much like the coax stub in Fig. 15–4.

In Fig. 15–6(c) we see a combination wavetrap that uses both series and parallel resonant elements. Several editions of the *ARRL Radio Amateur's Handbook* had construction projects for AM broadcast band wavetraps of this type. Although that publication is
for amateur radio operators, the techniques presented for interference suppression are
quite useful for the hospital-based technician and engineer.

If the interfering station is an FM broadcaster, then most video shops and electronic
parts stores carry FM traps for 75-Ω antenna systems. These traps cost only a few dollars
but provide a tremendous amount of relief from the interference produced by broadcast
stations.

The interfering signal might be entering the equipment on the ac power line. This situation is especially likely if you live very close to a high-power broadcasting station. If
there is space in the rig, then it might pay to install an ac line filter (Fig. 15–7). Electronic

**FIGURE 15–7**
ac line EMI filter.

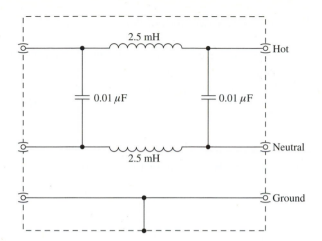

parts distributors sell EMI ac line filters suitable for equipment up to 2000 W or so (look at the ampere rating of the filter). Also, it is possible to install ferrite blocks on the line cord to serve as RF chokes. Some computer stores sell these to reduce EMI from long runs of multiconductor ribbon cable. It is sometimes possible to reduce the EMI pickup by either reducing the cord length or rolling it up into a tight coil. This solution works well where the cord is of a length nearly resonant (quarter wavelength) on the interfering signal's frequency.

## 15–4–3   Dealing with TVI/BCI

Television interference (TVI) is the bane of the radio communications and broadcasting communities. True TVI occurs when emissions from a transmitter interfere with the normal operation of a television receiver. Except for solo explorers on the North Slope of Alaska, and missionaries among the Indians of the Amazon Basin, all radio operators have a potential TVI problem as close as their own TV set . . . or their neighbors' sets.

One way to classify TVI is according to its cause. All electronic devices must perform two functions:

**1.** Respond to desired signals.
**2.** Reject undesired signals.

All consumer devices perform more or less in accordance with the first requirement, but many fail with respect to the second. Often a legally transmitted signal will be received on a neighbor's TV or hi-fi set even though the transmitter operation is perfectly clean. On some TV sets the high-intensity signal from a nearby transmitter drives the RF amplifier into nonlinearity, and that creates harmonics where none existed before. In other cases audio from the transmission is heard on the audio output of the set or is seen in the video as a result of signal pickup and rectification inside the set. Improper shielding can cause signals to be picked up on internal leads and fed to the circuits involved.

Remember two rules of thumb:

1. If transmitter emissions are not clean, then getting rid of the TVI/BCI is the radio operator's responsibility.
2. If the transmitter emissions are clean and the TVI/BCI is caused by poor medical equipment design, then it is the medical equipment owner's responsibility to fix the problem—not the radio operator's.

Unfortunately, in a society that depends too much on lawyers, the solution opted for by some ill-advised administrators is to hire lawyers to intimidate the radio owner into silent submission. Instead, wise managers will hire an engineer to solve the problem in a manner that will allow both parties to operate compatibly.

On the other hand, radio transmitter owners must respond responsibly to EMI complaints for several reasons. First, they have a legally imposed responsibility to keep transmitter emissions clean and free of spurious or unnecessary components. They must operate the transmitter legally. Second, it makes for good neighborhood relations if they attempt to help solve the problem. After all, we live in a society where more people seem willing to go running to lawyers than to engineers to solve technical problems. Even though the radio owners may win, a lawsuit can break them financially. Also, some neighbors who don't bring suit will, nonetheless, turn to local regulatory bodies even though these bodies don't have jurisdiction over the matter. (Only the Federal Communications Commission may legally regulate radio stations, a position that the courts have affirmed.)

The following steps should be taken to locate and eliminate the source of TVI:

1. Make sure that the transmitter emissions are clean. Although a spectrum analyzer must be used to ensure that the harmonics are down 40 dB or more below the carrier, a few simple checks will tell the tale in many cases. An absorption wavemeter can spot harmonics that are way too high. Also, listening on another receiver from a long distance (1 mi or more away) will give some indication of problems: at that distance, if you can hear the 2nd or 3rd harmonic, then so can the next door neighbor. Finally, if you own an RF wattmeter, or a forward-reading VSWR meter, and a dummy load, then you can make a quick check by measuring the output power with and without a low-pass filter installed in the line. If the power varies between the two readings by appreciably more than the insertion loss of the low-pass filter, then you should suspect that harmonics are present.

    The techniques for making the transmitter clean are simple (Fig. 15–8): a low-resistance earth ground and adequate filtering. The station shown in Fig. 15–8 has three frequency-selective elements after the transmitter: a low-pass filter, an antenna tuning unit (in lower-frequency stations), and a resonant antenna. All of these will help in reducing whatever harmonics the transmitter puts out.
2. Determine that you really have a TVI problem. Check for TVI with the transmitter both turned on and turned off. Also, make sure that the TV is properly adjusted.

    Also make sure that the interference is not coming from another source. Once an antenna goes up, all of the neighbors will assume the worst and blame every flicker of

**FIGURE 15–8**
Low-pass filter and antenna tuning
unit help suppress spurious radiation.

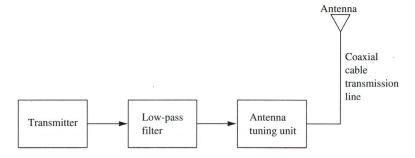

the set on the radio operations. Local broadcast stations, landmobile radio stations, and all manner of electronic devices can generate TVI—it's not just an amateur radio problem. In my own house I changed the dimmer switches the former owners installed; the switches produced "salt-and-pepper" snow on the TV and made 75/80 m on my ham radio receiver almost unusable.

3. Try adding a high-pass filter to the TV set. These filters will pass only those signals with a frequency greater than about 50 MHz, and they will severely attenuate HF amateur or CB signals. As shown in Fig. 15–9, the high-pass filter must be installed as closely as possible to the antenna terminals of the TV set. Make the connection wire as short as possible to prevent it from acting as an antenna in its own right.

   Counsel the owner of the affected TV to install an antenna that uses a coaxial cable transmission line, rather than 300-Ω twin lead. Although theory tells us that there should be no difference, it is nonetheless true that 75-Ω coaxial systems are less susceptible to all forms of noise, including TVI. Install a high-pass filter and a 75-to-300-Ω, TV-type BALUN impedance transformer at the TV's antenna terminals.

   Many, perhaps most, TV receivers have very poor internal shielding, so signals can bypass the high-pass filter and get picked up on the leads between the TV tuner (inside the set) and the antenna terminals on the rear of the set. That situation makes

**FIGURE 15–9**
Installation of high-pass filter.

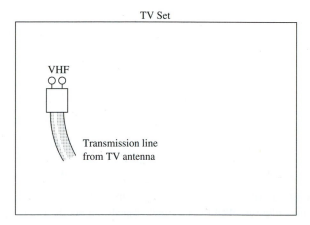

the high-pass filter almost useless. A solution for this dilemma is to mount the filter directly on the tuner, inside the set, making the lead length essentially zero.

## 15–4–4      Measuring Equipment and EMI

Medical instruments seem especially sensitive to picking up interfering signals. For example, the electrode wires of an ECG are exposed at the ends. In addition, the patient's body is basically an electrical conductor, so it makes a relatively decent "antenna" to pick up signals that exist in the air. The solution to that problem is a low-pass filter (Fig. 15–10) that blocks radio signals but not the ECG. These filters are not like the 60-Hz filter that is used in the ECG machine to eliminate power line artifact. Those filters take out a segment of the 0.05-to-100-Hz spectrum that is normally part of the ECG waveform, so they will always affect the shape of the waveform presented to the medical person using the machine. The RF low-pass filter has such a high cutoff frequency that the ECG waveform is not affected. Note that other low-level, low-frequency analog signal acquisition systems are also likely to experience high levels of EMI in the presence of RF fields.

While some neighbors will blast you for every little problem, others are not so bad and may actually respond in a positive manner if you point the way to a TVI/BCI solution even when the offending station is not yours. Let's revisit Intermod Hill.

A common problem for TV viewers in the area is FM interference. That form of TVI forms a herringbone pattern, which characterizes the interference as FM, not

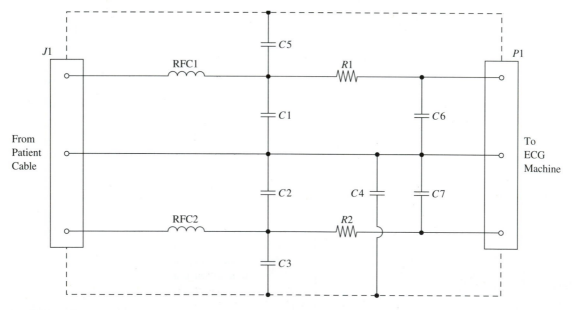

**FIGURE 15–10**
Low-pass filter installed to block radio signals but not the ECG.

AM/SSB. This form of interference often succumbs to either a bandstop filter or a shorted half-wave stub (refer again to Fig. 15–4). FM broadcast band bandstop or "notch" filters are available from most electronics parts distributers, as well as from many video stores and TV outlets. Both 75-$\Omega$ and 300-$\Omega$ forms are available (use the right kind). The shorted stub of Fig. 15–4 can be built from either twin lead or coax, depending upon the type of transmission line used. It is connected directly to the TV antenna terminals and has a length of

$$L_{(in)} = \frac{2952V}{F_{(MHz)}} \qquad\qquad \textbf{(15–2)}$$

where $\qquad L_{(in)} =$ the length in inches
$\qquad\quad F_{(MHz)} =$ the offending transmitter frequency in MHz
$\qquad\qquad\quad V =$ the velocity factor of the line (Use 0.82 for TV-type coax.)

■ **Example 15–2**

Find the length required for a shorting stub to suppress an FM broadcast signal on 101.1 MHz. Assume a 75-$\Omega$ coaxial transmission line (with a velocity factor of 0.82).

*Solution*

$$L_{(in)} = \frac{2952V}{F_{(MHz)}}$$

$$= \frac{(2952)\,(0.82)}{(101.1\ MHz)} = \frac{2421}{101.1} = \textbf{23.9 in}$$

## SUMMARY

1. Low-level signal acquisition systems are often subject to electromagnetic interference (EMI). The cures include shielding, filtering, and the use of differential input amplifiers.
2. Under some circumstances in a shielded differential input amplifier, a common mode signal can be accidentally unbalanced to create a valid differential input signal. The solution is to use a guard shield system for the input wiring.
3. When two or more radio signals combine in a nonlinear input circuit, or in a linear input circuit that was driven into nonlinearity by the presence of extraneous signals, intermodulation can occur, producing new frequencies according to the rule $mF1 \pm nF2$.
4. Various forms of EMI from RF sources can be eliminated with various combinations of bandpass filters, high-pass filters, low-pass filters, and notch filters.

## RECAPITULATION

Now go back and try to answer the questions at the beginning of the chapter. When you are finished, answer the questions and work the problems given below. Place a mark beside each problem or question that you cannot answer, and then go back to the text and reread appropriate sections.

## QUESTIONS

1. Describe the process by which a common mode signal can be made into a differential signal in shielded differential input amplifiers.
2. Draw a schematic diagram of a guard shield input differential amplifier, and explain in your own words how it works.
3. A radio transmitter at 36.8 MHz is interfering with a television receiver. Discuss ways of eliminating this interference and restoring proper operation of the TV set.
4. A radio telemetry system operates at 456.3 MHz. A local FM broadcaster is interfering with the reception of the telemetry, and intermodulation is suspected. (a) Calculate possible frequencies for the other interfering signal (assume only 2nd-order intermodulation). (b) Propose a solution to eliminate the interference.

## PROBLEMS

1. Calculate the length of a half-wavelength shorted stub to suppress a signal at 144.91 MHz when foam-filled coaxial cable ($V = 0.82$) is used.
2. Calculate the length of a half-wavelength shorted stub to suppress an FM broadcast station signal at 107.1 MHz if regular polyethylene coaxial cable ($V = 0.66$) is used.
3. Two signals are present in a locality: 106.4 MHz and 165.55 MHz. Find the first two orders of intermodulation product frequencies.
4. Find the 2nd-order intermodulation products for two signals: 52.3 MHz and 88.5 MHz.

# 16

# Data Converters

16–1 **OBJECTIVES**

1. To learn the operation of digital-to-analog converters (DACs).
2. To learn the operation of analog-to-digital converters (ADCs).
3. To learn the principal applications of data converters.
4. To learn the key design parameters used to select data converters.

16–2 **SELF-EVALUATION**

Before studying the material in this chapter, try to answer the questions given below. These questions test your knowledge of the subject. If you cannot answer a particular question, place a check mark beside it, and then look for the answer as you read the text.

1. Most binary DAC circuits use either the _____ _____ or the _____ resistance ladders.
2. List three types of ADC circuits.
3. How many different *states* can be represented on an 8-bit DAC? What is the highest *number* that can be represented on an 8-bit DAC?
4. Each step in a 10-bit DAC with a full-scale output potential of 10.00 V represents an output voltage change of _____ mV.

16–3 **WHAT ARE DATA CONVERTERS?**

Analog circuits and digital instruments occupy mutually exclusive realms. In the analog world a signal may vary between upper and low limits and may assume any value within the range. Analog signals are continuous between limits. But the signals in digital circuits may assume only one of two discrete voltage levels, one each for the two binary digits, 0 and 1. A **data converter** is a circuit or device that examines signals from one of these realms and then converts it to a proportional signal from the other.

A **digital-to-analog converter (DAC),** for example, converts a digital (i.e., binary) "word" consisting of a certain number of **bits** into a *voltage* or *current* that represents the binary number value of the digital word. An 8-bit DAC, for example, may produce an output signal of 0 V when the binary word applied to its digital inputs is $00000000_2$, and (say) 2.56 V when the digital inputs see a word of $11111111_2$. For binary words applied to the inputs, then, a proportional output voltage is created.

In the case of an **analog-to-digital converter (ADC, or A/D),** an analog voltage or current produces a proportional binary word output. If an 8-bit ADC has a 0-to-2.56-V input signal range, then a 0-V input could produce an output word of $00000000_2$, while the +2.56-V level seen at the input would produce an output word of $11111111_2$.

Data converters are used primarily to interface transducers (most of which produce analog output signals) to digital instruments or computer inputs, and to interface digital instrument outputs to analog-world devices such as meter movements, chart recorders, motors, and so on.

We have already discussed one form of ADC circuit, the dual-slope integrator, in Chapter 7. It is the primary ADC used to make digital voltmeters. In this chapter we will consider common DAC circuits and the following ADC circuits: **servo** (also called the **binary counter** or **ramp ADC**); **successive approximation; parallel converter;** and **voltage-to-frequency converter.**

## 16–4    DAC CIRCUITS

Figure 16–1 shows a **binary weighted resistance ladder** and operational amplifier used as a binary DAC. The operation of this circuit can be deduced from operational amplifier theory given in Chapter 12.

The resistors in the ladder are said to be binary weighted because their values are related to each other by powers of 2. If the lowest-value resistor is given the value *R,* then the next in the sequence will have a value 2*R,* followed by 4*R,* 8*R,* 16*R,* all the way up to the *n*th resistor (the last one in the chain), which has a value of $(2^{(n-1)})R$.

The switches ($B1$ through $B_n$) represent the input bits of the digital word. Although shown here as mechanical switches, they would be transistor switches in actual practice. The switches are used to connect the input resistors to either ground, or to voltage source *E,* to represent binary states 0 and 1, respectively. Switches $B1$ through $B_n$ create currents $I1$ through $I_n$, respectively, when they are set to the "1" position.

We know from Ohm's law that each current $I1$ through $I_n$ is equal to the quotient of *E* and the value of the associated resistor:

$$I1 = E/R1 = E/R$$
$$I2 = E/R2 = E/2R$$
$$I3 = E/R3 = E/4R$$
$$\bullet \qquad \bullet \qquad \bullet$$
$$\bullet \qquad \bullet \qquad \bullet$$
$$\bullet \qquad \bullet \qquad \bullet$$
$$I_n = E/R_n = E/(2^{(n-1)})R$$

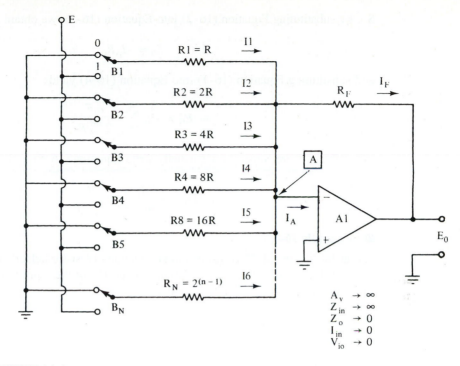

**FIGURE 16–1**
Binary weighted resistor ladder DAC circuit.

The total current $I_A$ into the junction (point A in Fig. 16–1) is expressed by the summation of currents $I1$ through $I_n$:

$$I_A = \sum_{i=1}^{n} \frac{a_i E}{2^{(i-1)} R} \tag{16–1}$$

where   $I_A$ = the current into the junction (point A) in amperes (A)
$E$ = the reference potential in volts (V)
$R$ = the resistance of $R1$ in ohms ($\Omega$)
$a_i$ = either "1" or "0," depending upon whether the input bit is "1" or "0"
$n$ = the number of bits, that is, the number of switches

From operational amplifier theory we know that

$$I_A = -I_f \tag{16–2}$$

and

$$E_0 = I_f R_f \tag{16–3}$$

So, by substituting Equation (16–2) into Equation (16–3), we obtain

$$E_0 = -I_A R_f \tag{16–4}$$

and substituting Equation (16–1) into Equation (16–4) yields

$$E_0 = -R_f \sum_{i=1}^{n} \frac{a_i E}{2^{(i-1)}R} \tag{16–5}$$

Since $E$ and $R$ are constants, we usually write Equation (16–5) in the form

$$E_0 = -\frac{E R_f}{R} \sum \frac{a_i}{2^{(i-1)}R} \tag{16–6}$$

■ **Example 16–1**

A four-bit (i.e., $n = 4$) DAC using a binary weighted resistor ladder has a reference source of 10 V dc, and $R_f = R$. Find the output voltage $E_0$ for the input word $1011_2$. (*Hint:* For input 1011, $a_1 = 1$, $a_2 = 0$, $a_3 = 1$, and $a_4 = 1$.)

*Solution*

$$E_0 = \frac{-E R_f}{R} \sum \frac{a_i}{2^{(i-1)}R} \tag{16–6}$$

$$= \left[ \frac{(-10 \text{ V})(R_f)}{R} \right] \left[ \frac{1}{2^{(1-1)}} + \frac{0}{2^{(2-1)}} + \frac{1}{2^{(3-1)}} + \frac{1}{2^{(4-1)}} \right]$$

$$= (-10 \text{ V}) \left[ \frac{1}{2^0} + \frac{1}{2^2} + \frac{1}{2^3} \right]$$

$$= (-10 \text{ V}) \left[ \frac{1}{1} + \frac{1}{4} + \frac{1}{8} \right]$$

$$= (-10 \text{ V}) \left[ \frac{1}{1.375} \right] = -7.27 \text{ V}$$

Although not revealed by the idealized equations, the binary weighted resistance ladder suffers from a serious drawback in actual practice. The values of the input resistors tend to become very large and very small at the ends of the range as the bit length of the input word becomes longer. If $R$ is set to 10 kΩ (a popular value), then $R8$ will be 1.28 MΩ. If we assume a reference potential $E$ of 10.00 V dc, then $I8$ will be only 7.8 *microamperes*. Most common nonpremium-grade operational amplifiers will not be able to resolve signals that low from the inherent noise. As a result, the bit length of the binary weighted ladder is severely limited. Few of these types of converters are found with more than 6- or 8-bit word lengths.

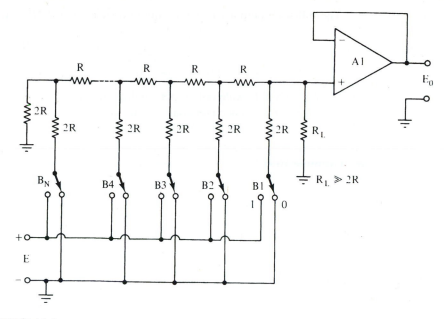

**FIGURE 16–2**
*R*-2*R* resistor ladder DAC circuit.

In commercial DACs, all of the resistors have a value of either $R$ or $2R$ (Fig. 16–2). The gain of the amplifier is unity, so $E_0$ can be expressed as

$$E_0 = E \sum_{i=1}^{n} \frac{a_i}{2^i} \qquad (16\text{–}7)$$

(provided that $R_L \gg R$, so that the voltage divider effect between the ladder and $R_L$ can be safely neglected).

■ **Example 16–2**

A 4-bit DAC using the $R$-2$R$ technique has a 5.00-V dc reference potential. Calculate $E_0$ for the input word $1011_2$.

*Solution*

$$E_0 = E \sum_{i=1}^{n} \frac{a_i}{2^i} \qquad (16\text{–}7)$$

$$= (5 \text{ V}) \left[ \frac{1}{2^1} + \frac{0}{2^2} + \frac{1}{2^3} + \frac{1}{2^4} \right]$$

$$= (5 \text{ V}) \left[ \frac{1}{2} + 0 + \frac{1}{8} + \frac{1}{16} \right]$$

$$= (5 \text{ V})(0.688) = \textbf{3.44 V}$$

The full-scale output voltage for any DAC using the $R$-$2R$ resistor ladder is given by

$$E_{fs} = \frac{E(2^{n-1})}{2^n} \tag{16–8}$$

where   $E_{fs}$ = the full-scale output potential in volts (V)
        $E$ = the reference potential in volts (V)
        $n$ = the bit length of the digital input word

■  **Example 16–3**

Find the full-scale output potential for an 8-bit DAC with a reference potential of 10.00 V dc.

*Solution*

$$E_{fs} = \frac{E(2^{n-1})}{2^n} \tag{16–8}$$

$$= \frac{(10\ V)(2^{8-1})}{2^8}$$

$$= \frac{(10\ V)(2^7)}{(2^8)}$$

$$= \frac{(10\ V)(255)}{(256)} = \mathbf{9.96\ V}$$

The output of a DAC cannot change in a continuous manner because the input is a digital word; that is, it can exist only in certain discrete states. Each successive binary number changes the output an amount equal to the change created by the least significant bit (LSB), which is expressed by

$$\Delta E_0 = \frac{E}{2^n} \tag{16–9}$$

So, for the DAC in Example 16–3, $E_0$ would be

$$\Delta E_0 = \frac{(10\ V)}{2^8}$$

$$= \frac{(10\ V)}{(256)} = \mathbf{40\ mV}$$

Note that $\Delta E_0$ is often called the 1 LSB value of $E_0$, and it is the smallest change in output voltage that can occur. It is interesting that if we let 0 V represent $00000000_2$ in our 8-bit system, then the maximum value of $E_0$ at $11111111_2$ will be 1 LSB less than $E$ (confirmed by the result of Example 16–3).

Numerous commercial DACs are on the market in IC, function module block, and equipment form. You should consult manufacturers' catalogues for appropriate types in any given application.

**16–5**     **SERVO ADC CIRCUITS**

The **servo** ADC circuit (also called the **binary counter** or **ramp** ADC circuit) uses a binary counter to drive the digital inputs of a DAC. A voltage comparator keeps the clock gate to the counter open as long as $E_0 \neq E_{in}$.

An example of such a circuit is the 8-bit ADC in Fig. 16–3(a), and the relationship of $E_0$ and $E_{in}$ relative to time is shown in Fig. 16–3(b).

Two things happen when a *start* pulse is received by the control logic circuits: the binary counter is reset to $00000000_2$, and the gate is opened to allow clock pulses into the counter. This will permit the counter to begin incrementing, thereby causing the DAC output voltage $E_0$ to begin rising [Fig. 16–3(b)]; $E_0$ will continue to rise until $E_0 = E_{in}$. When this condition is met, the output of the comparator drops *low,* turning off the gate. The binary number appearing on the counter output at this time is proportional to $E_{in}$.

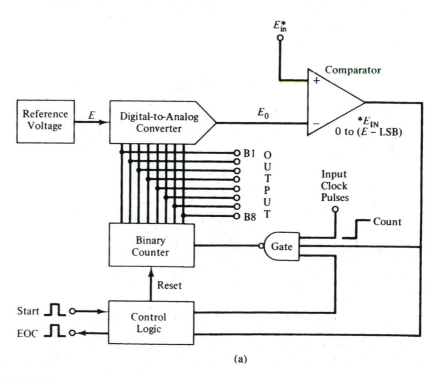

(a)

**FIGURE 16–3**
(a) Servo ADC circuit.

**FIGURE 16–3,** *continued*
(b) Operation of the servo-type
ADC circuit.

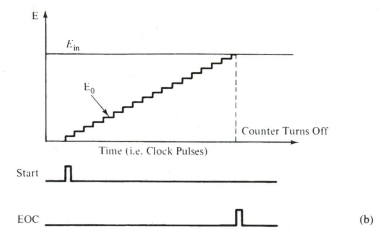

The control logic section senses the change in comparator output level and uses it to issue an **end-of-conversion (EOC)** pulse. This EOC pulse is used by instruments or circuitry connected to the ADC to verify that the output data are valid.

The **conversion time** $T_c$ of an ADC such as this depends upon the value of $E_{in}$, so when $E_{in}$ is maximum (i.e., full scale), so is $T_c$. Conversion time for this type of ADC is on the order of $2^n$ clock pulses for a full-scale conversion.

## 16–6        SUCCESSIVE APPROXIMATION (SA) ADC CIRCUITS

The conversion time of the servo ADC is too long for some applications. The successive approximation (SA) ADC is much faster for the same clock speed; that is, it takes $n + 1$ clock pulses instead of $2^n$. For the 8-bit ADC that has been our example, the SA type of ADC is 28 times *faster* than the *servo* ADC.

The basic concept of the SA ADC circuit can be represented by a platform balance, such as in Fig. 16–4, in which a full-scale weight $W$ will deflect the pointer all the way to the left when pan 2 is empty.

Our calibrated weight set consists of many separate pieces, which weigh $W/2$, $W/4$, $W/8$, $W/16$, and so on. When an unknown weight $W_X$ is placed on pan 2, the scale will deflect to the right. To make our measurement, we place $W/2$ on pan 1. Three conditions are now possible:

$$W/2 = W_X \qquad \text{(scale is at zero)}$$
$$W/2 > W_X \qquad \text{(scale is to the left of zero)}$$
$$W/2 < W_X \qquad \text{(scale is to the right of zero)}$$

If $W/2 = W_X$, then the measurement is finished, and no additional trials are necessary. But if $W/2 < W_X$, then we must add more weights in succession ($W/4$, then $W/8$, etc.) until we find a combination equal to $W_X$.

**FIGURE 16–4**
Successive approximation ADC circuits are like the platform balance.

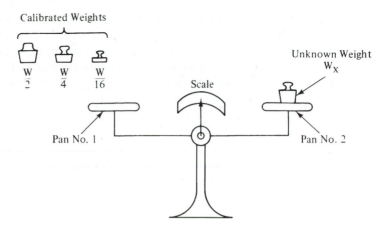

If, on the other hand, $W/2 > W_X$, then we must *remove* the $W/2$ weight and in the second trial start again with $W/4$. We will continue this procedure until we find a combination equal to $W_X$.

The SA ADC circuit uses not a scale but a shift register, as shown in Fig. 16–5. A
successive approximation register (SAR) contains the control logic, a shift register, and a
set of output latches, one for each register section. The outputs of the latches drive a
DAC.

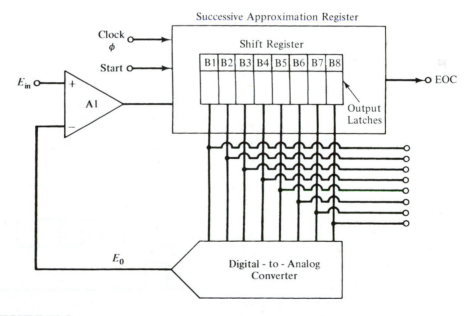

**FIGURE 16–5**
Successive approximation ADC circuit.

A *start* pulse sets the first bit of the shift register *high*, so the DAC will see the word $10000000_2$ and therefore produce an output voltage equal to one-half the full-scale output voltage. If the input voltage is greater than $1/2E_{fs}$, then the $B1$ latch is set *high*. On the next clock pulse, register $B2$ is set high for trial 2. The output of the DAC is now 3/4-scale. If, on any trial, it is found that $E_{in} < E_0$, then that bit is reset *low*.

Let's follow a 3-bit SAR through a sample conversion. In our example the full-scale potential is 1 V and $E_{in}$ is 0.625 V. Consider Fig. 16–6.

Time $t_1$: The *start* is received, so register $B1$ goes *high*. The output word is now $100_2$, so $E_0 = 0.5$ V. Since $E_0 < E_{in}$, latch $B1$ is set to "1," so at the end of the trial the output word remains $100_2$.

Time $t_2$: On this trial (which starts upon receiving the next clock pulse), register $B2$ is set *high*, so the output word is $110_2$. Voltage $E_0$ is now 0.75 V. Since $E_{in} < E_0$, the $B2$ latch is set to "0," and the output word reverts to $100_2$.

Time $t_3$: Register $B3$ is set *high*, making the output word $101_2$. The value of $E_0$ is now 0.625 V, so $E_{in} = E_0$. The $B3$ register is latched to "1," and the output word remains $101_2$.

Time $t_4$: Overflow occurs, telling the control logic to issue an EOC pulse. In some cases the overflow pulse *is* the EOC pulse.

Note that this example had a 3-bit SAR, so by our $(n + 1)$ rule four clock pulses were required to complete the conversion. The SA type of ADC was once regarded as difficult to design because of the logic required. But today, IC and function blocks are available that use this technique, so the design job is reduced considerably. The SA technique can be implemented in software under computer control using only an external DAC and

**FIGURE 16–6**
Timing diagram for Fig. 16–5.

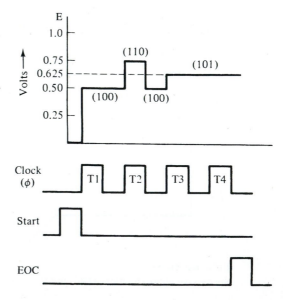

comparator. All register functions are handled in the software (program). (See the first three references in the Bibliography at the end of this chapter.)

## 16–7    PARALLEL CONVERTERS

The **parallel** ADC circuit (Fig. 16–7) is probably the fastest type of ADC known. In fact, some texts call it the "flash" converter in testimony to its speed. It consists of a bank of $(2^n - 1)$ voltage comparators biased by reference potential $E$ through a resistor network

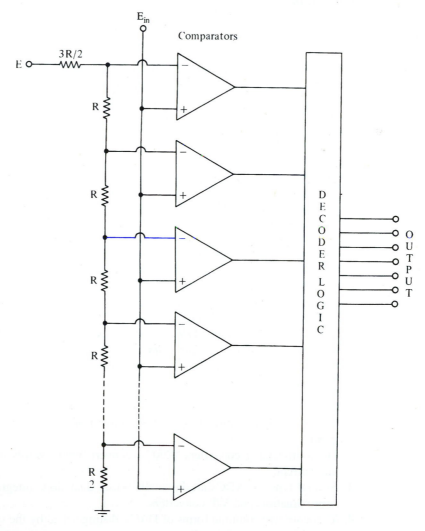

**FIGURE 16–7**
Parallel or "flash" ADC circuit.

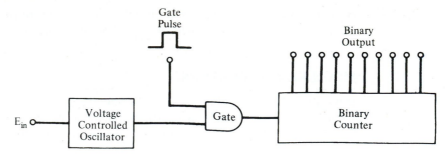

**FIGURE 16–8**
Voltage-to-frequency converter ADC.

that keeps the individual comparators 1 LSB apart. Since the input voltage is applied to all of the comparators simultaneously, the speed of conversion is essentially the slewing speed of the *slowest* comparator in the bank and the decoder propagation time (if logic is used). The decoder converts the output code to binary code or possibly BCD in some cases.

## 16–8        VOLTAGE-TO-FREQUENCY CONVERTERS

A voltage-to-frequency (V/F) converter is a voltage-controlled oscillator (VCO) in which an input voltage $E_{in}$ is represented by an output frequency $F$. An ADC using the V/F converter is shown in Fig. 16–8. It consists of no more than the VCO and a frequency counter. The display, or output states, on the counter gives us the value of $E_{in}$.

Voltage-to-frequency converters are used mainly where economics dictate serial transmission of data from a remote collection point to the instrument. Such data can be transmitted by wire or radio communications channels. Another application is the tape recording of analog data that is, in itself, too low in frequency to be recorded.

The inverse procedure, F/V conversion, is a form of DAC in which an input *frequency* is converted to an output *voltage*.

## SUMMARY

1. Analog-to-digital converters (ADCs) convert analog voltages or currents to binary words.
2. Digital-to-analog converters (DACs) convert binary words to proportional voltages and currents.
3. Several types of ADCs are in common use: dual-slope integrators, servo, successive approximation, and V/F converters.
4. There are two common forms of DACs, distinguished by the type of resistance ladder used: binary weighted or *R-2R*.

# RECAPITULATION

Now go back and try to answer the questions at the beginning of the chapter. When you are finished, answer the questions and work the problems given below. Place a mark beside each problem or question that you cannot answer, and then go back to the text and reread appropriate sections.

# QUESTIONS

1. Define in your own words the purposes of (a) the DAC and (b) the ADC.
2. Draw a circuit for a binary weighted resistor ladder DAC.
3. Draw a circuit for an R-2R resistor ladder DAC.
4. List two types of ADCs that use a DAC as a critical element.
5. The output of a DAC is a ———————— function of voltage or current.
6. Describe in your own words the operation of a successive approximation ADC.
7. Describe in your own words the operation of a servo ADC.
8. A very fast ADC circuit that uses a bank of voltage comparators is called the ———————— or ———————— ADC.

# PROBLEMS

1. An 8-bit DAC using a binary weighted resistor ladder has a +7.5-V dc reference potential. Find the output potential for the input word $10110111_2$ if $R_f = R$.
2. An 8-bit DAC using an R-2R ladder has a +2.56 V dc reference potential. Find the full-scale output potential.
3. Find the 1-LSB output voltage for the DAC in Problem 2.
4. Find the conversion time for a full-scale input potential for (a) a servo ADC and (b) an SA ADC if the clock speed is 2.5 mHz.

# BIBLIOGRAPHY

Aldridge, Don. "A/D Conversion Systems: Let your µP do the Working." *EDN Magazine* (May 5, 1976), p. 75.

Freeman, Wes, and Ritmanich, Will. "Cut A/D Conversion Costs by Using D/A Converters and Software." *Electronic Design* (April 26, 1977), p. 86.

Sheingold, Daniel H. *Analog-to-Digital Conversion Handbook.* Norwood, MA: Analog Devices, Inc., n.d.

"Software Controlled Analog to Digital Conversion Using the DAC-08 and the 8080A Microprocessor." Precision Monolithics, Inc., Application Note AN-22, 1976.

# 17

# Testing Electronic Components

## 17–1    OBJECTIVES

1. To learn how to test pn junction diodes.
2. To learn how to test transistors.
3. To learn how to test capacitors.
4. To learn the types of test equipment on the market.

## 17–2    SELF-EVALUATION

Before studying the material in this chapter, try to answer the questions given below. These questions test your knowledge of the subject. If you cannot answer a particular question, place a check mark beside it, and then look for the answer as you read the text.

1. What instrument can be used to make a simple qualitative test of a pn junction diode?
2. The same instrument as in Question 1 can also be used to test bipolar transistors. True or false?
3. List three types of transistor testers.
4. Capacitors and inductors can be tested on a ——————— meter.

## 17–3    TESTING COMPONENTS

It frequently becomes necessary to check individual electronic components either for value (i.e., μF, ohms, beta, etc.) or for quality. Service technicians must often use test procedures that determine which component is causing a malfunction. Design engineers often test components to assure critical parameters. The quality control departments of manufacturing plants often inspect incoming components purchased from other manufacturers in order to minimize failures on the assembly line, or they test their own equipment to minimize product warranty service after the equipment has been sold to the customer.

Some of the tests that we will consider are qualitative in nature, whereas others are quantitative. Most of the qualitative tests are simple and very easy to perform in a short period of time. Such a test is used mostly in troubleshooting, where we look for gross defects, rather than subtle variations, in performance. These tests are valid because of the time they save; why use a curve tracer to determine that a power transistor has a *c–e* short when an ohmmeter will do just as well?

Quantitative tests, on the other hand, are needed to determine actual device performance or to select out those in a large group that meet tighter specifications than are normally available or guaranteed.

## 17–4    SIMPLE SEMICONDUCTOR TESTS

A qualitative tester for diodes and bipolar transistors may consist of a simple ohmmeter, while a tester that actually measures the device parameters may tend to be both complex and costly. A solid-state diode consists of a pn junction of semiconductor material. The pn junction will pass current in only one direction, and this forms the basis of a simple test procedure. Figure 17–1 shows the use of an ordinary dc ohmmeter to test diodes. The ohmmeter uses a dc voltage source (usually a dry cell battery) and some calibrating resistors, and measures resistance via the current that is drawn when the unknown resistance is in the circuit. The ohmmeter probes, then, are polarized because the reference voltage is from a dc source; one probe is positive with respect to the other (i.e., "negative") probe. When the ohmmeter is connected as in Fig. 17–1(a), the pn junction of the diode will be forward biased, so the ohmmeter will register a low resistance, in the range 50 to 1000 Ω in most cases. When the ohmmeter probes are reversed, the pn junction becomes reverse biased, so the ohmmeter registers a high value of resistance.

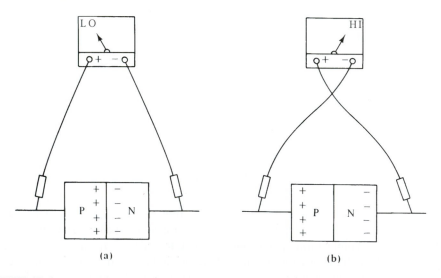

(a)                (b)

**FIGURE 17–1**
(a) Diode shows low resistance when forward biased by ohmmeter. (b) Diode shows high resistance, or open circuit, when reverse-biased by ohmmeter.

The terms **high** and **low** in this context are used in a relative sense. The low resistance is typically in the under-1000-$\Omega$ range. The high value tends to be from 5× the low reading in older germanium diodes to greater than 50× the low reading in most modern silicon diodes.

It is important to use the correct ohmmeter scale in this type of test. The use of a too-low scale (i.e., ×1 or ×10) can burn out small-signal diodes, while too-high scales tend to give false results. As a general rule, when using an ohmmeter with a 1.5-V battery, use the ×1 and ×10 scales for rectifier diodes that are rated at 500 mA or greater forward current. The ×100 and ×1000 scales are suited to testing small-signal diodes (e.g., 1N60, 1N914, etc.).

A problem develops when you are using the ohmmeter sections of the modern digital multimeter (DMM) in this test; the voltage source in the DMM ohmmeter is not a battery but an electronically regulated power supply. In fact, many manufacturers of these instruments tout these instruments for use in solid-state circuits because the ohmmeter will not forward-bias pn junctions or damage anything. This feature is an advantage in that it reduces the chance of false resistance readings in-circuit that are often caused by pn junctions' becoming inadvertently forward biased. But the low-voltage ohmmeter cannot be used as a tester for pn junctions for the same reason. A few DMM models are equipped with a "high-low power" switch (often labeled "diode" or with the diode circuit symbol) for the ohmmeter specifically so that the user may test pn junctions, while retaining the advantages of low-voltage operation.

A transistor is essentially a pair of pn junctions connected back to back, or at least this is a reasonable model of a transistor for our purposes. We may, therefore, use the same test procedure to test transistors. An ohmmeter, then, may be used to qualitatively evaluate bipolar transistors. Figure 17–2 details the test procedure.

It is necessary to check each pn junction of the transistor separately and to check the collector-emitter leakage current. Figures 17–2(a) and 17–2(b) show the test to the *b–e* junction, and Figures 17–2(c) and 17–2(d) show the test applied to the *b–c* junction. Both pnp and npn transistors are tested in exactly this same manner, but expect the polarity of the high-low readings to be reversed.

Check leakage between the collector and emitter terminals also by using the ohmmeter, as shown in Fig. 17–2(e). It is a good practice to measure this resistance by using the reversed probes method—not to discern the quality of a junction, but to keep from forward-biasing one of the other junctions. Use the higher of the two readings as the true reading.

The same advice as to ohmmeter scales must also apply when you are testing transistors. Use the ×1 and ×10 ranges for testing power transistors, and the ×100 or ×1000 scales when testing small-signal transistors.

In testing either diodes or transistors, you should never use an ohmmeter that has a voltage source greater than 1.5 V. Some older models used batteries such as 4.5, 22.5, or even 45 V. These voltages *will* burn out the pn junction being tested. You can determine both the voltage and the polarity of the probes by using a voltmeter to measure the open-terminal voltage that appears across the ohmmeter probes.

The pn junction tests tell us the relative condition of the junctions in the transistor, but they do not tell us if the base is able to control the collector current. Those tests are

**FIGURE 17–2**
(a) and (b) Checking the *b–e* junction of a transistor; (c) and (d) checking the *c–e* junction of a transistor; (e) checking *c–e* leakage.

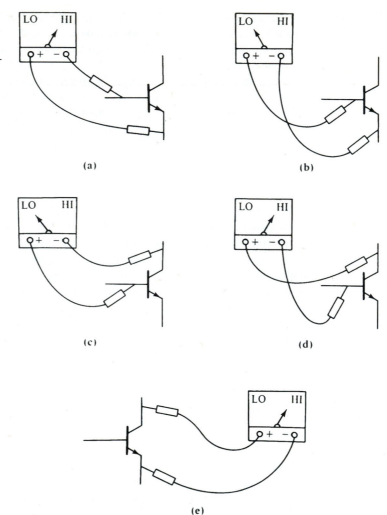

fairly accurate; there are only a few false positives. But a simple test that also uses the ohmmeter, shown in Fig. 17–3, does tell us whether the base terminal is able to control *c–e* current flow. The ohmmeter is connected across the *c–e* terminals in such a way that the *c–e* polarity is proper for normal operation. With the base open there will be a high resistance [Fig. 17–3(a)], but as shown in Fig. 17–3(b), the *c–e* resistance drops low when the base is shorted to the collector. Some authors, including this one, feel that it is better to connect the base to the collector through a resistor of a value that would normally be used to bias the device (i.e., 30–150 $\Omega$ in power transistors, 10–100 $\Omega$ in moderate-beta, small-signal transistors, and 100 to 1000 $\Omega$ in high-beta devices). This will reduce the dependence of the test on interpreting how much of the low-resistance reading is due to a forward-biased *b–e* junction.

**FIGURE 17–3**
Checking a transistor to see if the
base terminal can control $I_{ce}$.

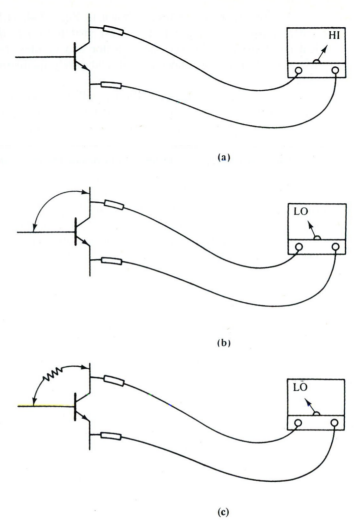

(a)

(b)

(c)

The ohmmeter tests are generally suited to transistors in the dc-to-UHF range, with the exception of high-voltage types, and VHF power transistors.

## 17–5          TRANSISTOR TESTERS

Several varieties of transistor testers are on the market: **conduction** (also called **leakage-gain**), **oscillator, beta,** and **curve tracer.** These different types of testers vary in their relative validity, but all prove useful under the correct circumstances. With the exception of the conduction type, all are considered a step above the simple ohmmeter tests of the previous section. The conduction type of tester is merely a formal version of the ohmmeter tests.

A simple conduction tester, shown in Fig. 17–4, is little more than an ohmmeter and a transistor socket. When the **test switch** (*S*1) is open, the meter will read the leakage resistance between the collector and emitter terminals of the transistor. But when *S*1 is closed, we have a rough and purely qualitative measure of the transistor's beta.

The meter in the conduction tester will have two scales; one each for *leakage* and *gain*. Only a few models bother with numbers, which in this case are essentially meaningless, but use color scales red-yellow-green to indicate bad-unknown-good (respectively).

The oscillator type of tester simply places the transistor in an audio or RF oscillator circuit to see if it will oscillate. An ac-coupled detector drives a meter to indicate when oscillations are present. The detector output is zero if the transistor is not oscillating. The oscillator type of tester works not only with bipolar transistors but also (if appropriate circuitry is provided) with both types of field effect transistors.

A true beta transistor tester gives both qualitative and quantitative information about the device under test. Some models are partially automatic, while others, costing very little money, are strictly manual.

The simplest method for measuring the beta of a transistor is to observe that beta is the ratio $I_c/I_b$. Figure 17–5 shows one of the simpler circuits used in this type of tester; it

**FIGURE 17–4**
Conduction-type transistor tester.

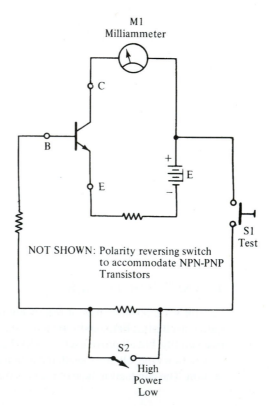

**FIGURE 17–5**
Beta tester.

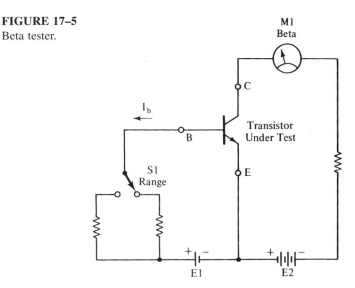

is similar to the circuit of the conduction tester. The difference between this and the other type of tester is that the base current is fixed, constant, and known. We may then place a milliammeter in the collector circuit to measure the collector current. We are justified in calibrating the collector meter in beta units, because the base current is known, because beta is a fixed property of the device, and because the collector current is the variable measured. In some models a small ac signal is applied to the base, and the meter is ac coupled to the collector. This will read ac beta, that is, $h_{fe}$ instead of $H_{FE}$.

Numerous instruments on the market accurately test transistors for beta. It is generally agreed that a good tester will do the following:

1. Measure $I_{cbo}$.
2. Measure $I_{ceo}$.
3. Measure beta in several ranges.
4. Measure super-beta up to 20,000 or more for testing Darlington devices.
5. Test pn junction diodes (both rectifiers and small-signal devices).
6. Test both types of FETs.

In-circuit transistor testers are capable of approximating the performance of out-of-circuit transistor testers. Most of these are ac beta testers, though a few are oscillator types. These testers are sometimes guilty of substantial error and thus are used for screening of several transistors in the circuit while troubleshooting. A majority of the errors are "false-bad" readings, so you are advised to remove a suspected bad transistor from the circuit for retesting out of circuit. The in-circuit tester appeals mostly to servicers, for whom it saves a lot of time.

The **curve tracer** is an instrument that uses an oscilloscope to trace the characteristic curve of a transistor or diode. These instruments plot the $I_c$-vs.-$V_{ce}$ curve of the device under test at anywhere from 1 to 10 different base current levels. Most of these instruments use step-switched circuits to automatically select the $I_b$ levels, and a variable voltage collector supply to change $V_{ce}$.

A simple curve tracer for testing diodes is shown in Fig. 17–6(a). In this circuit an ac source is connected between the vertical and horizontal inputs of an oscilloscope. A load is connected across the vertical input, and the diode being tested is connected across the horizontal input of the oscilloscope. If the diode is open, then only horizontal deflection occurs[Fig. 17–6(b)]. A shorted diode allows only vertical deflection [Fig. 17–6(c)]. But a good diode will appear shorted on one-half of the ac cycle, and open on the other half of the ac cycle, because the alternating current alternately forward- and then reverse-biases the diode. This action creates the L-shaped curve of Fig. 17–6(d); it is essentially a combination of the first two traces. A poor diode has a lot of leakage current flowing across the diode even in the reverse-bias direction. Such a diode will produce the trace shown in Fig. 17–6(e).

Zener diodes can be tested on a similar curve tracer, as shown in Fig. 17–7(a). The major difference between these two test circuits is the series resistance $R2$. The oscilloscope trace for a good diode is shown in Fig. 17–7(b).

An example of a commercial curve tracer is shown in Fig. 17–8. This unit, a Tektronix model 576, contains all of the needed circuitry, as well as the oscilloscope, in one cabinet.

**FIGURE 17–6**
(a) Diode curve tracer, (b) pattern for open diode, (c) pattern for shorted diode, (d) pattern for good diode, (e) pattern for poor diode.

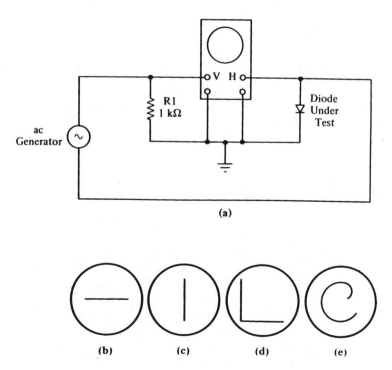

(a)

(b)        (c)        (d)        (e)

**FIGURE 17–7**
(a) Zener diode curve tracer,
(b) curve for good zener diode.

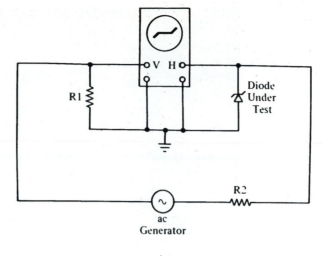

**(a)**

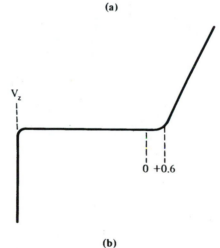

**(b)**

## 17–6    CHECKING CAPACITORS

Capacitors are simple devices, so they seemingly should be very easy to test. But this is not always true, especially at very low or very high capacitance values. Fortunately, the qualitative tests tend to be very simple.

## 17–7    TESTING CAPACITORS WITH AN OHMMETER

When you connect a discharged capacitor across an ohmmeter, the capacitor will charge because of the ohmmeter current. The initial current flow is very large, so the ohmmeter will show a low resistance. But as the capacitor charges, the resistance reading will

**FIGURE 17–8**
Commercial curve tracer (courtesy of Tektronix, Inc.).

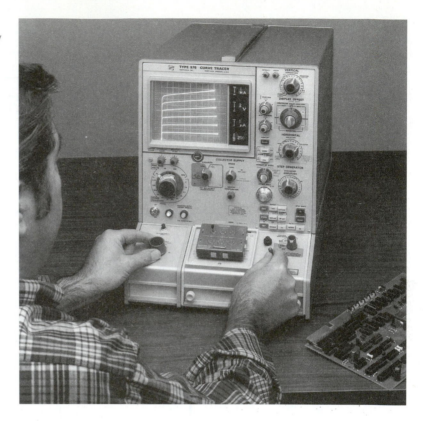

become progressively higher until it shows an open circuit. At that point, the charge voltage across the capacitor will be equal to the open-terminal voltage of the ohmmeter, so no further current flows. Use the highest ohmmeter range available, or the capacitor will charge too fast for a heavily damped analog meter pointer to deflect properly. This problem becomes especially critical at low capacitance values where the capacitance and the looking-back resistance of the ohmmeter form a short RC time constant. It takes some practice to become adept at judging capacitors of various values by using this technique, so some practice is in order for the student.

## 17–8     TESTING CAPACITORS WITH A VOLTMETER

We may use a voltmeter in two different ways to evaluate the leakage across the dielectric of a capacitor. Both of these methods are generally used in troubleshooting electronic problems.

In one method, useful in circuits at points where there is a potential of at least several volts across the capacitor, we check for the existence of current flow through the capacitor by looking for a dc voltage at one end. The "cold" end of the capacitor is disconnected from the circuit, and a voltmeter is used to measure the voltage from the disconnected end to ground. If there is a problem with leakage, then a voltage will be present.

The second method involves charging the capacitor to some voltage with a power supply. The supply is then disconnected, and a very high impedance voltmeter (i.e., an FET or a VTVM type) is used to measure the voltage across the capacitor. A perfect capacitor will hold the voltage indefinitely, but practical capacitors will leak, so the charge diminishes. Again, some practice is needed before an individual can correctly judge whether any given capacitor is "bad" by noting the rate at which the charge decays, so some more practice is in order.

## 17–9    MEASURING CAPACITANCE

Two traditional methods have been popular for measuring capacitance. One method is to place the capacitor in one leg of an ac bridge (see Chapter 4). A properly constructed bridge circuit can be calibrated to yield very accurate measurements. The other technique is to use a "Q-meter." This instrument places an unknown reactance ($L$ or $C$) in a series-tuned tank circuit and then measures the voltage drop across the unknown element. At any given frequency, the Q-meter can yield an accurate determination of either inductance or capacitance, depending upon how it is configured.

In recent years the bridge and the Q-meter have been somewhat eclipsed, except at the highest accuracy end, by the simple digital capacitance meter. A block diagram of such a meter is shown in Fig. 17–9. The circuit consists of a period counter in which the *on* time of the main gate is set by a monostable multivibrator (i.e., a one-shot). The capacitor under test sets the timing of the one-shot stage. In the past, RC one-shot circuits were subject to a lot of error, but today IC one-shot circuits are capable of 1% or better accuracy.

The period of the one-shot output pulse is equal to the time constant RC. The clock rate in this example is 1 mHz. The count that is accumulated in the counter will be numerically equal to the capacitance in picofarads. For example, if $R = 1$ M$\Omega$, and if

**FIGURE 17–9**
Digital capacitance checker.

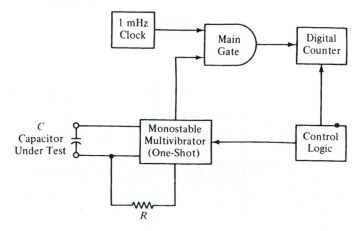

**TABLE 17–1**
Timing values vs. counts for Fig. 17–9.

| C (pF) | R (Ω) | T = R × C (s) | Count |
|---|---|---|---|
| 10 | $1 \times 10^6$ | $1 \times 10^{-5}$ | 10 |
| 50 | $1 \times 10^6$ | $5 \times 10^{-5}$ | 50 |
| 100 | $1 \times 10^6$ | $1 \times 10^{-4}$ | 100 |
| 500 | $1 \times 10^6$ | $5 \times 10^{-4}$ | 500 |
| 1000 | $1 \times 10^6$ | $1 \times 10^{-3}$ | 1000 |
| 5000 | $1 \times 10^6$ | $5 \times 10^{-3}$ | 5000 |
| 10,000 | $1 \times 10^6$ | $1 \times 10^{-2}$ | 10,000 |

we place a 10-pF capacitor across the test terminals, the gate time will be $(10^{-11}$ F$)$ $(10^6$ Ω$)$, or $10^{-5}$ s. During this time the 1-mHz clock will input 10 pulses to the counter, so the count is 10. If a 100-pF capacitor were used instead, the gate's *on* time would be $10^{-4}$, and thus 100 counts of the 1-mHz clock would be accumulated. Table 17–1 shows an appropriate timing scheme.

Another approach, less widely used, is shown in Fig. 17–10. In this case a voltage comparator circuit is used to gate pulses into an operational amplifier integrator (see Chapter 12). The output voltage of the integrator increments a small amount for each pulse received. One input of the comparator is biased to 0.636$E$, while the other is biased by the capacitor charge voltage in the RC timing network. When $E_c$ rises to 0.636$E$ (at the end of one RC time constant), the comparator will shut off the flow of pulses to the integrator.

When the integrator input no longer sees pulses, the output voltage will cease to increase. The voltage at this instant is proportional to the capacitance. This technique was popular in predigital instruments because an analog voltmeter, with a scale calibrated in capacitance, could be used as a readout device.

**FIGURE 17–10**
Simple analog capacitance tester.

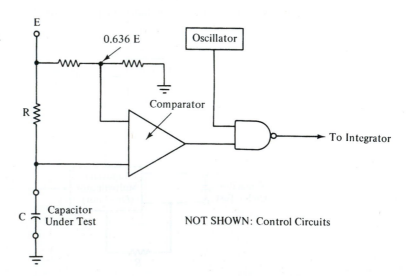

# SUMMARY

1. Qualitative tests of diodes and transistors can be made by using an ohmmeter.
2. Four different types of bipolar transistor testers are **conduction** (i.e., **leakage-gain**), **oscillator, beta,** and **curve tracer.**
3. Capacitors may be checked qualitatively by using an ohmmeter.
4. Capacitance can be measured in an ac bridge circuit by a Q-meter or with a digital counter that is controlled by a one-shot in which the capacitor in question sets the timing.

# RECAPITULATION

Now go back and try to answer the questions at the beginning of the chapter. When you are finished, answer the questions and work the problems given below. Place a mark beside each problem or question that you cannot answer, and then go back to the text and reread appropriate sections.

# QUESTIONS

1. An ohmmeter can be used to test a pn junction diode by —————————— the ohmmeter probes and comparing the —————————— and —————————— resistances across the junction.
2. Describe in your own words a procedure for testing bipolar transistors with an ohmmeter.
3. Which ohmmeter scales are appropriate for testing small-signal transistors?
4. Why are some digital multimeter ohmmeter sections *not* suitable for testing pn junctions?
5. List four types of transistor testers.
6. Describe how an ohmmeter can be used to make a qualitative test of a capacitor.
7. Which ohmmeter range should be used for the test discussed in Question 6?
8. List two ways to measure capacitance.

# PROBLEMS

1. A comparator allows pulses from a 1000-kHz clock into a digital counter for one RC time constant. If the resistor is 1 M$\Omega$ and the count is 5625, the capacitance is —————————— .
2. Find the RC time constant of a 0.001-$\mu$F capacitor and a 1-M$\Omega$ resistor.
3. A bipolar transistor creates a 10-mA collector current when 86 $\mu$A flows into the base terminal. Find the beta.

# 18

# Measurement of Frequency and Time

18–1

## OBJECTIVES

1. To learn the different types of equipment used to measure frequency.
2. To learn how to measure time in electronic circuits.
3. To learn frequency measurement techniques.
4. To learn the different types of frequency standards.

18–2

## SELF-EVALUATION

Before studying the material in this chapter, try to answer the questions given below. These questions test your knowledge of the subject. If you cannot answer a particular question, then look for the answer as you read the text.

1. How do frequency counters and period counters *differ* from each other?
2. What are **Lecher wires?**
3. How can frequency be measured with a WWVB comparator receiver?
4. List three types of absorption frequency meters.
5. List two techniques for using a low-frequency counter to measure a VHF frequency.

18–3

## FREQUENCY AND TIME

**Frequency** is the measure of **events per unit of time.** In the context of electrical circuits the "events" are cycles of alternating current or pulses, and the unit of time is the second or a subunit such as millisecond (1 ms = $10^{-3}$ s), microsecond (1 μs = $10^{-6}$ s), or nanosecond (1 ns = $10^{-9}$ s).

The unit of frequency is the hertz (Hz), and 1 Hz is equal to 1 cycle per second. In many applications where the Hz is too small a unit, kilohertz (1 kHz = $10^{3}$ Hz) and megahertz (1 MHz = $10^{6}$ Hz) are used.

**FIGURE 18–1**
Time relationships on a pulse.

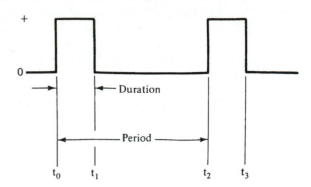

Time that people are accustomed to is measured in hours, minutes, and seconds; it is keyed to the rotation of the earth. In fact, most early astronomical observatories were constructed not to conduct scientific research therein but to provide the local monarch with the means to keep time. It was the furtherance of commerce and navigation that prompted governments to build observatories; the scientific researches of the astronomer were a benefit.

When dealing with non-dc electrical waveforms, we must often consider frequency, period, and duration, all of which are related to time. We have already defined frequency as events per unit of time. Figure 18–1 shows a pulse waveform. The frequency of the pulse train is the number of pulses that occur in 1 second.

Period and duration are both time measurements, but they are slightly different. **Period** is defined as the time interval between identical features or points on successive waveforms. The concept of period assumes **periodicity;** that is, the waveform is not a single event but repeats itself with predictable regularity. We could then, specify the period of the waveform in Fig. 18–1 as interval $t_2 - t_0$ or, identically, as $t_3 - t_1$. The **duration,** on the other hand, is the time interval during which the pulse is high, that is, interval $t_1 - t_0$ or interval $t_3 - t_2$ in Fig. 18–1. Note that duration is generally not considered a feature of sine waves, in which case we consider only period, but is a property of pulse waveforms.

Frequency and period in repetitive wave trains are related by the expression

$$F = \frac{1}{T}$$

(18–1)

where    $F$ = the frequency in hertz (Hz)
$T$ = the period in seconds (s)

■ **Example 18–1**

Calculate the frequency of a waveform that has a 225-μs period.

*Solution*

$$F = 1/T \tag{18-1}$$

$$= \frac{1}{225 \ \mu s \times \dfrac{1 \ s}{10^6 \ \mu s}}$$

$$= \frac{1}{2.25 \times 10^{-4} \ s} = \textbf{4444 Hz}$$

---

## 18–4        PERIOD MEASUREMENT

Period is measured by comparing the event being tested with a known clock or time base. In an earlier chapter we studied the digital counter in which the precision pulses from a time base section were passed through a gate to a decimal counting assembly under command from the event being measured. Counter input circuitry can be adjusted to trigger the gate open and closed, using identical points on successive waveforms.

Alternatively, you can often use an oscilloscope time base to graphically measure the period of a waveform displayed on a CRT screen. Figure 18–2 shows two pulses displayed on an oscilloscope screen. The period is 4.4 horizontal divisions, and since the time base was set to 2 ms/div, it represents a time interval of

$$4.4 \ \text{div} \times \frac{2 \ ms}{\text{div}} = 8.8 \ ms$$

The limitation on the accuracy of this technique is the accuracy of the oscilloscope time base. If greater precision is required, then an external time base marker generator can be used on the second vertical channel. This generator will place precisely timed, marked spikes on the CRT screen, by which time can be measured.

**FIGURE 18–2**
Measuring period on an oscilloscope screen.

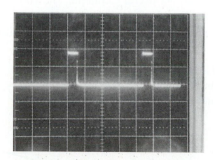

## 18–5    FREQUENCY MEASUREMENT

Frequency measurements are part of many different electronic and scientific activities. In communications, for example, the Federal Communications Commission (FCC) requires that transmitters be kept on frequency, within a certain error or tolerance band. An AM band broadcast station, for example, must operate on an actual frequency that is within ±20 Hz of its assigned frequency. FM band broadcast stations, on the other hand, must stay within ±2000 Hz of their assigned frequency. Class-D citizens' band transmitters must be within ±0.005% (i.e., 50 parts per million, 50 ppm). Some two-way radio services must maintain transmitter operating frequency to within ±2.5 ppm.

In most noncommunications electronics, frequency measurement requirements may be greater or less than those in any given communications service. Alternatively, there may be a special measurement problem. An example of a "problem" measurement is sometimes encountered in servicing electronic musical instruments; we may be required to measure a tone to a resolution of two or three decimal points of a hertz, for example, 680.24 Hz. Many instruments will measure 680 Hz with little difficulty but will not be able to resolve the fractional cycle (i.e., the 0.24) portion. A **high-resolution counter** will do the job by actually measuring the *period* and then taking its reciprocal in calculatorlike circuits before displaying the result on the digital readout. The period of 680.24 Hz, for example, is $1.47 \times 10^{-3}$ s, or 1.47 ms. A period counter will accurately measure this period, to the required resolution, if it has a 1-mHz time base.

## 18–6    ROUGH FREQUENCY MEASUREMENTS

Several techniques may be used to make *approximate* frequency measurements and are useful in those cases where there is no pressing requirement for great precision.

One popular rough measurement is to use the oscilloscope time base to find the period and then take the reciprocal of period [Equation 18–1] to find the frequency. In Fig. 18–2, for example, the period was 8.8 ms, so the frequency is

$$F = 1/T$$
$$= 1/(8.8 \times 10^{-3} \text{ s}) = 114 \text{ Hz}$$

Another useful technique, shown in Fig. 18–3, is to connect a calibrated oscillator to an oscilloscope input (the horizontal input is used in Fig. 18–3) and apply the unknown signal to the other input. When the oscillator frequency is adjusted to be an integer (i.e.,

**FIGURE 18–3**
Using Lissajous figures to measure frequency.

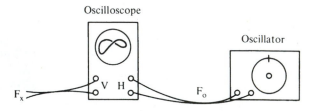

1, 2, 3, . . . , $n$) multiple or submultiple of the unknown frequency, then a stable Lissajous pattern will be seen on the screen of the CRT.

There are two practical limitations on the use of Lissajous patterns. One, of course, is the precision with which the oscillator is calibrated. In many instances the oscillator frequency calibration is quite good, while in others it is poor. It is often recommended that the oscillator frequency be measured on a frequency counter, but if a counter is available, then the use of Lissajous patterns is an exercise in the absurd.

The other limitation involves the inability to lock the oscillator on the unknown. The Lissajous pattern will probably *rotate* at a frequency equal to the *difference* between the two frequencies. Trying to make the pattern stand still long enough to make a measurement is a lot like trying to nail Jello to the wall.

It is, however, possible to use the rotation frequency of the trace, and its direction of rotation, to interpolate the correct frequency. The direction of rotation will tell you whether the unknown frequency is above or below the oscillator frequency. You can obtain this information experimentally by rocking the dial of the oscillator back and forth slowly across the correct frequency while observing the Lissajous pattern on the CRT screen. For example, let's say that we have a 1:1 Lissajous pattern rotating in the direction that indicates that the unknown frequency $F_x$ is *less* than $F_0$, the known frequency. The pattern rotates 26 times per minute. Thus, the error is

$$\text{Error} = \frac{26 \text{ cycles}}{\text{min}} \times \frac{1 \text{ min}}{60 \text{ s}}$$

$$= \frac{26 \text{ cycles}}{60 \text{ s}} = \textbf{0.43 Hz}$$

Since the unknown frequency is less than the indicated frequency, we subtract the error from the indicated frequency:

$$F_x = 562 \text{ Hz} - 0.43 \text{ Hz}$$
$$= 561.57 \text{ Hz}$$

The usefulness of Lissajous patterns diminishes rapidly as the frequencies involved increase. This problem is due to the severity of certain practical matters, such as the resolution of the RF generator frequency dial.

Another technique for making rough frequency measurements is to use a communications receiver to tune in the unknown signal. We may then obtain a rough indication of this frequency by reading the receiver's dial.

The dial markings of some professional-grade high-frequency (3-to-30 mHz) communications receivers can be quite good. Some can boast dial calibration that is accurate to within ±100 Hz. Certain modern digital models currently on the market not only match the resolution and accuracy of the older models but will surpass it by an order of magnitude.

Most communications receivers, however, have relatively poor frequency dial calibrations. Even these receivers provide a rough indication, and they can be made substantially more accurate if the dial markings can be compared with an external frequency standard or with one of the National Institute of Standards and Technology (NIST) standard frequency-time radio broadcast stations. These stations, using the call letters WWV and WWVH, operate on frequencies of 2.5, 5.0, 10.0, 15.0, and 20.0 mHz. Alternatively, the dial markings may be compared with a 100- or 1000-kHz crystal oscillator that was calibrated against WWV and WWVH.

The crystal oscillator method can be made very accurate, as will be demonstrated in a later section of this chapter, but only when the receiver dial markings are used only as a guide. Another source must be the standard of measurement.

## 18–7     ABSORPTION WAVEMETERS

Figure 18–4 shows two elementary forms of absorption wavemeters. These instruments provide only the roughest indication of frequency but are useful for some purposes, such as determining whether or not the signal a transmitter is producing is a harmonic, a parasitic, or the fundamental.

Both types of absorption frequency meters consist of a resonant tank circuit—that is, a fixed inductance and a variable capacitor. The capacitor is equipped with a dial that is calibrated in units of frequency. The coil in both cases is mounted in a manner that allows it to be coupled closely with the circuit or component being tested. The capacitor, however, must be mounted inside a shielded enclosure so that it will not be detuned by stray capacitances (including the operator's own hand). As a result, it is common to find the capacitor mounted inside an aluminum shield box. The inductor, then, is mounted on an insulated socket on the shield box.

The circuit in Fig. 18–4(a) is the classic absorption wavemeter. The capacitor, the inductor, and a lamp form a series-resonant tank circuit, so it will have a very low impedance at the resonant frequency. Energy from the tank circuit of the radio transmitter being tested is induced into the wavemeter circuit through the coil, which is placed in close proximity to the transmitter's tank. This will set up an oscillating current in the wavemeter tank circuit that causes the lamp to light up. It takes a relatively large amount of current to light the lamp, so it is prudent to use low-current types in this application.

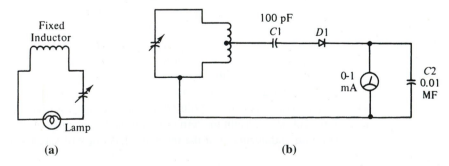

**(a)**                                                    **(b)**

**FIGURE 18–4**

(a) Simple absorption wavemeter, (b) rectified absorption wavemeter.

In some cases, such as vacuum tube transmitters that have a plate or cathode milliammeter, the lamp may be omitted. If the inductor of the wavemeter is coupled to the inductor in the transmitter's *plate* tank, then some of the plate energy will be absorbed by the wavemeter. When the two tank circuits are tuned to exactly the same frequency, the amount of energy absorption increases dramatically. If the dial of the wavemeter is tuned slowly back and forth across the transmitter tank's resonant frequency, there will be a momentary inflection of the milliammeter caused by the dramatic change in energy absorption level.

A more sensitive variation on the wavemeter circuit is shown in Fig. 18–4(b). In this case the tank of the wavemeter is used in the parallel resonant mode. Signal from an impedance-matching tap on the inductor is coupled through capacitor $C1$ to the diode, where it is rectified. The pulsating dc waveform is filtered and smoothed by capacitor $C2$. This produces the steady direct current required by the meter.

The milliammeter in Fig. 18–4(b) is deflected an amount that is roughly proportional to the strength of the applied signal. This wavemeter, then, will also function as a *relative* field strength indicator to aid in tuning the transmitter.

The use of an impedance-matching tap on the inductor is necessary because of the low impedance of the diode rectifier. Ordinarily, tank circuits should have a high impedance so that the $Q$ is increased. Also, recall that maximum power transfer between two circuits, or between portions of a single circuit, occurs only when the source and load impedances are equal. Furthermore, the low impedance of the diode would load the tank circuit too much, thereby lowering the $Q$. The effect of lowered $Q$ is to broaden the tuning, thereby obscuring the resonant point. Some models have a link-coupled secondary coil going to the diode circuit, but in this case the main tuning inductor is used as a tapped autotransformer.

The absorption wavemeter is a very crude device, no matter how well it is constructed. It is also limited in sensitivity. A form of active wavemeter is the **dip oscillator.** If an LC tank circuit is used as the frequency-determining network in a variable oscillator, then a dip oscillator type of wavemeter is made. Older models used vacuum tubes as the oscillating active element, so they were called "grid dip oscillators." Modern dip instruments are called either **base** or **gate dip oscillators,** depending upon whether a bipolar or JFET transistor is used as the active element. One model, using a tunnel diode, was sold under the name of "tunnel dipper."

A typical dip oscillator is shown in Fig. 18–5. This circuit is a variable-frequency oscillator (VFO) of the Colpitts type. The inductor from the resonant tank circuit is mounted externally to the cabinet housing the rest of the circuit. This is done so that the coil may be easily coupled to the tank circuit or component being tested. The gate signal of the oscillator is monitored by microammeter $M1$.

Any dip oscillator works because energy from its tank circuit is absorbed by any closely coupled conductor. If that conductor is a resonant tank circuit, however, the amount of energy absorbed will be greatest when the oscillator frequency and the resonant frequency of the tank are the same.

Capacitor $C1$ is coupled to a dial calibrated in units of frequency. The operator holds the inductor on the dip meter close to the circuit being tested and *slowly* tunes $C1$ until a pronounced dip in signal is noted. The resonant frequency of the tank circuit is then read from the dial of the dip oscillator.

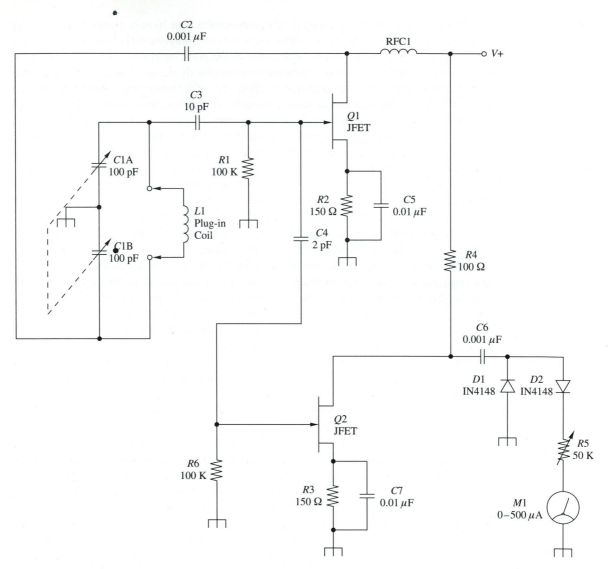

**FIGURE 18–5**
Typical dip oscillator.

The chief advantage of the dip oscillator over the simple absorption wavemeter is that it is active and thus more sensitive (i.e., it will detect even weak signals). The dip oscillator may also be used on turned-off circuits in which no power is present, because the active element provides the power needed for the test.

The dip oscillator may also be used with energized circuits in a manner that is very similar to the techniques using absorption wavemeters. In that case we are using the dip oscillator as an **oscillating detector.** A pair of earphones is plugged into jack $J1$, and it takes the place of the meter movement. The coil is coupled to the circuit under test, and $C1$ is adjusted. When the dipper frequency gets close to the frequency of the signal in the tank circuit being tested, an audio tone is heard in the headphones. This tone is a heterodyne between the oscillator's frequency and the unknown frequency in the tank circuit. Whenever two frequencies mix together in a nonlinear circuit, extra frequencies are generated equal to their sum and difference. In other words, if two frequencies are mixed together, *at least* four frequencies will be produced. If we call one frequency $A$ and the other $B$, then the four output frequencies will be $A$, $B$, $A + B$, and $A - B$. Of these, the first three will be higher than audio, but the difference frequency will be an audio tone. The exact pitch of the tone, then, is equal to the difference between the two frequencies, so when the two frequencies are equal (i.e., when $A = B$), then the difference frequency ($A - B$) will be zero. The two frequencies are said to be in "zero-beat" when they are equal.

None of the absorption techniques is a precise method for measuring frequency. They are used only in cases where simplicity and low cost count for more than accuracy. In transmitter work, for example, we might want to know whether an output signal is due to fundamental frequency oscillation of an amplifier, harmonic generation (i.e., $2F$, $3F$, etc.) or to a parasitic that is not harmonically related to the fundamental frequency $F$.

## 18–8    LECHER WIRES

**Lecher wires** are a mechanical device used for determining the *wavelength* of signals in the range of frequencies between approximately 50 and 500 mHz. Above that limit, similar techniques involving slotted lines (Section 18–9) are used. The slotted line works at frequencies well over 1000 mHz.

A set of Lecher wires consists of a pair of large-diameter parallel conductors, such as copper tubing or copper wire of #12 size or larger. The length of the conductors must be at least one-half wavelength at the lowest frequency to be tested. This partially explains the lack of popularity for Lecher wires below 50 mHz; that is, lengths become too great.

Parallel to the conductor pair will be a measuring scale calibrated in units of length, usually meters, centimeters, and millimeters. A few units might be found that use English units of length.

Some Lecher wires have an RF voltmeter (i.e., a diode detector and a microammeter) such as that shown in Fig. 18–6. This is not strictly necessary, however, if the RF source being tested is a vacuum tube model equipped with a dc milliammeter in the plate or cathode circuits.

A shorting bar between the two conductors is designed so that it can be moved along the length of the conductor from one end to the other. The effective electrical length of the Lecher wires is the distance between the input end and the shorting bar. When one is performing a frequency measurement, the shorting bar is placed initially at the 0-cm mark and is then moved slowly down the length of the parallel conductors while the RF voltmeter is being monitored. When the bar is moved to the point representing one-half wavelength of the applied frequency, then a pronounced dip is noted in the RF voltmeter reading. The distance between this point and the 0-cm point is equal

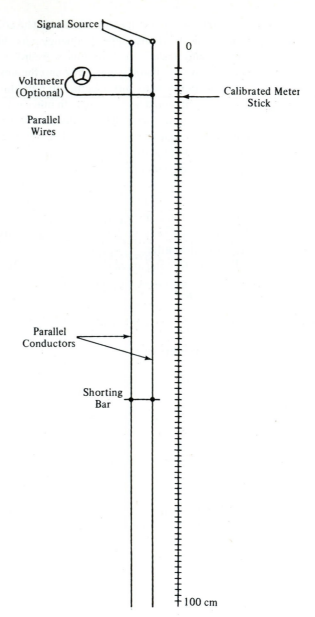

**FIGURE 18–6**
Lecher wires for measuring
VHF/UHF frequencies.

Signal Source

Voltmeter
(Optional)

Parallel
Wires

Parallel
Conductors

Shorting
Bar

Calibrated Meter
Stick

0

100 cm

to the half wavelength of the applied signal. Some authorities claim that better accuracy results from noting this distance and then continuing the short further down the wires until another dip is noted. Make an average of the two distances (i.e., 0 cm to first dip and first dip to second dip), and then use the resultant average distance in the standard formula:

$$F = \frac{C}{2\lambda} \tag{18-2}$$

where    $C$ = the speed of light ($3 \times 10^8$ m/s)
        $F$ = the frequency in hertz (Hz)
        $\lambda$ = the wavelength in meters (m)

This equation is also written as

$$F_{(Hz)} = \frac{3 \times 10^8}{2\lambda_{(m)}} \tag{18-3}$$

Since most frequency measurements using Lecher wires are in the over-50-MHz range, hertz units are too cumbersome, especially given the lack of accuracy inherent in the Lecher wire technique. We may rewrite Equation (18–3) to use megahertz instead of hertz, as follows:

$$F_{(MHz)} = \frac{300}{2\lambda_{(m)}} = \frac{150}{\lambda_{(m)}} \tag{18-4}$$

and if English units are used (i.e., *inches*),

$$F_{(MHz)} = \frac{11,811}{2\lambda_{(in)}} = \frac{5906}{\lambda_{(in)}} \tag{18-5}$$

In either case, however, we may find an expression for using the full distance between 0 cm and the second dip by eliminating 2 from the denominator in Equations (18–4) and (18–5).

Accuracies on the order of 1000 ppm are possible, if the operator of the Lecher wire set takes great care and makes several measurements that are averaged together. Great care in observing the nulls, care in recording the distance data, loose coupling to the source (tight coupling broadens the null), and averaging several measurements are the key to accuracy in Lecher wire use.

18–9       **SLOTTED LINE MEASUREMENTS**

A technique similar to the Lecher wires is the slotted line detector of Fig. 18–7. The slotted line is a special section of coaxial transmission line that has a slot in its outer conductor running most of its length. An RF detector probe is fitted through the slot and is

**FIGURE 18–7**
(a) Slotted line, (b) close-up of
detector probe.

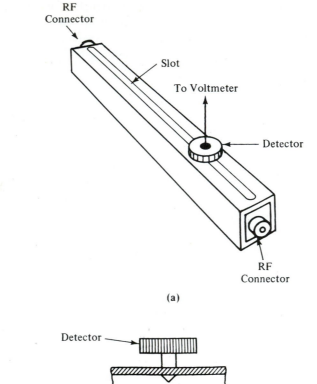

**(a)**

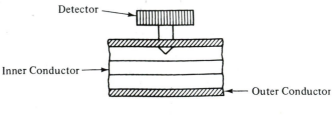

**(b)**

coupled to the center conductor by mutual inductance. The output of the detector is a dc
voltage proportional to the intensity of the signal in the line at that point.

Some slotted line assemblies have a knob and worm-gear mechanism that allows precise adjustment of the detector position to any position along the slot. A metric scale calibrated in millimeters tells the operator how far along the line the probe is located. A
micrometer dial is usually provided, because it is necessary to have a more precise resolution of distances at the higher frequencies covered by the slotted line, compared with
the Lecher wire set.

A slotted line is used in a manner that is nearly identical to that of the Lecher wires.
Reasonably accurate wavelength (and hence frequency) determinations are possible at
any frequency above the frequency at which the slot is one-half wavelength. The detector will locate nodes and antinodes along the length of the line, and nodes are one-half
wavelength apart, as are antinodes. The nodes, incidentally, are usually more sharply

defined and so are best used to make the measurement. The formulas given earlier for Lecher wires also apply to the slotted line.

## 18–10    NATIONAL INSTITUTE OF STANDARDS AND TECHNOLOGY (NIST) RADIO BROADCAST TIME/FREQUENCY SERVICES

The National Institute of Standards and Technology (NIST), formerly the National Bureau of Standards (NBS), an element of the U.S. Department of Commerce, operates precise time and frequency radio broadcast stations. Table 18–1 summarizes the radio station services.

Beginning with broadcasts in 1923, the NIST time/frequency services now include three radio broadcast stations (WWV, WWVH, and WWVB); the GOES satellite; Loran-C navigation signals; and a telephone time service available to computer users via modem.

### 18–10–1    NIST Radio Stations WWV, WWVH, and WWVB

Radio stations WWV and WWVH are high-frequency (short-wave) stations operating on 2.5, 5.0, 10, and 15 MHz. (Radio station WWV also operates on 20.0 MHz.) The operating frequencies of these radio stations are held to a **received accuracy** of $1 \times 10^{-7}$ and a time accuracy of 1 to 10 ms. The **transmitted accuracy** of the short-wave stations are about $1 \times 10^{-11}$, with day-to-day deviations normally on the order of $1 \times 10^{-12}$ (i.e., 1 part in 1000 billion!). Reasons for the difference between received and transmitted accuracies are various radio propagation effects.

The short-wave stations (WWV and WWVH) operate on a wide range of frequencies in order to accommodate the varied radio propagation conditions over the course of a single day, from day to day and from season to season. Sunspots play a role in short-wave propagation, so a spread of frequencies also helps accommodate differing conditions over the (approximately) 11-year sunspot cycle. Most users will be able to find at least one WWV/WWVH frequency that will yield good reception, depending on when and where the reception is attempted. In most locations away from the transmitter sites, the 10-, 15-, and 20-MHz frequencies will yield best results during the day, while the frequencies 2.5, 5.0, and 10.0 MHz might be better at night.

Radio WWVB operates in the very low frequency (VLF) band at a frequency of 60 kHz. The accuracy of the WWVB frequency is $1 \times 10^{-11}$, and the frequency accuracy is 0.100 to 1.0 ms. Reception is good at most points in the continental United States; European and South American users can usually obtain good results from this station as well.

Radio WWV, located at Fort Collins, Colorado (40°40′49.0″ N, 105°46′27.0 W), has been on the air since March 1923 (at first from Beltsville, Maryland). WWVH has been on the air since November 1948 and is located near Kekaha on the island of Kauai, Hawaii (21°59′26.0″N, 159°46′00.0 W). The VLF station, WWVB, has been on the air since July 1956 and is also located at Fort Collins. These stations operate at power levels from 2.5 to 13 kW. The short-wave stations operate using amplitude modulation, while WWVB operates with a special binary AM in which each cycle will shift from a normal level for LOW to a level +10 dB higher for HIGH.

**TABLE 18–1**
Summary of radio broadcast services

| Characteristics and services | WWV | | | WWVH | | WWVB |
|---|---|---|---|---|---|---|
| Date service began | March 1923 | | | November 1948 | | July 1956 |
| Geographical coordinates | 40°40′49.0″ N 105°02′27.0″ W | | | 21°59′26.0″ N 159°46′00.0″ W | | 40°40′28.3″ N 105°02′39.5″W |
| Standard carrier frequencies | 2.5 and 20 MHz | 5, 10, and 15 MHz | | 2.5 MHz | 5, 10, and 15 MHz | 60 kHz |
| Power | 2500 W | 10,000 W | | 5000 W | 10,000 W | 13,000 W |
| Standard audio frequencies | 440 (A above middle C), 500, and 600 Hz | | | | | |
| Time intervals | 1 pulse/s, minute mark, hour mark | | | | | s, min |
| Time signals: voice | Once per minute | | | | | |
| Time signals: code | BCD code on 100-Hz subcarrier, 1 pulse/s | | | | | BCD code |
| UT1 corrections | UT1 corrections are broadcast with an accuracy of ±0.1 s | | | | | |
| Special announcements reports | OMEGA reports, geoalerts, marine storm warnings, global positioning system status reports | | | | | |

The range of services provided by WWV and WWVH include the following:

Standard frequencies

Standard time intervals

Time announcements

UT1 time corrections

Binary coded decimal (BCD) time code for correcting automated and computer-based clocks

Geophysical alerts (including propagation reports)

Marine storm warnings

OMEGA (a VLF navigation system operating in the 10-to-16-kHz range) navigation system status reports

Global Positioning System (GPS) status reports

The hourly broadcast formats for WWV and WWVH are shown in Figs. 18–8 and 18–9, respectively. Station identification is performed during minute 1 on WWV and during minute 59 on WWVH (about 15 seconds before the WWV announcement commences). So that listeners can readily identify each station's voice reports, WWV uses a male announcer, and WWVH uses a female announcer.

The normal transmissions from WWV and WWVH differ slightly during the course of those minutes not reserved for other services. At WWV, the minute format (see inset to Fig. 18–8) includes an initial 500-Hz tone or special announcement for 45 seconds, followed by a "silent" period in which only a 1-second "tick" is heard. At the end of 60 seconds (52.5 to 60.0 seconds) a voice time announcement gives the time in Universal Coordinated Time (UTC).

[At one time, UTC was known as Greenwich Mean Time (GMT) because the zero reference was the Royal Observatory at Greenwich, England. In military and scientific circles, UTC is sometimes called "Z" or "Zulu" time for reasons that are a bit obscure but that are based on the need for radio phonetic and telegraphic shorthand. Local time varies from UTC by a fixed number of hours (see Fig. 18–10). Eastern Standard Time (EST) in the United States is −5 hours relative to UTC (except during Daylight Savings Time months when it is −4 hours). Time zones to the west of Greenwich are negative (i.e., earlier) a certain number of hours (depending on location), while locations to the east of Greenwich are at positive hours (i.e., later).]

During the following minute, WWV transmits a 600-Hz tone for the first 45 seconds, followed by the "silent tick" and a UTC voice announcement from 52.5 to 60.0 seconds.

WWVH follows a slightly different protocol. It transmits a 600-Hz tone from 00 to 45 seconds in minute 1, followed by a UTC voice announcement (45 to 52.5 seconds) and a "silent tick" (52.5 to 60.0 seconds). The second minute places a special announcement or 500-Hz tone from 00 to 45 seconds, a UTC time announcement from 45 to 52.5 seconds, and a "silent tick" from 52.5 to 60.0 seconds.

Additional signals include a tone at the beginning of each hour (a 0.8-s, 1500-Hz tone from both WWV and WWVH). At the beginning of each minute, WWV transmits a

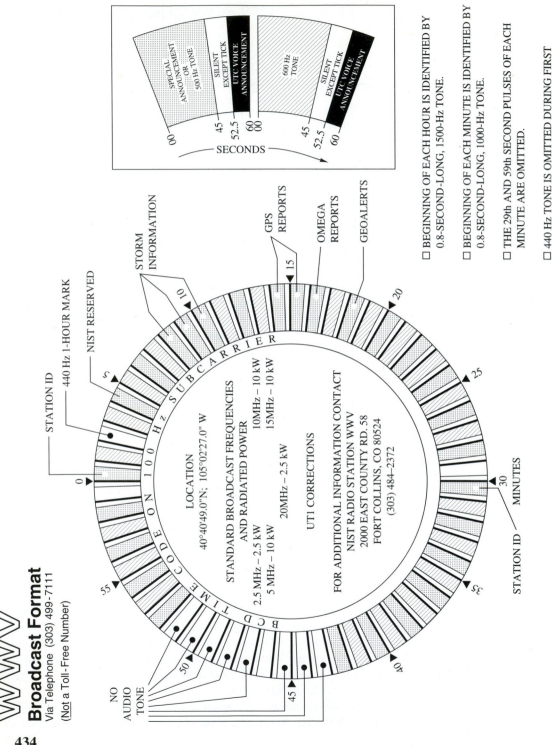

## WWV
**Broadcast Format**
Via Telephone (303) 499-7111
(Not a Toll-Free Number)

STATION ID

440 Hz 1-HOUR MARK

NIST RESERVED

STORM INFORMATION

GPS REPORTS

OMEGA REPORTS

GEOALERTS

STATION ID

MINUTES

NO AUDIO TONE

### LOCATION
40°4049.0"N; 105°0227.0" W

STANDARD BROADCAST FREQUENCIES AND RADIATED POWER

2.5 MHz – 2.5 kW    10MHz – 10 kW
5 MHz – 10 kW    15MHz – 10 kW
20MHz – 2.5 kW

UT1 CORRECTIONS

FOR ADDITIONAL INFORMATION CONTACT
NIST RADIO STATION WWV
2000 EAST COUNTY RD. 58
FORT COLLINS, CO 80524
(303) 484–2372

SPECIAL ANNOUNCEMENT OR 500 Hz TONE

SILENT EXCEPT TICK

UTC VOICE ANNOUNCEMENT

600 Hz TONE

SILENT EXCEPT TICK

UTC VOICE ANNOUNCEMENT

SECONDS

☐ BEGINNING OF EACH HOUR IS IDENTIFIED BY 0.8-SECOND-LONG, 1500-Hz TONE.

☐ BEGINNING OF EACH MINUTE IS IDENTIFIED BY 0.8-SECOND-LONG, 1000-Hz TONE.

☐ THE 29th AND 59th SECOND PULSES OF EACH MINUTE ARE OMITTED.

☐ 440 Hz TONE IS OMITTED DURING FIRST HOUR OF EACH DAY.

**FIGURE 18–8**
The hourly broadcast schedules of WWV.

434

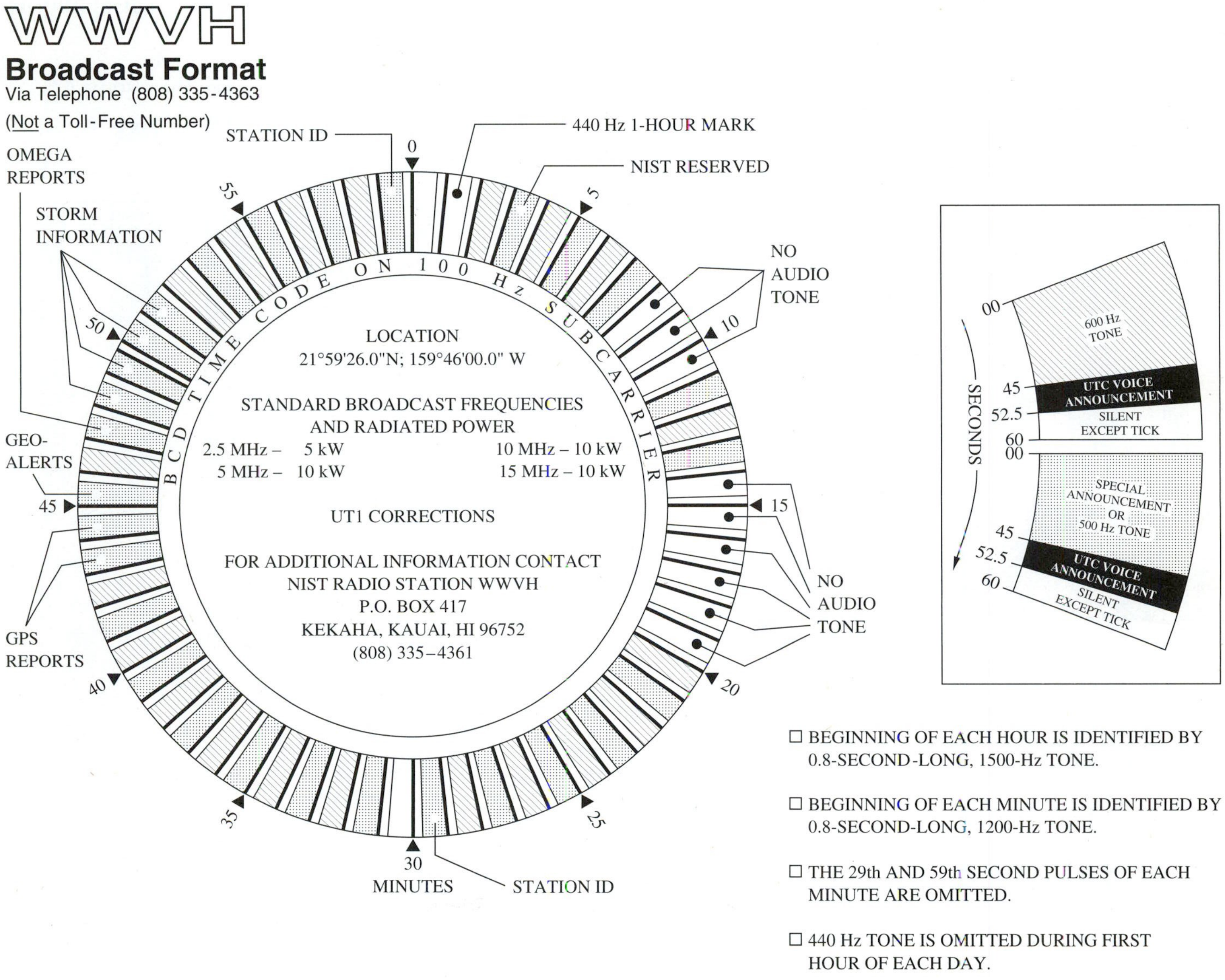

# WWVH

**Broadcast Format**

Via Telephone (808) 335-4363

(Not a Toll-Free Number)

OMEGA REPORTS

STORM INFORMATION

STATION ID

440 Hz 1-HOUR MARK

NIST RESERVED

NO AUDIO TONE

NO AUDIO TONE

STATION ID

MINUTES

GPS REPORTS

GEO-ALERTS

UT1 CORRECTIONS

FOR ADDITIONAL INFORMATION CONTACT
NIST RADIO STATION WWVH
P.O. BOX 417
KEKAHA, KAUAI, HI 96752
(808) 335–4361

STANDARD BROADCAST FREQUENCIES
AND RADIATED POWER

| | | |
|---|---|---|
| 2.5 MHz – | 5 kW | 10 MHz – 10 kW |
| 5 MHz – | 10 kW | 15 MHz – 10 kW |

LOCATION
21°59'26.0"N; 159°46'00.0" W

B C D   T I M E   C O D E   O N   1 0 0   H z   S U B C A R R I E R

SECONDS

00   45   52.5   60   00   45   52.5   60

600 Hz TONE

UTC VOICE ANNOUNCEMENT

SILENT EXCEPT TICK

SPECIAL ANNOUNCEMENT OR 500 Hz TONE

UTC VOICE ANNOUNCEMENT

SILENT EXCEPT TICK

☐ BEGINNING OF EACH HOUR IS IDENTIFIED BY 0.8-SECOND-LONG, 1500-Hz TONE.

☐ BEGINNING OF EACH MINUTE IS IDENTIFIED BY 0.8-SECOND-LONG, 1200-Hz TONE.

☐ THE 29th AND 59th SECOND PULSES OF EACH MINUTE ARE OMITTED.

☐ 440 Hz TONE IS OMITTED DURING FIRST HOUR OF EACH DAY.

**FIGURE 18–9**

The hourly broadcast schedules of WWVH.

435

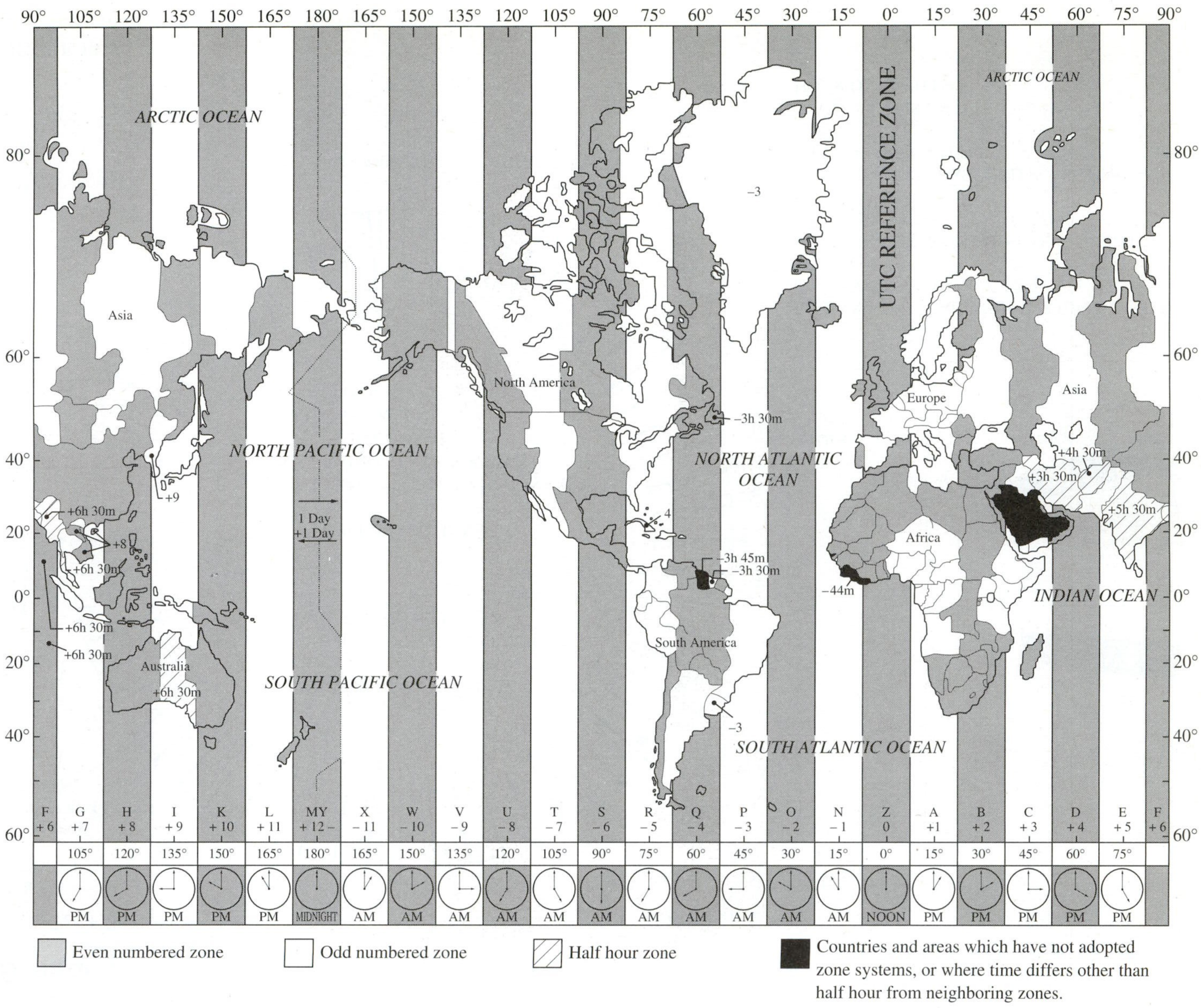

**FIGURE 18-10**

Standard time zones of the world and their relationship to UTC.

**FIGURE 18–11**
Format of WWV and WWVH seconds pulses.

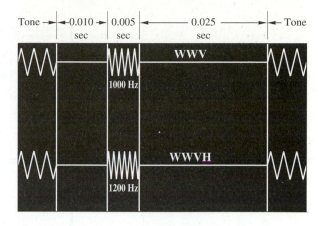

0.8-s, 1000-Hz tone, while WWVH transmits a 0.8-s, 1200-Hz tone. Both WWV and WWVH omit the 29th and 59th second pulses of each minute. The format is shown graphically in Fig. 18–11.

A standard audio tone of 440 Hz (musical "A above middle C") is broadcast at 2 seconds past the hour at WWVH and 3 seconds past the hour at WWV, except during the first hour of each broadcast day.

## 18–10–2     BCD Coded Signals on WWV and WWVH

All three NIST radio stations operate a binary coded decimal (BCD) coding scheme for transmitting time and time correction data to automated and computer-based clock systems. The code (Fig. 18–12) is broadcast on a 100-Hz subcarrier signal. (A few shortwave receiver models will not receive this signal because of inherent high-pass filtering in the AM detector circuits.) The pulse code is a modified IRIG H format. The UTC time code presents a 1-PPM frame reference marker ("R"), BCD year and time-of-year data, 6-PPM position markers, and 1-PPS index markers. The UTC data are linked to an atomic cesium beam clock. Navigators, astronomers, and certain others need a slightly less stable clock, so the UT1 scale, corrected from UTC, is provided. UT1 sign and numerical corrections are provided by the BCD transmission.

## 18–10–3     Geophysical Services from WWV and WWVH

Radio propagation forecasts are available over the NIST radio stations. At 18 minutes after the hour on WWV and at 45 minutes after the hour on WWVH, information about current radio propagation conditions and a forecast for the next 24 hours are broadcast. The information includes the 1700 UTC solar flux data from Ottawa, Ontario, Canada, and the Boulder "A" and "K" indexes. The A index is a number between 0 and 400 and is based on data taken over the previous 24 hours. The K index is a number generally less than 10 and is based on the current 3 hours data.

Quiet geomagnetic conditions are indicated by an A index of 10 or less. Higher values of the A index indicate high ionospheric absorption; the effect is especially severe in

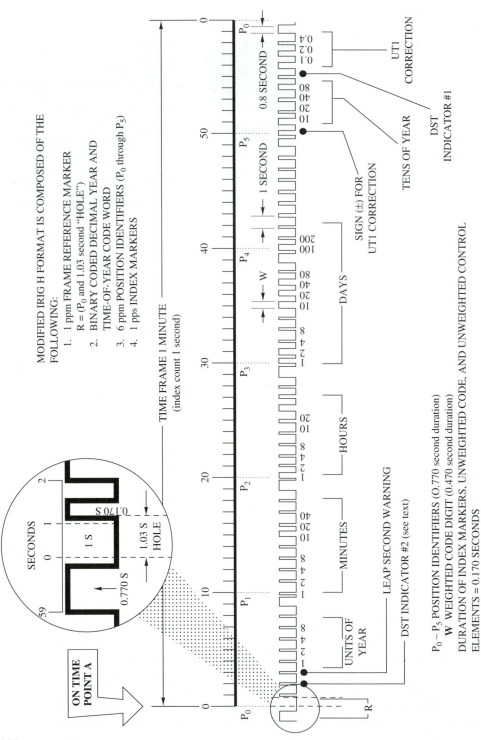

**FIGURE 18–12**
WWV and WWVH time code format.

438

high-latitude paths (such as the infamous North Atlantic path). When the A index reaches the vicinity of 100, there is severe disruption of short-wave communications, and the visible aurora borealis appears in high northern latitudes.

The K index is based on a smaller-range scale. When the geomagnetic field is quiet and normal, the value of K is 0. When K is about 1 or 2, the geomagnetic field conditions are unsettled, while values of $K \geq 3$ indicate possible auroral conditions. The K index is variable with geography. The value given is for Boulder, Colorado; locations to the south of Boulder have lower K values, and locations north of Boulder have higher K values.

The NIST stations WWV and WWVH also broadcast somewhat subjective observations of solar activity and geomagnetic conditions using terms such as "very low," "low," "moderate," "high," or "very high" for solar activity, and "quiet," "unsettled," and "active" for the geomagnetic conditions.

At one time WWV and WWVH used a radio conditions scale that featured a letter and number combination. The letters "W", "U," and "N" were used for "warning," "unsettled," and "no-warning," while conditions were broadcast on a scale of 1 to 9 (with better propagation conditions being indicated by higher numbers). An N9 reading meant good long distance ("DX") listening, while W2 meant nonexistent DX. The information was broadcast in 400-Hz modulated CW at 19 and 49 minutes after the hour. These scales are sometimes found in older texts and papers on propagation, so their meanings should be understood by anyone who reads back-literature in their studies.

### 18–10–4    Using WWV and WWVH Broadcasts for Frequency Measurement

Within the limits of received accuracy, the WWV and WWVH signals can be used to compare local frequency standards to a moderate degree of accuracy. Marker oscillators (operating at decade frequencies from 100 kHz to 10 MHz) and the time base oscillators for lower-priced digital frequency counters can be calibrated using the WWV and WWVH broadcasts off-the-air. However, for higher-accuracy applications a better standard is usually needed.

Crystal marker oscillator output frequencies are usually subharmonics of one or all WWV and WWVH frequencies. When the two signals (i.e., WWV/WWVH and the marker oscillator) are both present at the receiver antenna input stage, a heterodyne beat note is created, and it has a frequency equal to the difference between the two input frequencies. The frequency-setting trimmer capacitor on the marker oscillator (usually a small variable capacitor in series or parallel with the piezoelectric crystal) is then adjusted to minimize the difference between two frequencies, as evidenced by the reduction of the beat note tone to near 0 Hz. (This process is called "zero beating.")

The meaning of "zero beat" differs from one observer to another, and these differences are a source of measurement error. A normal human adult, with average hearing acuity, can hear tones down to 25 to 30 Hz. This fact, coupled with lower-end bandwidth limits in the receiver audio and detector circuits, limit the aural zero beat to about ±30 Hz, and possibly to ±100 Hz in many people.

If the receiver used to pick up WWV and WWVH is equipped with a signal strength meter (S-meter), then it can often be used as a zero beat detector down to difference frequencies below human hearing acuity. As the two input frequencies come closer together, the beat note drops in frequency and diminishes to zero when the two input frequencies are identical. It is nearly impossible to attain an exact 0-Hz beat note, but it is quite possible to obtain beat notes of less than 0.5 Hz, as indicated by a slow "bobbing" of the S-meter back and forth.

The accuracy of the zero beat method of calibrating oscillators is improved considerably by using the *highest* WWV/WWVH frequency that produces a *strong* signal in your locality. Weak signals are not as useful because certain other (mostly propagation) errors become more important. The differences between results obtained using weak and strong signals suggest that it is prudent to wait to make a calibration until one of the higher frequencies becomes loud in your area—a function of radio propagation conditions between the transmitter site and your reception site.

Let's assume, for purposes of an example, that a receiver with an S-meter or an external phase detector is used to zero-beat a 500-kHz time base oscillator against WWV and that a ±2-Hz accuracy in adjustment is obtained. If we zero-beat against 2.5 MHz WWV, then we are using the 5th harmonic of the 500-kHz oscillator signal to create the beat note. The accuracy of the crystal oscillator frequency is the beat note accuracy divided by the harmonic factor, so a ±2-Hz error translates to ±2/5 Hz, or ±0.4 Hz. Thus, the actual oscillator frequency will probably lie between 499,999.6 and 500,000.4 Hz. As striking as this accuracy seems, if the ±2-Hz error results were obtained by beating against the 15-MHz WWV signal, then the *30th* harmonic of the 500-kHz oscillator is used, so the actual error is ±2/30 Hz or ±0.0667 Hz, or nearly a full order of magnitude better.

## 18–10–5     The VLF (60-kHz) WWVB Broadcasts

NIST radio station WWVB offers what is the best frequency comparison possible because its 60-kHz operating frequency is less subject to propagation effects that induce errors into the short-wave broadcast signals. The VLF station is less susceptible to fading, ionospheric disturbances, and phase shift errors than the HF stations. In addition, the propagation phase shifts for WWVB are more predictable and less variable than those for WWV and WWVH. The transmitted accuracy of the WWVB 60-kHz frequency is better than 1 part in 100 billion ($1 \times 10^{-11}$), with a day-to-day deviation of 5 parts in 1000 billion ($1 \times 10^{-12}$). The BCD time code accuracy is approximately 0.1 ms. These accuracies can be improved by the use of statistical averaging techniques such that the transmitted and received accuracies approach each other.

Modulation of the WWVB 60-kHz signal is a modified amplitude modulation that allows only two levels to account for the two-level binary codes being transmitted. A 10-dB downward amplitude (Fig. 18–13) occurs at the start of each second, creating a negative-going leading edge for each pulse. The full amplitude is restored 0.2 second later if the first bit is a 0, 0.5 second if it is a 1, and 0.8 second later if a position marker occurs. A 45° carrier phase shift occurs from 10 minutes to 15 minutes after the hour and serves to identify WWV (other stations outside the United States also operate on 60 kHz). The same basic data are available on WWVB as on WWV and WWVH, as detailed in Fig. 18–14.

**FIGURE 18–13**
10-dB downward amplitude.

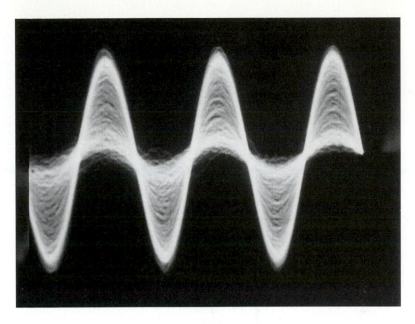

## 18–10–6    WWVB Comparator Receivers

Figure 18–15 shows the block diagram for a 60-kHz WWVB **frequency comparator receiver,** several brands of which are on the market. The antenna for the receiver is a shielded loop tuned to 60 kHz. In most cases the loop antenna includes either an untuned or a tuned 60-kHz preamplifier stage. The receiver consists of a 60- to 100-dB, 60-kHz tuned amplifier. Although LC tuning predominates, the 60-kHz frequency also allows the use of RC active filters and gyrator circuits for frequency selection. The raw output signal of the amplifier chain is usually made available to the user through a front or rear panel connector on the receiver.

Following the high-gain amplifier is a level detector, or similar stage, that produces a square wave output for each cycle of the 60-kHz received signal. This signal will then be used to drive a **phase detector** of some sort that will compare the WWVB signal to the signal being tested. In older receivers LC phase detector circuits are used, but in all modern receivers digital coincidence detectors followed by integrators are used for phase detection.

The input frequency being measured is usually passed through a divide-by-$N$ frequency counter before being applied to the other input port of the phase detector. In some cases a second divide-by-$N$ counter is also used between the 60-kHz output of the receiver and the primary phase detector input.

The output of the phase detector coincidence gate is a series of pulses of widths that are proportional to the differences in frequency or phase between the 60-kHz WWVB and test signal. If the coincidence gate pulses are integrated, then a dc level is produced that

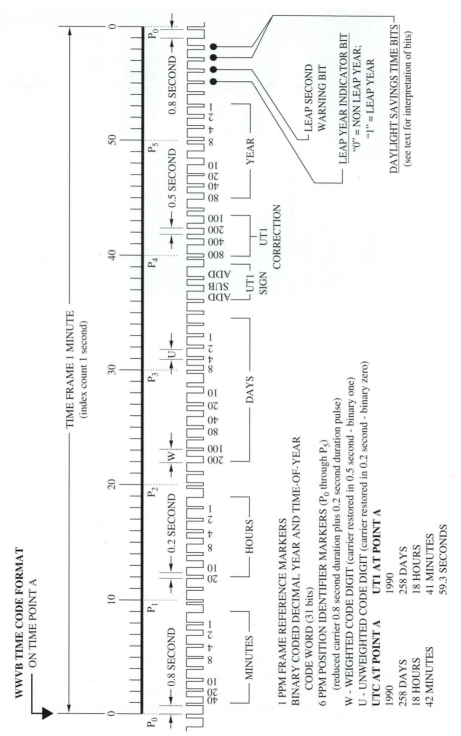

**FIGURE 18–14**
WWVB time code format.

442

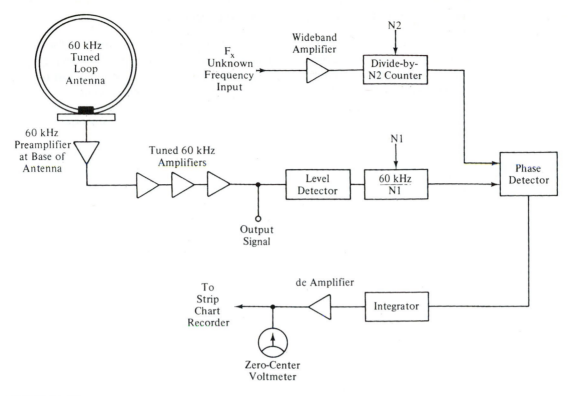

**FIGURE 18–15**
Block diagram of a 60-kHz comparator receiver.

is proportional to the frequency or phase difference, including sign. This voltage is displayed on a zero-center meter, a strip chart recorder, or a computer.

The unknown frequency can be measured by adjusting the divide-by-$N$ counters until a zero difference frequency indication is obtained. The frequency of the unknown signal is given by

$$F_X = \frac{(N2)\ (6 \times 10^4\ \text{Hz})}{N1} \qquad \textbf{(18–2)}$$

where    $F_X$ = the frequency of the unknown signal
$N1$ = the division ratio applied to the 60-kHz WWVB signal
$N2$ = the division ratio applied to the unknown signal

The use of the comparator receiver is a little clumsy for ordinary, short-term frequency measurements, but it is ideal for long-term measurements such as the frequency stability of local frequency standards.

**18–11**      ## ATOMIC FREQUENCY AND TIME STANDARDS

Until recent decades the earth's rotation was used as the reference standard for the measurement of frequency and time. Frequency was sometimes measured by having the signal run a synchronous electric clock, so that by observing how the clock lost or gained time we would know whether or not the frequency was precisely accurate.

But the earth's rotation is not the constant it was once believed to be. In fact, there are even seasonal variations thought to correlate highly with the waxing and waning of the polar ice caps.

In 1964, the Twelfth General Conference of Weights and Measures redefined time and frequency in terms of the **atomic clock,** which contains a phase locked loop (PLL) in which a reference frequency source is either a rubidium gas cell device or an atomic beam of thallium or (more frequently) cesium atoms. These clocks lock the frequency of a 5- or 10-mHz crystal oscillator to the oscillation frequency of the atoms. It has been found that accuracies of $\pm 7 \times 10^{-12}$ are possible.

The atomic resonator works on the natural resonant frequency of the atom used, which when excited will oscillate (i.e., wobble) at an extremely stable fixed frequency. Examples of oscillation frequencies of some atoms often used in atomic clocks are as follows:

| | |
|---|---|
| Hydrogen maser | 1420.405751 mHz |
| Cesium beam | 9192.631770 mHz |
| Rubidium gas cell | 6834.682608 mHz |

**18–12**      ## BASIC FREQUENCY METERS

Two basic types of frequency meters are in common use: **heterodyne** and **counter.** Although each class has numerous different examples, there are some commonalities among them all.

Figure 18–16 shows the elementary heterodyne frequency meter. In all heterodyne techniques, the unknown signal is mixed with a known signal from a precisely cali-

**FIGURE 18–16**
Heterodyne frequency meter.

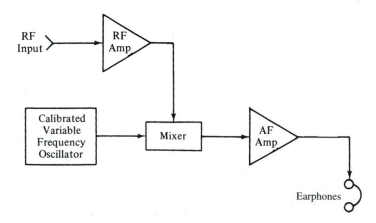

**FIGURE 18–17**
Frequency counters. (a) Hand-held,
(b) laboratory and workbench
model [(a) courtesy of MFJ).

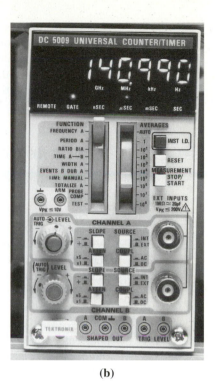

**(a)**                                   **(b)**

brated frequency source. The frequency of the beat note gives the frequency difference
between the two sources. If the source is zero-beat with the unknown signal, then the
unknown will be equal to either the known frequency or one of its harmonics or subharmonics.

Frequency counters, examples of which are shown in Figs. 18–17(a) and 18–17(b),
were discussed in detail in an earlier chapter. You may find it prudent to go back and
review appropriate sections at this time. Additional commercial examples are given in
Chapter 24, "Radio Transmitter Measurements."

The example shown in Fig. 18–17(a) is a small, hand-held digital frequency counter
intended for measurement of radio transmitter frequencies in the field. It features two frequency ranges (200 MHz and 600 MHz) in order to maximize the usefulness of the digital display resolution, as well as two gate times (0.1 and 1.0 s). The example in Fig.
18–17(b) is a laboratory and workbench model that will measure two frequencies or periods independently, as well as the ratio and difference between the two frequencies. It will
also measure time interval and event totals.

## 18–13    USING FREQUENCY METERS

In the sections to follow we will demonstrate common techniques used to measure frequency. Most of these techniques are found in common usage, even though the use of the
digital counter has become predominant over other methods in the past few years. The

counter has always been the most convenient for many types of measurement, but until technological spinoffs from the space program reduced IC costs while improving performance, digital counters were too costly for most people. It was not unusual to see the 500-mHz counter that is needed by some two-way radio shops costing over $5000, yet today, models offering the same or better performance can be purchased for considerably less than $1000.

## 18–14     USING FREQUENCY COUNTERS

The simplest method for checking a transmitter output frequency with a digital counter is shown in Fig. 18–18. A short whip antenna is connected to the input of the counter through a short piece of coaxial, or directly through a coaxial connector mounted on the base of the whip. The transmitter is then keyed on, and the radiated signal is picked up by the whip antenna. If the counter is close enough to the transmitter (less than 500 ft, depending upon power level of the transmitter), sufficient signal will be applied to the input of the counter to cause triggering. The transmitter output frequency is then displayed on the counter readout. Caution is required to measure an unmodulated signal; be duly cautious per the instructions on triggering given in Chapter 6.

Because the signal radiated in this test may interfere with other users of the same or adjacent frequencies, it is a proper test only on the citizens' band and in the amateur (ham) radio bands. Even there, it is legal only after listening on the frequency to make sure that nobody is using it.

To prevent interference with normal communications traffic, it is necessary to measure the frequency by keying the transmitter into a nonradiating dummy load. One version of this technique is shown in Fig. 18–19. Here the transmitter output power is absorbed by the dummy load. A coaxial "tee" connector and a 50-$\Omega$ attenuator feed a portion of the signal to the frequency counter. The attenuator is used to reduce the transmitter's power level to a point where it will not harm the counter input stages. If insufficient attenuation is used for the power level applied, then damage to the counter *will* result.

**Caution: Never feed the output of a transmitter directly into a frequency counter unless instructed to do so by the manufacturer.**

Another legal method for measuring transmitter output frequency is shown in Fig. 18–20. This technique employs a "throughline" RF sampler that picks off a small portion of the transmitter's output signal and routes it to the digital frequency counter. Some models of counter are equipped with a throughline, and they are the *only* types that may be connected directly to the transmitter's output.

## 18–15     TRANSFER OSCILLATORS

A **transfer oscillator** is a stable, precisely calibrated, low-frequency oscillator that is loosely coupled to a communications receiver through a small-value capacitor or a wire "gimmick"—a few turns of insulated wire wrapped around the receiver antenna lead.

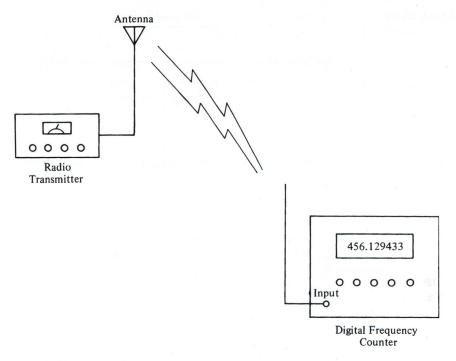

**FIGURE 18–18**

"On-the-air" method of frequency counting (*Note:* This technique is illegal except on the citizens' band and when monitoring on-the-air broadcasting stations.)

**FIGURE 18–19**

Correct method for measuring transmitter frequency with a counter.

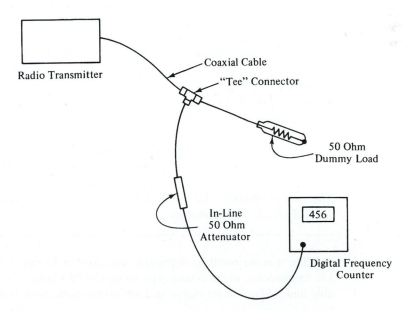

**FIGURE 18–20**
Shielded pick-off box.

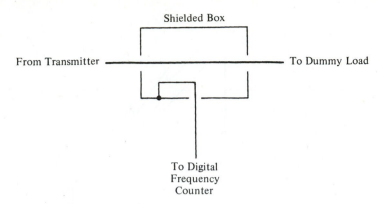

There are several different ways to use a transfer oscillator in making frequency measurements. It is assumed that we know the approximate frequency of the transmitter from information given as to its channel of operation.

In one transfer oscillator technique the oscillator is a precision device operating on a subharmonic of the transmitter frequency. The frequency of the transfer oscillator is adjusted until it zero-beats with the unknown signal in the receiver. In this case a harmonic of the transfer oscillator beats against the unknown signal. It is, therefore, important to know which harmonic is being used. We can then multiply the transfer oscillator frequency by the harmonic factor to obtain the actual transmitter frequency.

### ■ Example 18–2

It is necessary to measure the frequency of an AM broadcast station that operates on an assigned frequency of 680 kHz. A precision square wave generator is used as the transfer oscillator. The frequency resolution of the generator can be accurately interpolated to 0.1 Hz. It is found that the zero beat occurs between the broadcast signal and the generator when the generator frequency is 136,001.1 Hz. What is the transmitter's frequency?

### *Solution*

We know that 136 kHz is 1/5 of 680 kHz, so we must be using the 5th harmonic of the generator to zero-beat the transmitter signal. Therefore,

$$F_x = 5 \times 136{,}001.1 \text{ Hz}$$
$$= 680{,}005.5 \text{ Hz} = \textbf{680.0055 kHz}$$

The error is 680,005.5 Hz − 680,000 Hz, or 5.5 Hz, well within the ±20-Hz specification for AM broadcast transmitters.

The transfer oscillator technique described in Example 18–2 is limited to relatively low frequencies, where it is easy to obtain variable-frequency oscillators that have believable dial calibration. At higher frequencies the usefulness of the transfer oscillator deteri-

**FIGURE 18–21**
Transfer oscillator method of measuring frequency.

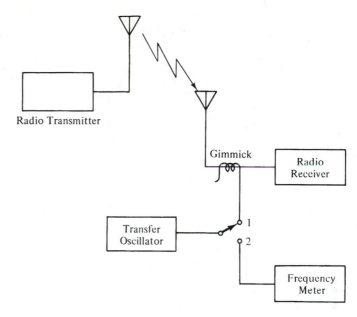

orates markedly, because the transfer oscillators are less well behaved. At these frequencies it might be more useful to use the method detailed in Fig. 18–21.

In this method the output signal from the transfer oscillator is fed through a switch. When the switch is in position 1, the oscillator is adjusted to zero-beat with the unknown signal. After this has been done, the switch is transferred to position 2, and a frequency meter is used to read the frequency of the transfer oscillator. Again, we may be using a subharmonic of the actual frequency; again, we must know which harmonic is being used. When the frequency is high, it is even more important to know exactly which harmonic is being used. If extremely low order (i.e., 1/10 or less) harmonics are used, then it is possible for two harmonics to be simultaneously audible in the receiver output.

This technique is also useful when the transfer oscillator operates on the same frequency as the transmitter, but the transmitter is at a distant location from the measurement site.

## 18–16     CRYSTAL MARKER OSCILLATORS

A technique that is related to the transfer oscillator is shown in Fig. 18–22. Instead of a transfer oscillator, which is a *variable-frequency* oscillator, we use a precision quartz crystal oscillator that operates on a single, standard frequency or on several harmonically related frequencies. This type of oscillator uses a piezoelectric quartz crystal as the frequency-control element. In general, crystal oscillators are more accurate and more stable than either RC or LC types. The primary requirement for a crystal oscillator, apart from accuracy and stability, is that the output waveform be rich in harmonics. A square wave meets this requirement, but any nonsinusoidal waveform will probably be useful up to the 15th or so harmonic, or even further.

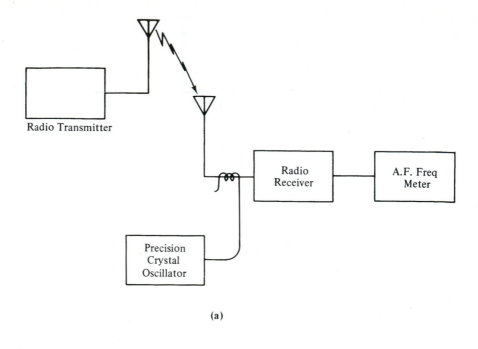

(a)

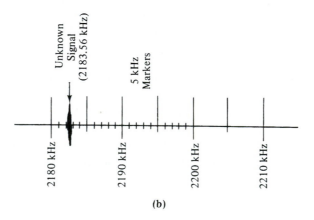

(b)

**FIGURE 18–22**
(a) Crystal oscillator method of frequency measurement, (b) receiver dial points during measurement.

Examples of crystal oscillators often used as marker oscillators are shown in Fig. 18–23. In both cases the frequency-determining network is crystal controlled. Also, the frequency dividers for both cases are the same, the only difference being the oscillator IC device. The frequency division consists of IC decade dividers that produce outputs of 1 mHz, 100 kHz, and 10 kHz, and a divide-by-2 that produces 5-kHz output.

Two different oscillator circuits are shown in Fig. 18–23. The main circuit [Fig. 18–23(a)] is preferred by the author. It uses a special IC device called the MC4024

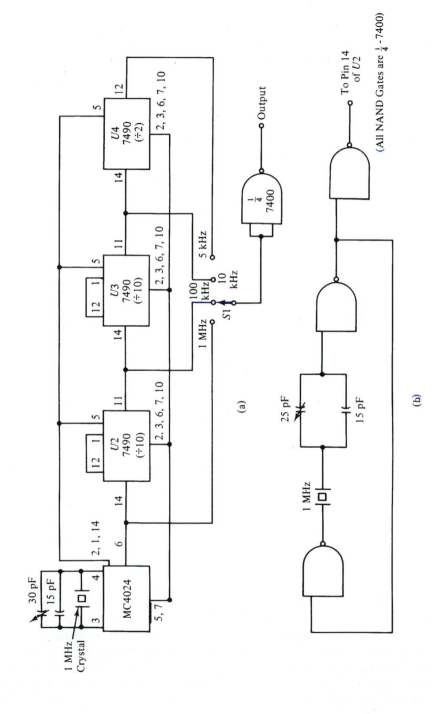

**FIGURE 18–23**

(a) Crystal marker oscillator using digital ICs for frequency division, (b) alternate oscillator circuit.

dual VCO. The variable capacitor across the crystal is used to set the oscillator frequency.

An alternate crystal oscillator is shown in Fig. 18–23(b). This circuit uses three TTL NAND gates that are part of a single type 7400 IC device. Again, a variable trimmer capacitor is used to fine-tune the oscillator frequency.

The crystal oscillators shown in Fig. 18–23 can be used for moderately precise applications. Where high precision is required, however, a commercially built crystal oscillator can be used instead. A wide variety of such oscillators are sold. Some are in a fancy cabinet complete with a regulated power supply, whereas others are less fancy, being built into a metal-cased, but epoxy-encapsulated, package.

Two designs are prevalent. In older models the crystal and often some other components are housed inside a temperature-controlled oven that usually operates at a temperature of 75°C. The later approach, and the one favored in many modern designs, is to temperature-compensate the circuit so that changes in temperature are not reflected as changes in output frequency. Such oscillators are called **temperature-compensated crystal oscillators,** or **TCXO.**

The output of a crystal oscillator is loosely coupled to the receiver's antenna circuitry through a low-value capacitor or a wire gimmick. This superimposes the marker signal on top of the other signals without overloading the receiver input.

There are two or more different ways to use the crystal oscillator markers to make frequency measurements with the aid of a communications receiver. The first, and most crude, is to use the marker oscillator to temporarily calibrate the receiver's dial over a small range near the unknown frequency. Most higher-quality communications receivers have a movable dial pointer to make this type of recalibration easier. For example, let's say that we want to calibrate a receiver to exactly 7.0 mHz. If we connect a 1-mHz crystal oscillator to the antenna input circuitry, then the marker signals will appear every 1 mHz up the band. We then locate the marker that falls very close to the 7-mHz point, and then move the dial cursor until it lies exactly over the 7.0 mark. For those frequencies close to 7 mHz, the dial will be relatively accurate. But one limitation to this method is that receiver dials rarely have better than 100-Hz resolution, and even that level is accomplished only in the higher-priced models.

Some authorities feel that the best technique for using the crystal marker oscillator is to allow it to heterodyne against the unknown signal and then to measure the frequency of the audio beat note. An audio frequency meter is used for this purpose in Fig. 18–22, and it is usually connected to the receiver's audio output jack or terminals. For very close resolution it may be connected directly to the output of the AM detector so that a near-dc beat note can be obtained.

The receiver is calibrated in the same manner as in the previous example. It is usually best to start with a higher-frequency marker and then work down, so that you can home in on the unknown frequency and thereby avoid mistakes. Figure 18–22(b) shows a simulated receiver dial. Also shown are the markers from the crystal oscillator and the location on the dial of the unknown signal. The generator is first adjusted to produce a 1-mHz output. It is found that the unknown falls between 2 and 3 mHz. The marker generator is then readjusted to produce a 100-kHz signal, and it is found that the marker lies between 2100 and 2200 kHz. Next, the frequency of the marker is set to 10 kHz, and it is found that the unknown is between 2180 and 2190 kHz. Since the unknown, in this case, falls closer to one of the 10-kHz markers, it is unnecessary to go to any lower marker frequency.

Recall that we are seeking a *beat note*. If the unknown is anywhere near the midpoint between two adjacent markers, there will be *two* beat notes of similar frequency. This ambiguity is cleared up only by going to a lower-frequency marker in order to gain some additional frequency resolution. In our present case, however, the unknown is close to one of the 10-kHz points, so no ambiguity will exist.

The beat note produced in the audio output is found to be exactly 3562 Hz, as measured on an audio frequency meter. This means that the unknown frequency is located at a point 3562 Hz away from the 2180-kHz marker. Since we are able to observe on the receiver dial that the unknown is located on the high side of the 2180-kHz point, we know that the actual frequency is 2180 kHz + 3562 Hz, or 2,183,562 Hz. Once again, it is desirable, and considered good engineering practice, to measure frequency three times and then average the results. Tune the receiver away from the unknown, and then go back again to the point where you believe the dial to be tuned exactly on the unknown frequency. Do this several times to take advantage of the possibility that errors will "add out."

The crystal marker oscillator may be calibrated by zero-beating its output against WWV, WWVH, WWVB, or a local atomic frequency standard or other precise source. If aural zero beats are used, be aware that the typical error, or uncertainty, is 30 Hz (more in observers with poor hearing acuity). This error is *additive*. If you have a 30-Hz uncertainty when you beat the oscillator to a primary frequency standard, and then another 30-Hz error when you beat the oscillator signal against an unknown, then the actual uncertainty may be as high as 30 + 30, or 60 Hz. And that is the best that will be possible for an aural zero beat immediately after calibration of the oscillator; it will deteriorate further as the oscillator ages. Use nonaural zero-beat methods for any but trivial cases.

## SUMMARY

1. Low-precision audio frequency measurements can be made by using Lissajous patterns on an oscilloscope screen.
2. Several types of frequency standards are used: atomic clocks, gas cell devices, crystal oscillators, and the NBS radio broadcast stations (WWV, WWVH, and WWVB).
3. Two basic categories of frequency meters are used in common practice: heterodyne and counter.
4. Three types of heterodyne frequency meters are used: direct, transfer oscillator, and crystal oscillator.
5. It is not legal to measure a transmitter's frequency off the air except in the citizens' band, amateur radio bands, and standard broadcast bands.

## RECAPITULATION

Now go back and try to answer the questions at the beginning of the chapter. When you are finished, answer the questions and work the problems given below. Place a mark beside each problem or question that you cannot answer, and then go back to the text and reread appropriate sections.

## QUESTIONS

1. Define **frequency**.
2. The *units* commonly employed in electronics to measure frequency are _____, _____, and _____.
3. Define **period** and **duration** as applied to electrical waveforms. Note the *difference* between these terms.
4. Units commonly employed in electronics to measure time are _____, _____, _____, and _____.
5. Give the "power-of-10" relationship between the units expressed in (a) Question 2 and (b) Question 4.
6. Period can be measured roughly on an _____ and more precisely on a _____ counter.
7. Describe a method whereby audio range frequencies can be measured to two or three decimal points of a single cycle.
8. List three devices that can be used to make rough frequency measurements in the high-frequency range.
9. List the operating frequencies of WWV and WWVH.
10. List the operating frequency of WWVB.
11. Draw the circuits for two types of absorption wavemeters.
12. List three types of absorption wavemeters.
13. Describe in your own words the phenomenon on which the grid dip oscillator depends.
14. List two types of transmission line devices capable of rough frequency measurements in the VHF, UHF, and microwave ranges.
15. What accuracy is possible if the devices of Question 14 are used?
16. WWV frequencies are maintained to an accuracy of _____%.
17. Audio tones broadcast by WWV and WWVH are _____ Hz, _____ Hz, and _____ Hz, and each tone has a duration of _____ s.
18. A female voice is heard making announcements on a 15-mHz standard time station. The call letters are _____.
19. When using WWV or WWVH to zero-beat a digital counter's time base oscillator, it is prudent to use the _____ frequency that produces a strong local signal.
20. The best accuracy possible when one is adjusting an oscillator frequency using the aural zero-beat technique is ± _____ Hz.
21. List two types of atomic frequency standards.
22. List a common gas cell type of frequency standard.

23. An atomic beam clock frequency standard has a potential accuracy of _____%.
24. List two basic forms of frequency meters.
25. What is the approximate resonant frequency of a cesium beam? (Use four significant figures.)
26. In the United States only three types of transmitters may have their frequency checked off the air: _____, _____, and _____.
27. _____ and _____ oscillators are examples of heterodyne frequency meters.
28. Only counters equipped for _____ monitoring can be connected directly to the transmitter. All others must be protected by connecting them to the transmitter through an _____.
29. The _____ technique of frequency measurement uses a local subharmonic oscillator to zero-beat the unknown frequency. The output of the local oscillator can be measured by an external frequency meter.
30. A frequency measurement device consisting of an inductor, a capacitor, and a lamp in series is called an _____ _____.
31. A _____ frequency meter zero-beats a calibrated signal source with an unknown signal.

## PROBLEMS

1. What is the frequency of a repetitive waveform that has a period of 856 ns?
2. What is the period of a 2.2250-mHz signal?
3. An oscilloscope is used to measure the *period* of a sine wave. The oscilloscope time base is set to 1 μs/div, and the sine wave is observed to occupy 8.6 horizontal divisions. What is the period?
4. What is the frequency of the signal measured in Problem 3?
5. A period counter reads 4.60211 μs. What is the frequency of the applied signal?
6. A 1:1 Lissajous pattern rotates in a direction that indicates that the unknown frequency is above the known frequency. Find the unknown frequency if the known frequency is 91 Hz and the pattern rotates 12 times per minute.
7. A 100-kHz crystal oscillator is zero-beat against 10 mHz WWV, the aural method being used. The resultant oscillator frequency will be between _____ Hz and _____ Hz.
8. A WWVB comparator receiver is adjusted so that $N1 = 64$ and $N2 = 1024$ when the output meter indicates the zero-beat condition. What is the unknown frequency?
9. A pair of Lecher wires is used to measure the operating frequency of a radio transmitter. The first null is noted at a distance of 28.2 cm, and the second null is at a distance of 56.45 cm. What is the probable frequency of operation?
10. A crystal oscillator and a communications receiver are used to measure a frequency near 9 mHz. A 1098-Hz beat note is heard and the unknown frequency is below the 9-mHz marker. Calculate the unknown frequency.
11. A transmitter is required to be within ±0.004% of its assigned frequency. Express this in ppm.

**12.** A transmitter must be operated on an actual frequency of 27,085 kHz ±0.005%. The actual measured frequency is found to be 27,086,023 Hz. Is the transmitter within the specified tolerance?

# BIBLIOGRAPHY

Carr, Joseph J. *Elements of Electronic Communications.* Reston, VA: Reston Publishing Co., 1977.

––––––. *Receiving Antenna Handbook.* San Diego, CA: HighText Publications, 1993.

––––––. *Practical Antenna Handbook,* 2nd ed. Blue Ridge Summit, PA: TAB/McGraw-Hill, 1994.

*NIST Special Publication 432 (Revised 1990): NIST Time and Frequency Services.* Boulder, CO: National Institute of Standards and Technology, June 1991.

*Reference Data for Radio Engineers,* 5th ed. Indianapolis, IN: ITT–Howard W. Sams, Inc., 1968.

# 19

# Measurements on
# Untuned Amplifiers

## 19–1      OBJECTIVES

1. To learn to make gain and power output measurements on untuned amplifiers.
2. To learn to make harmonic distortion measurements on amplifiers.
3. To learn intermodulation distortion measurement techniques.
4. To learn the operation of principal audio test instruments.

## 19–2      SELF-EVALUATION

Before studying the material in this chapter, try to answer the questions given below. These questions test your knowledge of the subject. If you cannot answer a particular question, place a check mark beside it, and then look for the answer as you read the text.

1. Total harmonic distortion (THD) is measured with a low-distortion oscillator, an ac voltmeter, and a _____ filter.
2. Define **voltage gain** in your own words.
3. Write the transfer function for a voltage amplifier.
4. In frequency response tests of a voltage amplifier, we plot _____ vs. _____.

## 19–3      UNTUNED AMPLIFIERS

Many electronic circuits and instruments contain untuned amplifiers. These amplifiers are generally thought of as *audio* amplifiers, but that assumption is erroneous. The audio amplifier is merely one type of untuned amplifier. A video amplifier is also untuned, but its frequency response may extend to 50 or 100 mHz, certainly far above "audio." At the other end of the spectrum, an ECG or EEG amplifier used in medical electronic instruments is certainly untuned, but its −3-dB frequency response extends only to 100 Hz on the high end and as low as 0.05 Hz (decidedly subaudio) on the other end of the range.

## 19–4     CONVERTING TO DECIBEL (dB) NOTATION

Many measurements require power, voltage, or current data but are best expressed in a unit called **decibels,** denoted **dB.** The dB is a *ratio* (and nothing more), so it is well suited for expressions of amplifier gain, attenuator loss, frequency response, and anything else where we must *compare* two levels. For electrical power we express decibels as

$$dB = 10 \ \log_{10} \frac{P1}{P2} \qquad \qquad (19\text{–}1)$$

where          db = the gain or loss in decibels
       $P1$ and $P2$ = the two power levels
          $\log_{10}$ refers to the base-10 logarithms

### ■  Example 19–1

Calculate the power gain of an amplifier if 100 mW (0.100 W) applied to the input produces an output power of 14 W.

*Solution*

$$\begin{aligned}
dB &= 10 \ \log_{10}(P1/P2) \qquad \qquad (19\text{–}1)\\
&= 10 \ \log_{10}(14/0.100)\\
&= 10 \ \log_{10}(140)\\
&= (10)(2.15) = \textbf{21.5}
\end{aligned}$$

Similarly, we can express voltage and current gains in decibels using Equations (19–2) and 19–3), respectively:

$$dB = 20 \ \log_{10} \frac{E1}{E2} \qquad \qquad (19\text{–}2)$$

$$dB = 20 \ \log_{10} \frac{E1}{E2} \qquad \qquad (19\text{–}3)$$

### ■  Example 19–2

Calculate the voltage gain of an amplifier if 100 mV causes an output potential of 14 V.

*Solution*

$$\begin{aligned}
dB &= 20 \ \log_{10}(E1/E2) \qquad \qquad (19\text{–}2)\\
&= 20 \ \log_{10}(14/0.1)\\
&= 20 \ \log_{10}(140)\\
&= (20)(2.15) = \textbf{42.9}
\end{aligned}$$

Examples 19–1 and 19–2 show the use of decibel notation for power and voltage *gains;* that is, the numerator was made the higher number. But what about *losses?* A convention that has been adopted is to place the *output* figure in the numerator so that the lesser number is in the numerator in the case of a loss. This will result in exactly the same number of decibels, but the fact that it is a loss shows up because the result is negative. If the "output/input" protocol is followed, then it is always possible to distinguish losses and gains in making system calculations.

■ **Example 19–3**

An attenuator drops a 10-V signal to 50 mV. Calculate the loss in decibels.

*Solution*

$$
\begin{aligned}
\text{dB} &= 20 \ \log_{10}(E1/E2) \qquad\qquad\qquad\qquad \textbf{(19–2)} \\
&= 20 \ \log_{10}(0.05 \ \text{V}/10 \ \text{V}) \\
&= 20 \ \log_{10}(0.005) \\
&= (20)(-2.30) = \textbf{–46}
\end{aligned}
$$

Note that in Example 19–3 the *loss* caused the result of the decibel equation to be negative, where it had been *positive* for *gains.* The same relationship is also found in current and power ratios.

**19–5**　　**SPECIAL dB-BASED SCALES**

Decibel scales are merely *ratios* between two sets of numbers. Sometimes, however, it is convenient to reference all measurements within a system, or even an entire industry, to a standard level that is usually referred to as "0 dB." If this is done, then we may simply add up all the gains and losses and then note how the result compares with the amount of signal required to do the job at hand. We may then use the data to predict signal levels and certain aspects of performance and to make changes in the design to take care of any problems.

In several areas of electronics we see these special dB scales. Some of the most common are as follows:

**Common dB Scales**

1. **dBm** (used in RF measurements). This scale defines 0 dBm as 1 mW dissipated in a 50-$\Omega$ load.
2. **VU** (audio **volume units**). The VU scale is a decibel-based scale in which 0 VU is defined as 1 mW dissipated in a 600-$\Omega$ load at 1000 Hz.
3. **dB** (obsolete). This scale defined 0 dB as 6 mW dissipated in a 500-$\Omega$ load at 1000 Hz.
4. **dBmv** (TV coaxial cable antenna systems). This scale, the only one of the four that is not defined in terms of power, defines 0 dBmv as 1 mV across 75 $\Omega$.

The user of a special dB-based scale may never be required to convert a measurement back to the basic units, that is, milliwatts or millivolts. The dB data will usually suffice. If a conversion is necessary, then it is a simple matter to substitute in the standard power or voltage level, as given above, and then solve the appropriate decibel equation for the unknown, such as it may be.

**19–6**     **VOLTAGE GAIN MEASUREMENT**

The gain of a voltage amplifier is defined by the amplifier transfer function:

$$A_v = \frac{E_{out}}{E_{in}} \tag{19–4}$$

where     $A_v$ = the voltage gain (dimensionless)
$\quad\quad\quad E_{out}$ = the output signal voltage
$\quad\quad\quad E_{in}$ = the input signal voltage

We can *measure* the gain of an amplifier by measuring the input signal voltage and the output signal voltage and then taking the ratio expressed by Equation (19–4). Figure

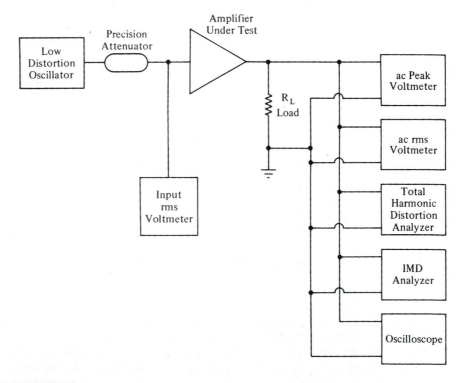

**FIGURE 19–1**

Test equipment setup for performing most audio tests and measurements.

19–1 shows the equipment configuration used to perform many of the tests that we will discuss in this chapter. We are assuming that ac (i.e., audio) amplifiers are being tested, but the same procedures also hold true for dc amplifiers if a dc signal source and dc voltmeters are substituted for their ac counterparts.

### Procedure

1. Adjust the attenuation, oscillator output level, and amplifier gain controls for a convenient output level. (In some cases the input voltage, total harmonic distortion point, or output level may be further specified.)
2. Measure the input and output signal voltages.
3. Make the gain calculation using Equation (19–4).
4. If dB gain figures are desired, then use a modified version of Equation (19–2):
   $dB = 20 \log_{10}(A_v)$

The voltage gain measurement is usually made with the amplifier gain at maximum (if it is variable) and the input signal level either at a point well below the level where output clipping is noted on the oscilloscope or at the point where the specified THD level is reached.

## 19–7  OUTPUT POWER MEASUREMENT

Hi-fi output power was one measurement that was often used to confuse buyers with unrealistically high numbers, rather than to convey anything truly coherent and meaningful about the product being sold. We used to mutter something about "taffy being distributed . . ." when salespeople talked of hi-fi amplifier power levels. There were seemingly innumerable abuses of measurement procedure that produced amplifier ratings in "watts" but that were essentially meaningless.

It would seem that such a "basic" measurement would be very cut and dried. But we often found the following types of "wattage" ratings: (a) rms watts at 1% THD, (b) rms watts at 5% THD, (c) peak watts, (d) peak-to-peak watts (!?!), and (e) something called "music power." Of these, only the rms watts rating accurately reflects the ability of the amplifier to do work, that is, move a loudspeaker cone. But note well that most amplifiers produce considerably more output power if the maximum permissible THD is 5% than they will if only 1% or 2% THD is permitted.

Let's consider these various types of "power measurement" with respect to a proper measurement of rms watts at 1% THD. We have a hypothetical power amplifier connected to an 8-$\Omega$ resistive load in a circuit such as that in Fig. 19–1. It is nominally a 25-W rms amplifier. At a THD of 1% it produces an rms output potential of 14 V ac.

1. rms watts at 1% THD
   Equation: $P = (E_{rms})^2/R$
   Solution: $P = (14)^2/8 = 24.5$ W
2. Peak watts
   Equation: $P = (E_p)^2/R$
   Solution: $P = (1.414 \times 14)^2/8 = 49$ W

3. Peak-to-peak watts
   Equation: $P = (2.83E_{rms})^2/R$
   Solution: $P = (2.83 \times 14)^2/8 = 196$ W
4. "Music power"
   Equation: $P = 2 \times$ (rms power)
   Solution: $P = (2)(14)^2/8 = 49$ W

Note the wide range of numerical results (and all of these have been advertised!). However, only rms watts at 1% THD is particularly meaningful to the user, although some public address applications can tolerate 5% THD, so there you are able to use the rms watts at a 5% THD rating. Modern equipment is able to offer THD figures that are less than 0.01%. The advertised figure should be used for any meaningful power measurement. Perhaps you can see why some people used to refer to "watts IYL" (If You're Lucky) to describe some audio amplifier power specifications.

### Procedure

1. Adjust the amplifier for maximum gain and increase the level from the oscillator either (a) until clipping begins (and then back off until clipping disappears) or (b) until the THD increases to 1% (or the manufacturer's specified level).
2. Measure the rms voltage across the load resistor.
3. Calculate the rms power from the expression $P = (E_{rms})^2/R_L$.
4. In most cases this measurement is performed at 1000 Hz or some other frequency in the 200-to-2000-Hz range specified by the manufacturer. Some feel, however, that a somewhat more stringent test is to measure the power at 1000 Hz and the upper and lower −3-dB points. The results of the three measurements are then averaged.

Audio power measurements can be deceiving, even when the rms rating is used. For example, in a *stereo* amplifier, the power supply must be capable of delivering sufficient power to produce maximum output from both channels. Some manufacturers will specify the output power with *both channels driven*. This is to take into account the regulation of the power supply, which must be able to handle twice the power load than when only one channel is driven. The output power from either channel should be nearly the same under both conditions; if the "both channels driven" power is significantly less than the power measured with only one channel driven, then suspect a weak power supply.

## 19–8     INPUT SENSITIVITY

The sensitivity rating of an amplifier tells us how much signal is required to drive the amplifier to a specified output level. Two methods are generally applied: drive level required for (a) full output power and (b) 1 W output.

### Procedure

1. Adjust the amplifier gain to maximum.
2. Increase the input signal level to produce the specified output level.

**3.** Measure the rms input voltage.

**4.** The sensitivity is usually written in the form "6 mV for 1 W output."

**19–9**      **FREQUENCY RESPONSE**

The **frequency response** of an amplifier refers to its ability to amplify signals of different frequencies. Ideally, a perfect amplifier will amplify all frequencies from dc to daylight equally. But in practical amplifiers this ideal is not met.

The response plot of an amplifier (Fig. 19–2) graphs **gain vs. frequency** on semilogarithmic paper. The relative gain is not actually measured but is inferred from the output voltage; the input signal voltage is kept a constant.

The gain, then, is expressed in terms of output volts, which are usually converted to decibel form. The 0-dB reference level will be specified as the output voltage at a given reference frequency, usually 400 or 1000 Hz. The reference frequency must be far enough removed from the upper or lower −3-dB points that the response is essentially flat.

In most cases the frequency response data are taken manually by adjusting the oscillator frequency and then reestablishing the input voltage to the amplifier using the attenuator. The output voltage and frequency are then recorded.

In some cases, however, modern voltage controlled (VCO) signal generators are used to *sweep* the spectrum. If a sawtooth sweep voltage is used for both the VCO and the oscilloscope time base, then the amplitude of the amplifier output signal (applied to the oscilloscope Y input) represents the frequency response data.

**19–10**      **SQUARE WAVE TESTING**

If a square wave input signal to the amplifier is used, then we may gain a qualitative indication of frequency response by examining the output waveform. Although quantitative

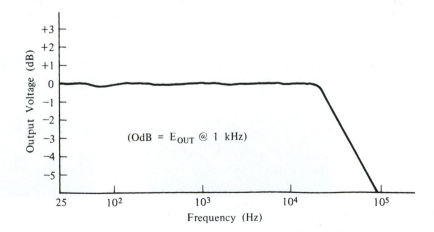

**FIGURE 19–2**
Frequency response curve.

data are lacking, this test gives a quick method for making a qualitative evaluation of the amplifier's response.

The square wave test is possible because of the nature of square waves. If you perform a mathematical analysis of a square wave to find its Fourier series, then you will find that it is made up of a fundamental sine wave at the frequency (some say repetition rate) of the square wave, plus an infinite number of *odd* sine wave harmonics. Each successive harmonic has less amplitude than the previous harmonic. Real square waves, that is, those that are produced in a signal generator, have up to approximately the 1000th harmonic in significant amplitude.

Figure 19–3 shows various types of square wave responses. In Fig. 19–3(a) we see the response of a very wide band amplifier, which is essentially the input square wave unaffected. In very wide band amplifiers only slight rounding of the upper corners may be expected, and this may not be visible on the oscilloscope trace. Figure 19–3(b) shows the results of passing a square wave through an amplifier that has a much more limited frequency response. Note the severe rounding of the corners. It is the *upper* harmonics that contribute to the rise time and "sharpness" of the waveform corners.

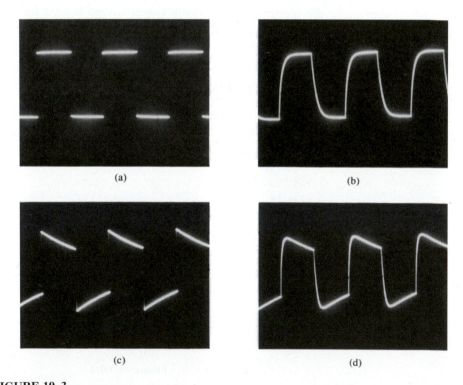

(a)                                             (b)

(c)                                             (d)

**FIGURE 19–3**
(a) Normal square wave, (b) loss of high frequencies, (c) loss of low frequencies, (d) loss of both high and low frequencies.

Poor low-frequency response produces a trace like that shown in Fig. 19–3(c). In this case the leading edges of both positive and negative excursions of the pulse are "overpeaked," and the flat portion is tilted.

## 19–11    TOTAL HARMONIC DISTORTION (THD)

All waveforms (and in fact all continuous mathematical functions) can be described in terms of a Fourier series of sine and cosines, and in most cases they bear harmonic relationships to the fundamental sine wave frequency. The specific harmonics that are present and their relative amplitudes and phases determine the actual shape of the waveform. Only the sine wave is pure; all other waveforms have harmonic components.

When an amplifier distorts a waveform, it is creating harmonic components that have not existed before. The measure of this type of distortion is the percentage of **total harmonic distortion (THD).**

Total harmonic distortion is measured with a harmonic distortion analyzer, blocked out in Fig. 19–4(a). The oscillator supplying the input signal to the amplifier under test must have a very pure (i.e., harmonic-free) sine wave output signal. The THD analyzer does not care where the distortion originates; it will measure the total harmonic distortion regardless of the source.

An rms ac voltmeter is used to measure distortion. First measure the rms voltage of the amplifier output potential $E$ and then measure the rms voltage due to harmonics $E_h$. This voltage is measured by passing the signal through a **notch filter** that is adjusted to pass all frequencies except the oscillator's output frequency, that is, the fundamental frequency. The response of a typical notch filter is shown in Fig. 19–4(b). The percentage THD is expressed by

$$\text{THD} = \frac{E_h}{E} \times 100\% \tag{19–5}$$

where    THD = the percentage of total harmonic distortion
$E_h$ = the rms output voltage measured after the notch filter
$E$ = the rms output voltage measured before the notch filter

### ■ Example 19–4

Calculate the percentage THD if the amplifier output voltage is 7 V rms, and the notch filter output voltage is 56 mV (the notch filter is tuned precisely to the oscillator frequency).

*Solution*

$$\text{THD} = (E_h/E) \times 100\% \tag{19–5}$$
$$= (0.056 \text{ V}/7 \text{ V}) \times 100\%$$
$$= (8 \times 10^{-3}) \times 100\% = \mathbf{0.8\%}$$

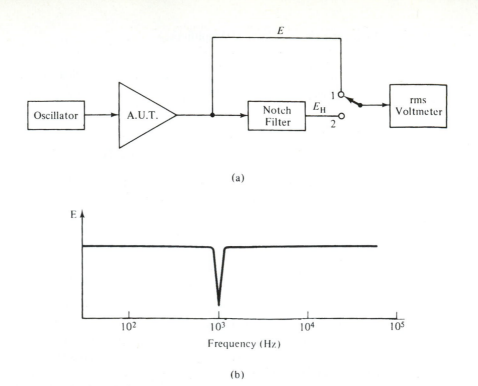

(a)

(b)

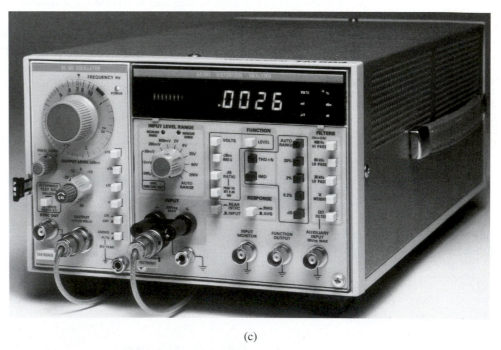

(c)

**FIGURE 19–4**
(a) Total harmonic distortion analyzer, (b) notch filter frequency response, (c) commercial THD analyzer (courtesy of Tektronix, Inc.).

Note that THD performance for any given amplifier varies with output power level. An amplifier in which THD is measured at 1 W output, for example, appears far more linear than one in which the THD was measured at maximum rated output power. An example of a commercial THD analyzer is shown in Fig. 19–4(c).

**19–12**

## INTERMODULATION DISTORTION (IMD)

When two signals are combined together in a linear network, they will both pass through the system without affecting each other. Figure 19–5(a) shows the result of mixing a high-frequency signal (i.e., 3–4 kHz) with a higher-amplitude, low-frequency signal (i.e., 40–100 Hz) in a linear network. If the output signal of the network were examined on a spectrum analyzer, only the two original frequencies would be present.

If these same signals are passed through a nonlinear network (i.e., an amplifier that produces distortion), then mixing or "heterodyning" will result, because the lower fre-

**FIGURE 19–5**
(a) Linear mixing of two tones; (b) nonlinear mixing of two tones produces AM waveform.

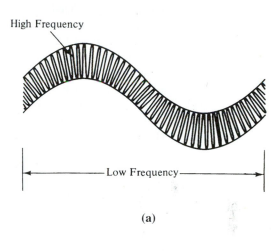

High Frequency

Low Frequency

**(a)**

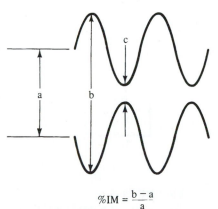

$$\%IM = \frac{b - a}{a}$$

**(b)**

quency will tend to modulate the higher frequency. If $F1$ and $F2$ are the two original frequencies, then the output spectrum would contain the following (at least): $F1$, $F2$, $F1 + F2$, and $F1 - F2$. There might also be present harmonics of these signals, plus the sum and difference frequencies caused by the harmonics.

If there is any significant nonlinearity in the circuit to cause distortion, then the effect on the two tones will be as shown in Fig. 19–5(b), the familiar amplitude modulation waveform. The percentage of intermodulation distortion is given by

$$\text{IMD} = \frac{b - a}{a} \times 100\% \qquad\qquad \textbf{(19–6)}$$

where   IMD = the intermodulation distortion in percent
        $b$ and $a$ refer to features in Fig. 19–5(b)

A simple IMD analyzer (Fig. 19–6) can be made by merely rectifying the amplifier output signal. A standardized test signal is used in this circuit. The higher frequency must be at least 50 times greater than the lower frequency. For a 60-Hz low frequency, then, the high frequency must be at least 3000 Hz. The two signals are combined in a linear resistor network, usually a Wheatstone bridge in which all four arms are equal.

The test circuit uses a high-pass filter to reject the low-frequency component of the output signal and pass only the high-frequency component plus the modulation products (i.e., the sum and difference frequencies), which lie near the high-frequency signal. A low-pass filter is used to average the rectifier output to remove any residual 3000-Hz signal. The remaining signal is a dc voltage that represents the 60-Hz component that

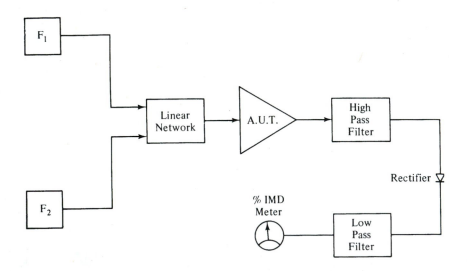

**FIGURE 19–6**
Test equipment needed for IMD measurement.

has been modulated onto the 3000-Hz "carrier." Since the original high-pass filter removed the main low-frequency component, it does not affect this reading. The output voltage, then, is proportional to the percentage of modulation or, in this case, intermodulation.

In untuned amplifiers (i.e., audio amplifiers), IM distortion is also sometimes called **cross-modulation.** When you are dealing with radio receivers, however, this terminology may be misleading. In receiver systems, intermodulation and cross-modulation have very similar meanings—but there is a slight difference. In speaking of untuned amplifiers, however, people use both terms even though the term **intermodulation** is preferred.

## 19–13    OTHER PARAMETERS

In this section we will discuss certain other aspects of untuned amplifiers. Many of the tests discussed so far in this chapter are particularly interesting when you are testing hi-fi and other audio amplifiers. But tests in this section are used primarily in other types of amplifiers, such as might be found in scientific, industrial, medical, or engineering instruments. Some of these tests are specific to *differential* amplifiers, whereas others are applicable to amplifiers in general. We will use the operational amplifier for our examples, but this is not to imply that the tests are not valid on other types as well.

## 19–14    SLEW RATE ($S_r$)

The **slew rate** $S_r$ of an amplifier is the measure of its maximum *change of output voltage* (*dE/dt*) while the amplifier is delivering its full rated output current. Figure 19–7 shows the test configuration used in this test.

The rated output current $I_m$ and the maximum output voltage $E_m$ are determined from the amplifier's specifications, and the value of $R_L$ is determined by Ohm's law: $R_L = E_m/I_m$.

The unity gain noninverting follower circuit is used in Fig. 19–7(a) because, in operational amplifiers especially, this configuration produces the worst-case results. An inverting follower, or noninverting follower *with* gain, can also be used, but the results cannot be expected to apply to all cases.

The input signal $E_{in}$ is adjusted to severely overdrive the amplifier in both positive and negative directions [see Fig. 19–7(b)]. The *slope* of the output voltage waveform, as viewed on an oscilloscope, represents the slew rate:

$$S_r = \frac{\Delta E_0}{\Delta t} \qquad\qquad (19\text{–}7)$$

The units of slew rate are *volts per unit of time.* In actual practice, the units commonly seen are *microvolts per second* (μV/s) or *volts per microsecond* (V/μs). Although both representations are valid, they can lead to some confusion unless you are aware of which is being used.

**FIGURE 19–7**
(a) Unity gain noninverting follower used in test circuit, (b)
definition of slew rate.

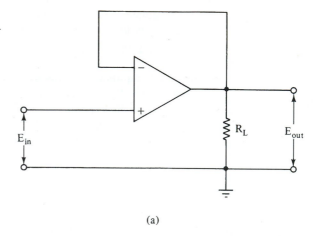

(a)

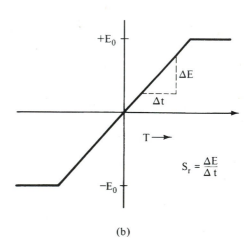

(b)

## 19–15     FULL-POWER BANDWIDTH

The **full-power bandwidth** of an amplifier is the frequency at which open-loop voltage gain $A_{vol}$ drops to unity while the amplifier is delivering its full rated output power. This frequency will usually be somewhat less than the gain-bandwidth product for small signals.

It is rarely feasible to measure this parameter in an open-loop configuration, because noise sources ordinarily present in all circuits will predominate in the output signal. Fortunately, we may approximate this parameter using a gain follower circuit, provided that the closed-loop voltage gain is set to something in the ×100 to ×500 range. The full-power bandwidth at these gains will approximate that of $A_{vol}$ to a negligible error.

The input signal amplitude is set to a level that drives the output to deliver full rated power to an external load resistor $R_L$. The frequency is then increased, with the input voltage maintained constant, until the output voltage $E_0$ is equal to the input voltage $E_{in}$:

$$A_v = \frac{E_0}{E_{in}} = 1 \qquad\qquad (19\text{–}8)$$

## 19–16          INPUT OFFSET VOLTAGE

The **input offset voltage** is defined as the potential across the inverting and noninverting inputs required to bring the output potential to zero at a time when $E_{in}$ is also zero. We would normally expect $E_0$ to be zero any time $E_{in}$ is zero, but certain circuit and internal IC defects often cause a small erroneous offset voltage.

The measurement circuit is shown in Fig. 19–8. It is an inverting follower with a gain of ×500 to ×1000. The "input" end of resistor $R1$ is grounded to ensure that $E_{in}$ is zero. Then a high-resolution, precision, dc power supply is applied directly across the amplifier inputs. The supply is adjusted until $E_0$ is zero. The voltage $E$ that reduces $E_0$ to zero is the input offset voltage of the amplifier.

## 19–17          COMMON MODE REJECTION RATIO (CMRR)

Common mode rejection is the property of a *differential* amplifier that rejects input signals that are *common* to *both* inputs. The **common mode rejection ratio (CMRR)** is expressed by

$$\text{CMRR} = \frac{A_{v(d)}}{A_{v(cm)}} \qquad\qquad (19\text{–}9)$$

$$\text{CMRR} = \frac{(E_0/E_{in})}{(E_0/E_{(cm)})} \qquad\qquad (19\text{–}10)$$

$$\text{CMRR} = \frac{E_{(cm)}}{E_{in}} \bigg|_{E_0 = \text{constant}} \qquad\qquad (19\text{–}11)$$

**FIGURE 19–8**
Test setup for measuring input off-set voltage.

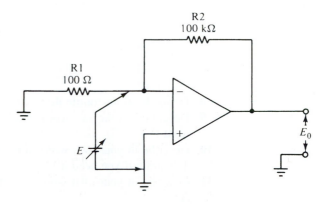

A convenient differential voltage is selected, and both the input voltage $E_{in}$ and the output voltage that it produces, $E_0$, are measured precisely. Next, the two inputs are tied together and are used as a single, common mode input. The input signal voltage is increased until the same output voltage as was obtained before is at the output. Equation (19–11) can then be used to calculate the CMRR.

## SUMMARY

1. Voltage gain is the ratio of output voltage to input voltage.
2. The *rms* power measurement is more closely related to the work that an audio power amplifier can do than are some of the other "power" measurements sometimes used.
3. The total harmonic distortion (THD) is found by measuring the rms output voltage with and without the fundamental present.
4. Intermodulation distortion is measured by noting the degree of modulation of a high-frequency signal by a low-frequency tone.

## RECAPITULATION

Now go back and try to answer the questions at the beginning of the chapter. When you are finished, answer the questions and work the problems given below. Place a mark beside each problem or question that you cannot answer, and then go back to the text and reread appropriate sections.

## QUESTIONS

1. Define in your own words what is meant by *decibel scale*.
2. Power or voltage gain is represented by _____ dB, while a gain or loss is represented by _____ dB.
3. Write the transfer equation if a voltage amplifier has an input signal $E1$ and an output signal $E2$.
4. Discuss in your own words the different power measurement techniques used on hi-fi amplifiers.
5. What is meant by **input sensitivity?**
6. Frequency response is a measure of _____ vs. _____.
7. A square wave test gives us a _____ view of an amplifier's frequency response.
8. Define **total harmonic distortion** in your own words.
9. Describe in your own words a technique for measuring THD. (Use a diagram if you want.)
10. Describe in your own words a method for measuring frequency response data. (Use a diagram if you need it.)
11. Describe in your own words a method for measuring ac voltage gain (i.e., at 1000 Hz).

12. Define **intermodulation distortion** in your own words.
13. Draw a block diagram of an IMD analyzer.
14. Define **slew rate** in your own words.
15. Define **full-power bandwidth** in your own words.
16. Define **input offset voltage** in your own words.
17. Draw circuits for measuring (a) slew rate and (b) full-power bandwidth—without looking back in the text.

## PROBLEMS

1. An amplifier produces 6 V output when 13 mV is applied to the input. Calculate the power gain in both scaler units and dB.
2. A wideband RF amplifier produces 650 W output when a 6-W signal is applied to the input. Calculate the power gain in scaler units and dB.
3. Calculate the *loss* in dB if an attenuator reduces a 1-V signal to 1 mV.
4. An RF signal source connected to a matching, nonreactive, load is set to produce −10 dBm. How much power is dissipated in the load?
5. A 600-$\Omega$ load is connected to an audio sine wave generator that is set to produce +3 VU output. How much *voltage* (rms) appears across the matched load?
6. In a voltage gain measurement, 7 V is produced at the output by a 100-mV signal at the input. Calculate the gain in dB.

# 20

# Measurement on Tuned Circuits

## 20-1    OBJECTIVES

1. To learn some of the problems encountered in making measurements on tuned RF circuits.
2. To learn sweep frequency techniques.

## 20-2    SELF-EVALUATION

Before studying the material in this chapter, try to answer the questions given below. These questions test your knowledge of the subject. If you cannot answer a particular question, place a check mark beside it, and then look for the answer as you read the text.

1. A parallel resonant circuit exhibits an impedance at resonance that is (high/low) compared to off-resonance impedance.
2. Calculate the resonant frequency of an LC network consisting of a 120-pF capacitor and a 5.6-μH inductor.
3. Find the capacitance needed to resonate a 1.2-μH coil to 5.5 MHz.
4. The sweep frequency used to test a high-Q resonant tank circuit must be kept low in order to prevent _____ in the circuit.

## 20-3    TUNED CIRCUITS

In this chapter we consider certain techniques and instruments used to measure various parameters in frequency-selective (i.e. tuned) circuits. Although any circuit that has a bandwidth less than "dc-to-daylight" may be considered to be frequency selective, we limit the discussion here to circuits of relatively narrow bandwidth. Generally, these circuits will be RF in nature—hence the emphasis on RF circuits.

You are also referred to Chapters 21 through 24 for additional study of this subject as it applies specifically to spectrum analyzers, antennas, receivers, and transmitters. Although an attempt is made at not presenting the same material more than once, there is,

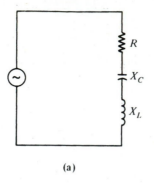

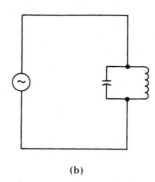

(a)                          (b)

of necessity, some overlap. For the broadest view of the subject of RF measurements, then, you are advised to study Chapters 21–24 in addition to this chapter.

Tuned circuits consist of R, L, and C elements in parallel or series combination. Figure 20–1 shows two forms of tuned circuits: the series resonant [Fig. 20–1(a)] and the parallel resonant [Fig. 20–1(b)].

A tuned circuit differs from an untuned circuit because of the property of resonance, which exists in LC circuits when the capacitive and inductive reactances are equal in magnitude; that is,

$$X_L = X_C \tag{20–1}$$

$$2\pi f L = \frac{1}{2\pi f C} \tag{20–2}$$

The capacitive and inductive reactances produce opposite effects on alternating currents, so when the reactance magnitudes are equal, the net effect is cancellation of the reactive component. The circuit becomes essentially resistive. The resonant frequency can be determined by solving Equation (20–2) for $f$, which results in the expression

$$f = \frac{1}{2\pi\sqrt{LC}} \tag{20–3}$$

where   $f$ = the resonant frequency in hertz (Hz)
        $L$ = the inductance in henries (H)
        $C$ = the capacitance in farads (F)

■ **Example 20–1**

Calculate the resonant frequency of a tank circuit consisting of a 100-μH inductor and a 100-pF capacitor.

*Solution*

$$f = \frac{1}{2\pi\sqrt{LC}} \tag{20–3}$$

$$= \frac{1}{(2)(3.14)\sqrt{(10^{-4}\ H)(10^{-10}F)}}$$

$$= \frac{1}{(2)(3.14)\sqrt{10^{-14}}} = \textbf{1.59 mHz}$$

There is a fundamental difference in the performance of these two types of resonant tank circuits. The graph plotting the **impedance vs. frequency** for these circuits is shown in Fig. 20–2. The series resonant circuit (curve A) presents a high impedance at frequencies removed from resonance and a low impedance at resonance. The behavior of the parallel resonant circuit is exactly the opposite. The parallel tank circuit offers a high impedance at the resonant frequency and a low impedance at frequencies removed from resonance.

Note that the response of these circuits falls off gradually from resonance in both cases. The "sharpness" of the fall-off depends upon the **quality factor, or figure of merit,** designated by the letter $Q$. The $Q$ of a tank circuit is given by

$$Q = \frac{f_r}{BW} \tag{20–4}$$

$$Q = \frac{X_L}{R} \tag{20–5}$$

where     $Q$ = the quality factor
          $f_r$ = the resonant frequency
          BW = the bandwidth
          $R$ = the circuit resistance

*Note:* Any units of frequency may be used provided that the *same* units are used for both $f_r$ and BW.

**FIGURE 20–2**
(a) Impedance in a series resonant circuit, (b) impedance in a parallel resonant circuit.

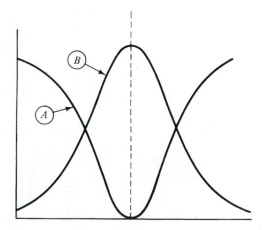

**FIGURE 20–3**
(a) Medium-$Q$, (b) high-$Q$, (c) low-
$Q$ resonant circuit frequency
response curves.

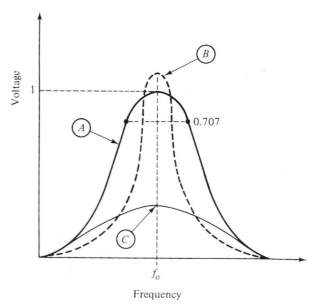

Figure 20–3 shows the frequency response curves for tank circuits with varieties of
$Q$ values. Curve $A$ shows the response given when $Q$ has a moderate value, while curve
$B$ shows the response obtained when the tank circuit has a high $Q$ value. A tank circuit
with a very low $Q$ will produce curve $C$.

The curve for a moderate $Q$ is marked at a point that is 70.7% of the maximum point.
This is known as the −6-dB, or "half-power," point. Bandwidth in tank circuits is measured with these points as the guidelines. When someone refers to a bandwidth of "so
many kilohertz," he is indicating the number of kilohertz between the points where the
response falls off 6 dB.

Tuned transformers (Fig. 20–4) are encountered frequently in RF circuits. In many
cases both primary and secondary windings of the transformer are resonated by capacitors, whereas in others only one winding is resonated.

The apparent $Q$ of a tuned transformer is affected by the degree of coupling between
the primary and secondary tank circuits. Figure 20–5 shows the three basic responses
obtained with differing degrees of coupling. Figure 20–5(a) shows the response obtained
when **subcritical** coupling is used. In this type of coupling few flux lines from the primary cut the secondary, so only the strongest signals close to the resonant frequency are
passed.

**FIGURE 20–4**
Tuned transformer.

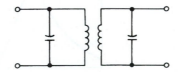

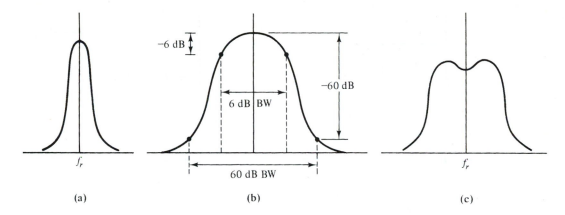

**FIGURE 20–5**
(a) Subcritical coupling, (b) critical coupling, (c) over-critical coupling.

The effect of **critical coupling** is shown in the moderate-$Q$ curve of Fig. 20–5(b). **Overcritical coupling** produces the low-$Q$, "double hump" curve in Fig. 20–5(c). In this last type of coupling most of the primary flux lines cut the secondary. This condition results in the classic response shown.

A filter is a special tuned circuit that passes frequencies in a certain band and rejects all others. The tuned transformer is a case of a **bandpass filter;** that is, it will pass only those frequencies in a band around the resonant frequency.

Figure 20–6 show the frequency responses for several different classes of filters. The response shown in Fig. 20–6(a) is for a **low-pass filter.** This type of filter passes frequencies *lower* than the cutoff frequency $f_c$ and rejects frequencies above $f_c$.

The response for a **high-pass filter** is shown in Fig. 20–6(b). This response is just the opposite of the low-pass filter; it passes frequencies *above* $f_c$ and rejects those below $f_c$.

When low-pass and high-pass filter elements with slightly overlapping cutoff frequencies are connected in cascade with each other, they form a **bandpass** filter such as in Fig. 20–6(c). A similar response is produced by a simple tank circuit, tuned transformers, and certain other RLC networks. Quartz and ceramic piezoelectric crystal networks also produce the bandpass response under the right conditions.

Two examples of **band-reject** filter responses are shown in Figs. 20–6(d) and 20–6(e). The principal difference between these two is in the bandwidth. The response shown in Fig. 20–6(d) rejects a stopband centered about $f_r$ but extending out in both directions quite a bit. The notch filter in Fig. 20–6(e) shows a response in which the stopband is very narrow around the resonant frequency.

**20–4        TESTING TUNED CIRCUITS**

Tuned circuits tend to be a little more critical of the techniques used than are many untuned circuits. While low-frequency tank circuits are relatively well behaved, the problems increase dramatically with increased frequency.

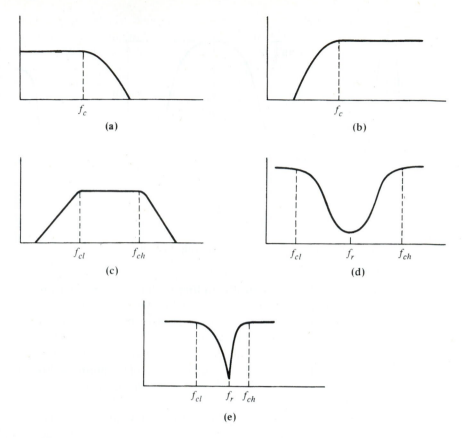

**FIGURE 20–6**

(a) Low-pass filter response, (b) high-pass filter response, (c) bandpass filter response, (d) band-reject filter response, (e) notch filter response.

Several problems will become immediately apparent once you begin to work with tuned circuits, but one of the most critical is loading of the circuit by the test equipment. Suppose, for example, that you want to measure $f_r$ by using a signal generator and an RF voltmeter. The cables to these instruments could easily cause a parallel capacitance across the tank circuit that has a value of several hundred picofarads. This value may be many times the actual tuning capacitance of the tank, but even if it is merely a significant fraction of the tuning capacitance, serious errors in the measurement will result. The solution to this problem is to isolate the tank circuit from the test equipment whenever possible.

One very gross type of measurement of resonant frequency is to use the dip meter, discussed in Chapter 18, but dip meter calibration is coarse, at best.

A signal generator and an RF voltmeter (or oscilloscope) can be used if proper isolation of the tank circuit is provided. If the tank is part of a cascade amplifier, then the normal input/output terminals may be used. Otherwise, it may prove wise to use a "gimmick" to excite the tank and a very low-capacitance probe for the voltmeter. It sometimes helps to connect the voltmeter or oscilloscope to a low-impedance tap on the tank circuit inductor or provide a transformer link for the indicator.

The response curve can be measured, if somewhat laboriously, by adjusting the signal generator to various points around $f_r$ and then manually plotting the data on graph paper. This technique "grows old" very quickly, however, because of the relative difficulty in obtaining the needed data. It is better to use a sweep generator to write the curve onto the oscilloscope CRT screen.

## 20–5   SWEEP TECHNIQUES

A **sweep generator** uses a voltage controlled oscillator (VCO) driven by a sawtooth or sine wave (in which case it is an FM generator). In either case we create a varying RF frequency that sweeps back and forth across the frequency set on the dial at a rate determined by the sawtooth or sine wave.

A block diagram of a sweep generator test configuration is shown in Fig. 20–7(a). The sweep generator signal and a signal from a crystal calibrator (i.e., a marker) are mixed in a linear network before being applied to the circuit under test. The output of the RF circuit is passed through a detector and a low-pass filter to the vertical input of an oscilloscope. The sweep generator will also send either the sweep voltage (i.e., the sawtooth) or a synchronization pulse to the oscilloscope. In the former case the sweep generator would be connected to the oscilloscope horizontal channel, whereas in the latter it would be connected to the *sync* or *trigger* input. Figure 20–7(b) shows a typical sweep generator-oscilloscope package.

When using sweep techniques, be careful not to use too high a sweep frequency, or **ringing** of the tank circuit will occur.

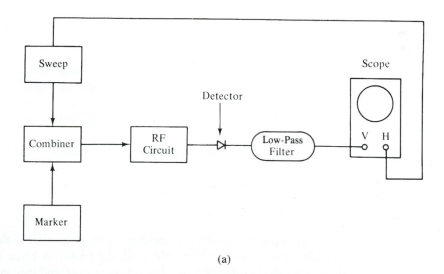

(a)

**FIGURE 20–7**
(a) Block diagram of equipment needed for sweep testing.

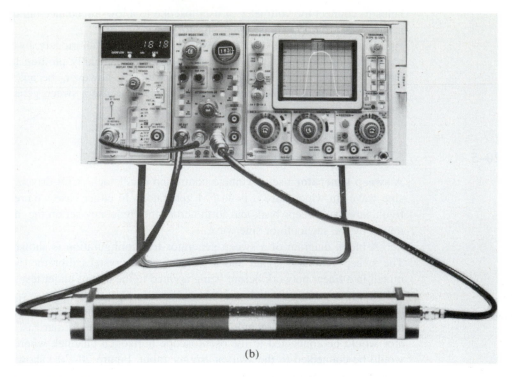

(b)

**FIGURE 20–7,** *continued*
(b) Sweep testing setup [(b) courtesy of Tektronix, Inc.].

## SUMMARY

1. The two basic types of resonant tank circuits are **series** and **parallel.**
2. Resonance exists in a tank at the frequency where the inductive and capacitive reactances are equal.
3. $Q$ is a measure of how fast the frequency response falls off at frequencies removed from resonance.
4. A sweep oscillator and detector produces an **amplitude-vs.-time** trace on an oscilloscope.

## RECAPITULATION

Now go back and try to answer the questions at the beginning of the chapter. When you are finished, answer the questions and work the problems given below. Place a mark beside each problem or question that you cannot answer, and then go back to the text and reread appropriate sections.

## QUESTIONS

1. List two types of resonant tank circuits.
2. The series resonant tank circuit exhibits a _____ impedance at resonance.
3. The parallel resonant tank circuit exhibits a _____ impedance at resonance.
4. At resonance, _____ _____ and _____
   _____ are equal, so they cancel, leaving only the resistive component.
5. Define $Q$ in your own words.
6. In tuned transformers, subcritical coupling causes the apparent $Q$ to be
   _____.
7. Draw the response curves for (a) critical, (b) subcritical, and (c) overcritical coupling.
8. Draw the typical response curves for (a) high-pass and (b) low-pass filters.
9. What is the principal difference between a stopband filter and a notch filter?
10. Describe how a sweep generator can be used to measure the frequency response of a
    circuit.
11. Why must a slow sweep frequency be used in sweep tests?

## PROBLEMS

1. Using the standard reactance equations, derive the equation for resonance in an LC
   tank circuit. Show your work.
2. Calculate the resonant frequency of a tank circuit that consists of a 33-pF capacitor
   and a 1.2-μH inductor.
3. Find the $Q$ of a 5-mHz tank circuit if the −6-dB points are 30 kHz apart.
4. Find the $Q$ of a 1000-kHz tank circuit if the coil has an inductance of 450 μH and a
   50-Ω series resistance (copper loss of the wire).

# 21

# Antenna and Transmission Line Measurements

21–1 **OBJECTIVES**

1. To learn the various parameters normally of interest in antenna/transmission line systems.
2. To learn what instruments are available for antenna measurements.
3. To learn antenna measurement techniques.
4. To learn how to adjust HF and VHF antennas, using *simple* instrumentation.

21–2 **SELF-EVALUATION**

Before studying the material in this chapter, try to answer the questions given below. These questions test your knowledge of the subject. If you cannot answer a particular question, then look for the answer as you read the text.

1. Describe two methods for measuring antenna impedance.
2. What are **standing waves,** and how are they formed in radio transmission lines?
3. How can a **noise bridge** be used to read antenna impedance?
4. What is a **slotted line?** What are **Lecher wires?**
5. **Nodes** and **antinodes** alternate with each other, repeating themselves every _____ wavelength along the line.

21–3 **ANTENNAS**

Radio antennas **radiate** an electromagnetic wave into space when excited by an oscillating electrical current. The radiating effect is present in some degree at all frequencies of oscillation, but the efficiency is extremely poor except at certain resonant frequencies. At the resonant frequencies the efficiency increases tremendously.

There are many forms of antennas, but we will limit this discussion to high-frequency (HF) and very-high-frequency (VHF) dipoles and monopoles. Brief excursions will be made into the UHF and low-microwave region and to directive arrays, but you are

**485**

advised to consult some of the referenced texts at the end of the chapter for more detail. This chapter is designed to discuss some of the measurements made on antennas, not the antennas themselves. Interestingly enough, for those who wish to pursue the subject, the amateur radio literature contains many high-quality articles from both theoretical and practical points of view; these articles are generally more available to the public than are the papers presented in professional journals.

Figure 21–1 shows two fundamental antennas: a half-wavelength dipole mounted horizontally to the earth's surface [Fig. 21–1(a)] and a vertical monopole that is quarter wavelength [Fig. 21–1(b)]. These antennas are similar in many ways, but a principal difference is in the nature of the field that is radiated by each. Both will radiate an electromagnetic field in which electric and magnetic field components radiate outward at right angles to each other, forming a plane. But the horizontal antenna radiates a signal in which the *electric* field is *horizontal* to the earth's surface, whereas the vertical antenna radiates a signal in which the electric field is vertical to the earth's surface; that is, it is perpendicular to the surface. Note that reception strength is greatest when the receiver and transmitter antennas have the *same* polarization.

**FIGURE 21–1**
(a) Half-wavelength antenna,
(b) quarter-wavelength antenna.

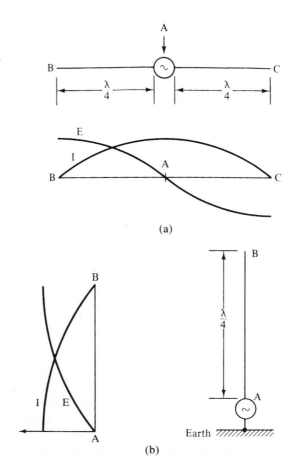

The dipole in Fig. 21–1(a) is one-half wavelength, which in free space is given by

$$L_{(ft)} = \frac{\lambda}{2} = \frac{492}{F_{(MHz)}} \tag{21–1}$$

Nearer the earth's surface the required length is a little less, on the order of

$$L_{(ft)} = \frac{468}{F_{(MHz)}} \tag{21–2}$$

Equation (21–1) holds true when the antenna is physically located many wavelengths from any conductive surface or object, and that includes the earth's surface. Equation (21–2) is an empirically derived approximation that *usually* works for antennas constructed at practical heights above the surface of the earth. Equations appropriate for quarter-wavelength antennas are the same and are merely Equations (21–1) and (21–2) with the velocity factor reduced by one-half, that is, 246 and 234 instead of 492 and 468, respectively.

Also shown in Fig. 21–1 are the normal voltage and current distributions along the lengths of the radiators. In both figures the **feed point** is a point of lowest impedance, that is, maximum current and minimum voltage. The feed-point impedance is given in the form

$$Z_L = R_0 \pm jX \tag{21–3}$$

where $Z_L$ = the antenna feedpoint impedance in ohms ($\Omega$)
$R_0$ = the antenna radiation resistance in ohms ($\Omega$)
$X$ = the reactive component of impedance
$j$ is the operator $(-1)^{1/2}$

At resonance—that is, when the oscillator frequency is the same as the antenna's natural resonant frequency—the antenna impedance is purely resistive, so $Z_L = R_0$. At frequencies higher than resonance, the phase angle between $E$ and $I$ becomes *inductive,* so $Z_L = R_0 + jX_L$; at frequencies lower than resonance it is capacitive, so $Z_L = R_0 - jX_c$.

## 21–4 TRANSMISSION LINES

It is rarely practical to locate the oscillator (i.e., the radio transmitter) exactly as shown in Fig. 21–1. In most cases the source will be located at some point that is remote from the antenna feed point, be it just a few feet or a few miles. A **transmission line** is used to transfer power between the source and the load, that is, from the transmitter output to the antenna feed point (see Fig. 21–2).

A transmission line may be a pair of parallel wires, as in Fig. 21–2, or a coaxial cable (which is most often the case in modern radio systems). Note, however, that the true transmission line has properties far more complex than might appear to be the case when

**FIGURE 21–2**
Parallel representation of a transmission line.

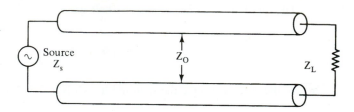

only a pair of wires is used. We will leave a complete discussion of transmission lines to a text on antenna systems or an engineering "fields and waves" (i.e., electromagnetics) course. Here we will consider only the **velocity factor** and the **characteristic impedance** (also called the **surge impedance**) of the line.

The velocity factor arises from the fact that radio waves do not propagate as fast in transmission lines as they do in free space. The velocity of radio signals in space is the speed of light, $3 \times 10^8$ m/s, but in a transmission line the velocity will be but a fraction of this speed. This fraction is the velocity factor and is usually expressed in decimal form (i.e., 0 to 1). Examples of the velocity factors for several popular transmission lines are as follows:

| Type of Cable | Value of $V$ |
|---|---|
| Twin lead | 0.82 |
| Foam-filled coaxial cable | 0.80 |
| Teflon | 0.70 |
| Regular dielectric coaxial cable | 0.66 |

The velocity factor $V$ is the fraction of the speed of light at which signals travel in the transmission line. For example, in foam-filled coaxial cable the signal velocity is 0.80 times the speed of light, or

$$V = (0.80)(3 \times 10^8 \text{ m/s})$$
$$= 2.4 \times 10^8 \text{ m/s}$$

The velocity factor affects the equations used to calculate electrical half and quarter wavelengths. The modified equations for cables are shown below.

*Half-wavelength case:*

$$L_{(ft)} = \frac{492V}{F_{(MHz)}} \tag{21–4a}$$

*Quarter-wavelength case:*

$$L_{(ft)} = \frac{246V}{F_{(MHz)}} \tag{21–4b}$$

where   $L$ = the length of the cable in feet (ft)
$\qquad$ $F$ = the frequency in megahertz (MHz)
$\qquad$ $V$ = the velocity factor (dimensionless)

■  **Example 21–1**

Calculate the length in feet of a half-wavelength section of foam-filled coaxial cable at a frequency of 20 MHz.

***Solution***

$$L = 492V/F_{(MHz)} \tag{21-4a}$$

$$= \frac{(492)(0.80)}{(20)}$$

$$= (394)/(20) = \mathbf{19.68\ ft}$$

---

The characteristic impedance of the transmission line is associated with the distributed inductance and capacitance per unit of length and is given in the most simple form by

$$Z_0 = \sqrt{L_0/C_0} \tag{21-5}$$

where   $Z_0$ = the characteristic impedance in ohms ($\Omega$)
$\qquad$ $C_0$ = the distributed capacitance per unit length
$\qquad$ $L_0$ = the distributed inductance per unit length

Practical values for $Z_0$ range from 150 to 1000 $\Omega$ for parallel conductor transmission lines and from 10 to 150 $\Omega$ for coaxial conductor transmission lines. Coaxial cable types RG-58/U and RG-59/U have characteristic impedances of approximately 52 and 75 $\Omega$, respectively. Television antenna twin lead has a characteristic impedance of 300 $\Omega$.

Stated another way, the characteristic impedance is equal to the value of the load *resistance* that will permit maximum transfer of power between the source and the load. If any other value of load is used, then not all of the power applied to the line by the source will be absorbed by the load. Some of the power will be reflected back down the transmission line toward the source. Reflected power results in **standing waves** (see Section 21.7).

## 21–5 ANTENNA SYSTEM MEASUREMENTS

The antenna system is potentially very complex, and its measurement is also sometimes complex. Yet, at the same time, it is possible to do much in the way of practical work on antennas using almost naive methods and instruments. This chapter will deal with simple but effective techniques widely used in industry.

Several key parameters are of interest, including **impedance, standing wave ratio, resonance, radiation pattern** (azimuthal and elevation), and **gain.**

## 21–6    ANTENNA IMPEDANCE

A number of different factors can affect the impedance of a practical antenna, including its physical proximity to the earth's surface and other conductive surfaces. In our simplified world that admits only half-wavelength horizontal dipoles and quarter-wavelength vertical monopoles, we find that "ideal" impedance values are 52 $\Omega$ for the vertical and 73 $\Omega$ for the horizontal antenna. But in actual practice, real-world antennas have feed-point impedances that vary considerably from these ideals. The horizontal dipole, for example, will exhibit a radiation resistance between 5 and 120 $\Omega$ for heights of 1/25 to 1 wavelength above ground. The 73 $\Omega$ figure is merely a theoretical impedance in free space. Similarly, the quarter-wavelength vertical radiator has a nominal or theoretical impedance of 52 $\Omega$ but exhibits impedances between 5 and 80 $\Omega$ (mobile installations, in which such antennas are particularly popular, usually show feed-point impedances of 10 to 30 $\Omega$). These values of impedance represent only the radiation resistance component of the overall impedance, which is given as

$$R_0 = \frac{E_0}{I_0} \qquad \text{(21–6)}$$

where    $R_0$ = the radiation resistance in ohms ($\Omega$)
$E_0$ = the feed-point potential in volts (V)
$I_0$ = the feed-point current in amperes (A)

The reactive component is zero in ideal, resonant antennas, but it usually has a value between 0 and ±100 $\Omega$ in real antennas.

## 21–7    STANDING WAVES AND SWR

**Standing waves** on antennas and transmission lines are simple phenomena but are often misunderstood. A certain mystique has built up over the matter of **standing wave ratio (SWR)** and what it means in terms of radio system performance. Many people spend a great deal of money and time attempting to idealize the SWR on an antenna without realizing that little improvement results in their ability to communicate with other stations.

Before discussing SWR, however, let's first determine the nature of standing waves. We may do this best by mechanical analogy, using a *model* that is easy to understand and interpret. Incidentally, modeling, although a valid teaching tool, is often poorly done. But in the case of standing waves, the model is another standing wave, and the behavior of the two types are nearly identical; the experiment using the model can be done by almost anybody who has a rope and a support to attach it to.

Our rope model, shown in Fig. 21–3, consists of a rope anchored at one end to a wall or other immovable, permanent structure. The rope is held *taut* by a person at the other end [Fig. 21–3(a)].

**FIGURE 21–3**
Concept of a reflected wave using a rope analogy.

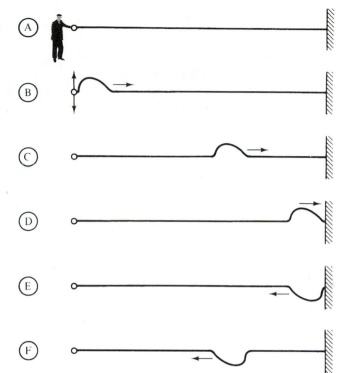

A *wave* is imparted to the rope by a quick up-and-down motion of the wrist [Fig. 21–3(b)]. This wave travels at a given velocity toward the wall. When the wave hits the wall end of the rope, it is *reflected* back toward the source end of the rope. Note that a phase reversal took place when the wave was reflected.

If the rope were ideal, and if both the wall and the operator's hand were perfectly rigid, then the wave would reflect back and forth along the line undiminished as time went on. But, like electrical circuits, there are losses that absorb energy and reduce the amplitude of the wave.

The situation in Fig. 21–3 is analogous to applying a *pulse* to an electrical transmission line. But what happens when a *continuous* oscillation is applied to the line? Consider Fig. 21–4. The source end of the rope is moved up and down in a precise, well-timed manner, thereby imparting sine waves to the rope. After the first wave has reflected from the wall end of the rope, we will actually have two waves: a *forward* wave traveling from left to right and a *reflected* wave that travels right to left. The wave amplitude at any given point along the rope will be the algebraic sum of these two waves. When viewed from the side, the rope will take on a blurred appearance, showing the nodes and antinodes of Fig. 21–4. These are known as **standing waves** because the nodes and antinodes are stationary; they do not move relative to the length of the rope.

Figure 21–5 shows three situations where there are standing waves on a radio transmission line (such as in Fig. 21–2). Figure 21–5(a) shows the situation when the forward wave is *totally* reflected back toward the source, that is, the case when the transmission

**FIGURE 21–4**

Standing waves on a line.

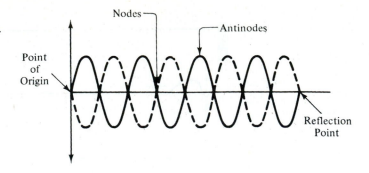

line is *shorted* ($Z_L = 0$) or totally *open* ($Z_L = \infty$). Interestingly enough, the pattern for both conditions is the same, except for the location of the nodes and antinodes. If the line is shorted, then the voltage at the load end is zero, so the nodes (i.e., the points of minimum voltage) are spaced at half-wavelength intervals along the line back from the short. But if the line is open, the voltage at the load end is maximum, so the antinodes (i.e., the points of voltage maxima) are spaced at half-wavelength intervals back from the load end of the cable. In other words, the two patterns are identical except for a 90° shift in the relative

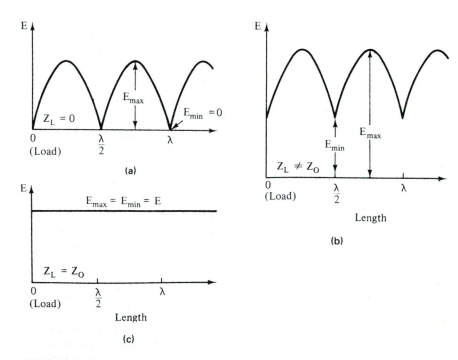

**FIGURE 21–5**

(a) Standing waves for $Z_L = 0$ or $Z_L = \infty$; (b) standing waves for $Z_L = Z_0$; (c) standing waves for $Z_L = Z_0$.

positions of the nodes and antinodes. The particular pattern selected for Fig. 21–5(a) is representative of the shorted condition.

Most antenna systems, however, do not totally reflect the forward wave; some of it is radiated into space. Thus, the reflected wave has a lower amplitude than the forward wave, so the cancellation is not total. The graph in Fig. 21–5(b) shows this situation; note that the voltage at the nodes is *not* zero.

The **standing wave ratio (SWR)** is given by

$$SWR = \frac{E_{max}}{E_{min}} \tag{21–7}$$

or, if current is measured instead of voltage,

$$SWR = \frac{I_{max}}{I_{min}} \tag{21–8}$$

Some people designate these measurements as VSWR and ISWR, respectively, but the numbers are the same. We may also calculate SWR from the following equations:

$$SWR = \frac{Z_0}{Z_L} \qquad (Z_0 > Z_L) \tag{21–9}$$

$$SWR = \frac{Z_L}{Z_0} \qquad (Z_0 < Z_L) \tag{21–10}$$

$$SWR = \frac{1 + \sqrt{P_r/P_f}}{1 - \sqrt{P_r/P_f}} \tag{21–11}$$

and

$$SWR = \frac{E_f + E_r}{E_f - E_r} \tag{21–12}$$

where  $Z_0$ = the characteristic impedance of the transmission line in ohms ($\Omega$)
  $Z_L$ = the impedance of the load in ohms ($\Omega$)
  $P_r$ = the power reflected from the load in watts (W)
  $P_f$ = the forward power applied to the transmission line by the source in watts (W)
  $E_r$ = the voltage component of the reflected wave in volts (V)
  $E_f$ = the voltage component of the forward wave in volts (V)

The last case, shown in Fig. 21–5(c), represents the situation where $Z_L = Z_0$. In this condition the forward power is totally absorbed by the load, and none is reflected. In this case there are no nodes or antinodes, and the line is said to be **flat.** Ideally, this is what is desired in a perfect antenna transmission line system; it represents the case where SWR is 1:1.

**21–8**        **MEASURING SWR**

Any of the equations for SWR presented so far can be used as the basis for an SWR measurement system. We could, for example, measure the impedance of the antenna and then use Equation (21–9) or (21–10). Or we could actually measure the power, voltage, or current in the forward and reflected waves.

All of the methods work, and you can find instruments that are based on just about any of them. In most low-cost VSWR indicators, however, either voltage or current is used, and that produces a problem that is the root of one of the most persistent myths about SWR adjustment and measurement. It is the *erroneous* belief of many people that SWR can be "adjusted" by trimming the transmission line length. This is a pernicious and vicious idea that ought to be consigned forthwith to the city dump. This idea is not true, but the belief that it is true persists because the nature of voltage- and current-based SWR meters make it *appear* true to the unknowing observer. When this type of instrument is used, the *numbers* obtained are valid *only* when SWR is measured at the antenna feed point or at half-wavelength intervals along the line back from the load. In other words, for the numbers to be valid, the transmission line between the SWR meter and the antenna must have a length $L$ of

$$L = \frac{492NV}{F} \tag{21–13}$$

where   $N$ = a convenient integer (i.e., 1, 2, 3, . . .)
        $V$ = the velocity factor (dimensionless)
        $F$ = the frequency in megahertz (MHz)

In addition to the problem just discussed, simple SWR meters also suffer greatly in accuracy when used in systems where the antenna is incorrectly built and has a grossly improper length. Such antennas are highly reactive, and so they will interfere with the correct measurement of SWR. Attempting to obtain accurate readings where SWR is very high, if low-cost instruments based on voltage or current are used, is a lot like trying to skin an amoeba.

The low-cost SWR meter is a valid instrument if its limitations are understood. If SWR is not too high, and if the user understands that it is a *relative* indicator, then it will prove very useful in correctly adjusting radio antenna systems. The correct use is to adjust the antenna length, or the L and C components in an antenna-matching network at its base, for a *minimum* SWR reading. The minimum will always be located at the same adjustment of the antenna components, even though we know not to take the numerical indication too seriously unless the line length is correct and the antenna is not terribly reactive.

If a true RF wattmeter is used to measure forward and reflected power, so that Equation (21–11) may be applied, SWR is totally independent of line length. Some RF wattmeter manufacturers publish an SWR nomograph based on Equation (21–11) to make this calculation easier (Fig. 21–7, to be discussed).

## 21–9     SWR METERS

In addition to the RF power meters discussed in Chapter 24, several low-cost meters are usually employed to measure SWR. Examples of some circuits are shown in Figs. 21–6(a) and 21–6(b). These instruments are voltage and current based, respectively.

The bridge SWR meter is shown in Fig. 21–6(a). It is a resistance Wheatstone bridge of the type discussed in Chapter 4. The bridge arms consist of resistors R1 through R3 and the radiation resistance of the antenna. The transmitter, or a low-power RF source such as a signal generator, produces the excitation potential. Note that this is a *resistance* bridge, so it will not measure the reactive component.

The circuit shown in Fig. 21–6(a) is set up to measure the radiation resistance of 50-Ω antenna systems because R3 is set to 51 Ω. If this resistor is changed to 75 Ω then

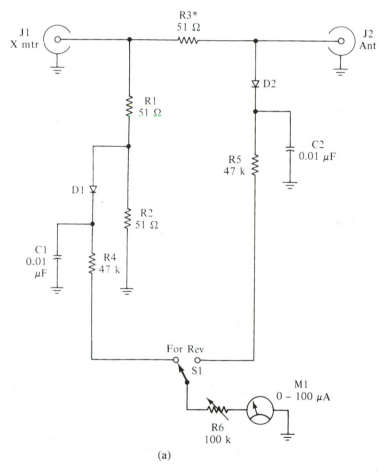

(a)

**FIGURE 21–6**

(a) Simple bridge-type VSWR meter.

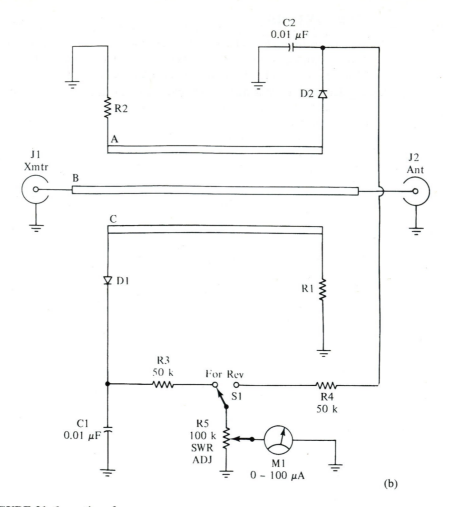

**FIGURE 21–6,** *continued*
(b) "Monimatch"-type VSWR meter.

the bridge will also work accurately in 75-Ω systems. The operation of the bridge is as follows:

## Procedure

1. Set *S1* to *forward.*
2. Adjust *R6* for a full-scale deflection.
3. Set *S1* to *reflected,* and read SWR from the special meter scale.

Another form of SWR meter, shown in Fig. 21–6(b), has the advantage that it may be left in the line while transmitting. The heart of this instrument is the pickup unit, consisting of the three conductors labeled *A* through *C.*

Conductor *B* is the center conductor of the transmission line; conductors *A* and *C* are identical to each other and spaced equidistant from *B*. In older units conductor *B* was a piece of coaxial cable, while conductors *A* and *C* were small-gage, enameled wires threaded between the braid and inner insulation of the coaxial cable. In most modern instruments, however, the pickup unit is made from foil patterns on a printed circuit board.

Resistors *R*1 and *R*2 are selected to have a resistance equal to the characteristic impedance of the transmission line (i.e., 50 or 75 Ω). In some units a trade-off value of 68 Ω is used so that the instrument may be used with either type of transmission line with but minimal error. Except for these resistors, the detector circuits and their operation are the same as on the bridge type of meter.

## 21–10  USING POWER METERS TO MEASURE SWR

SWR can also be measured with an RF wattmeter that is equipped with a direction coupler device so that forward and reflected power can be measured. We can then use the power levels obtained on the wattmeter in Equation (21–11) to determine SWR. Bird Electronics, a manufacturer of professional RF wattmeters, publishes the nomograph shown in Fig. 21–7; this chart is based on Eq. (21–11). Two different ranges are shown in Figs. 21–7(a) and 21–7(b).

It is also possible to use the radiation resistance of the antenna as measured on an antenna impedance bridge. Once this value is determined, then either Equation (21–9) or Equation (21–10) may be used, as appropriate, to compute SWR.

There are also some computing SWR meters on the market. Most of these use a pick-off unit such as in Fig. 21–6(b) or a related mechanism in which a toroidal (i.e., dough-nut-shaped) core current transformer replaces the pickup unit but also includes an analog circuit that automatically performs the three-step procedure given earlier.

## 21–11  ANTENNA IMPEDANCE BRIDGES

Several types of antenna impedance bridge are on the market, most of which are based on the Wheatstone bridge circuit, as shown in Fig. 21–8(a). Typically, impedance *Z*4 is the antenna impedance, *Z*3 is fixed, and both *Z*1 and *Z*2 are variable.

A simple resistance bridge of the form shown previously in Fig. 21–6(a) will often be used, but this type of circuit measures only the radiation resistance component of impedance and ignores altogether the reactive portion. But this is not necessarily disadvantageous, because the reactive component tends to disappear in simple antennas that are made resonant; a procedure for resonating the antenna can be performed without a complex impedance bridge.

Figure 21–8(b) shows another popular impedance bridge circuit in which *Z*1/*Z*2 is replaced by a differential capacitor, *C*1. This instrument is also limited to measuring the resistive component of impedance.

A low-cost commercial radiation resistance bridge is shown in Fig. 21–8(c), the Leader model LIM-870A. This simple instrument will determine the value of antenna resistances between 0 and 1000 Ω at frequencies up to 150 MHz.

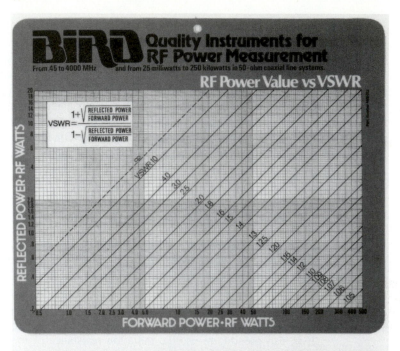

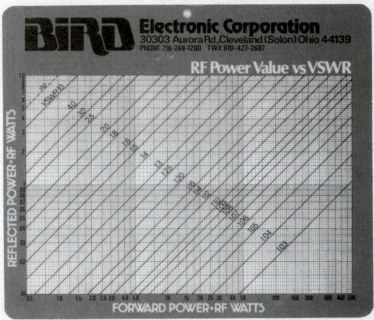

**FIGURE 21–7**
Nomograph for computing VSWR from RF power (courtesy of Bird Electronics Corp.).

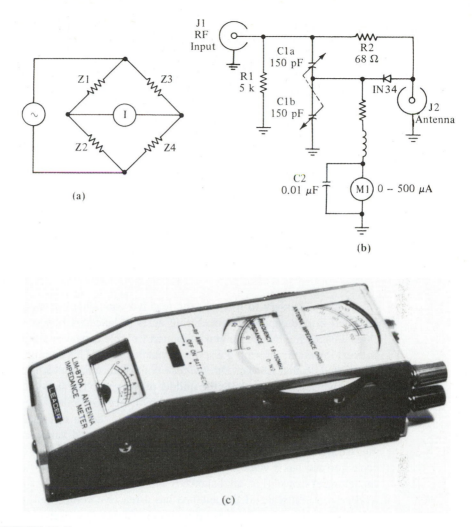

**FIGURE 21–8**
(a) General form of Wheatstone bridge, (b) RF impedance bridge for antenna measurements, (c) low-cost commercial antenna impedance bridge.

None of the bridges shown so far will measure the antenna's reactive component of impedance; only the antenna's radiation resistance can be actually measured numerically. But an experienced operator can usually *spot* highly reactive conditions by observing the depth and sharpness of the null. A resistive antenna shows a deep, narrow null, whereas a reactive antenna exhibits a shallower, broader null. This type of determination is, at best, subjective and requires skill in interpretation. But it can aid by showing that the antenna is not resonant.

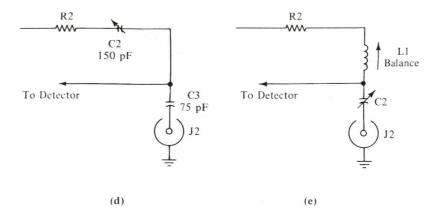

**(d)**                    **(e)**

**FIGURE 21–8,** *continued*
Modifications of resistance bridge to allow measurement of complex impedances: (d) inductive,
(e) capacitive.

If it is necessary to measure the reactive component of antenna impedance, then a
modified bridge capable of measuring ac impedances (Chapter 4) is required. In some
cases a balancing inductor and a calibrated variable capacitor are placed in series with the
antenna impedance, forming a series resonant circuit. This combination [see Fig. 21–8(e)]
is adjusted so that $L1$ (i.e., the *balance* control) resonates with $C2$ at the excitation frequency when $C2$ is exactly in the middle of its range. The impedance of $L1/C2$ under this
condition is entirely resistive and will have a very low value.

Another technique that is often employed is shown in Fig. 21–8(d) and consists of
two capacitors, $C2$ and $C3$. Capacitor $C2$ is a calibrated variable model and must be of
the "straight line capacitance" type of construction. Capacitor $C3$ is a precision fixed
capacitor and has a value that is equal to $\frac{1}{2}C2$. When $C2$ is set to midrange, then, these
capacitances cancel (i.e., are balanced) in the bridge circuit. The dial connected to $C2$ will
read zero at midscale. If the antenna has a reactive component, then $C2$ must be offset
higher ($X_L$) or lower ($X_c$), depending upon whether the reactance is inductive or capacitive.

Note that these measurements are also subject to the same constraints as SWR measurements: the bridge must be connected into the system at the antenna feed point or
through a piece of transmission line that is a one-half wavelength long.

## 21–12      NOISE BRIDGES

The impedance noise bridge is a low-cost method for testing antenna systems; a low-cost
commercial example is shown in Fig. 21–9.

A representative noise bridge circuit is shown in Fig. 21–10. The noise source is a
reverse-biased zener diode. It produces a wide-spectrum, pseudogaussian noise signal that
is amplified in a wideband (i.e., −3 dB at over 150 mHz) amplifier. The noise output from
the amplifier is applied to the primary of a special transformer, $T1$.

**FIGURE 21–9**
Low-cost RX noise bridge.

This transformer (*T*1) has a 1:1:1 ratio, and windings *L*1 through *L*3 are **trifilar wound** on a toroid core (see inset to Fig. 21–10). The dots in Fig. 21–10 indicate the same ends of *L*1, *L*2, and *L*3.

One end of *L*2 is connected to a noninductive potentiometer that has a dial calibrated in ohms, from zero to full scale (in this example, 250 Ω), and a variable capacitor connected to a dial marked in noncalibrated units of reactance from $X_c$–0–$X_L$. As in Fig. 21–8(d), *C*1 is balanced by *C*2 so that at $X = 0$, $C1 = C2$.

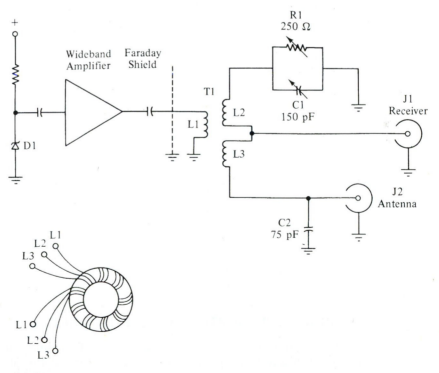

**FIGURE 21–10**
Circuit for noise bridge.

The detector used with a noise bridge is a general-coverage communications receiver, preferably a model equipped with an S-meter or at least an AGC voltage meter. The receiver will show a high noise level at all frequencies *except* the resonant frequency of the antenna.

The noise bridge can be used to find the resonant frequency of an antenna, to determine the correct physical length for a piece of transmission line that must be cut to an electrical half wavelength, and for certain other applications.

## 21–13    FINDING AN ANTENNA'S RESONANT FREQUENCY

There are two ways to use a noise bridge to find the resonant frequency of an antenna. For those bridges that do not use $C1$ and $C2$, the procedure is as follows:

**Procedure**

1. Set $R1$ to the anticipated radiation resistance of the antenna; tune the receiver to the anticipated resonant frequency.
2. Adjust the receiver tuning *very slowly* above and below the antenna's anticipated resonant point until a null is heard in the noise level or until a dip is noted on the receiver's S-meter.
3. Adjust the receiver dial and $R1$ for the deepest null in the noise level or for the minimum S-meter reading. These controls are slightly interactive. The receiver dial now indicates the resonant frequency, while the resistance of $R1$ is equal to the antenna's radiation resistance.

On models such as Fig. 21–10, which have $C1$ as a reactance control, use the following procedure:

**Procedure**

1. Set $R1$ to the anticipated radiation resistance; set $C1$ to $X = 0$ (i.e., midscale); and tune the receiver to the anticipated resonant frequency.
2. Adjust $C1$ for a null, or minimum, reading on the *S*-meter.
3. If the null occurs when $C1$ is in the $X_c$ region, the antenna is too long. But if the null is in the $X_L$ region, then it is too short.
   a. $X = X_c$: tune the receiver *downband* to locate $f$.
   b. $X = X_L$: tune the receiver upband to locate $f$.
4. Adjust $C1$, $R1$, and the receiver dial for best null, as in the previous example.

## 21–14    FINDING λ/2 LENGTH OF COAXIAL CABLE

We learned from Equation (21–13) that a piece of transmission line will vary from a physical half wavelength because of the velocity factor $V$. This phenomenon is due to the slower velocity of the radio signal in coaxial cable as compared with free space. The correct physical length to make an electrical half wavelength can be found by using a noise bridge and Equation (21–13) in the following procedure:

## Procedure

1. Use Equation (21–13) to calculate the required length, and then cut a length of cable that is slightly *longer.*
2. Set $X$ to 0 and $R1$ (in Fig. 21–10) to some resistance slightly above zero.
3. Connect one end of the cable to the noise bridge antenna port, and short the other end.
4. Tune the receiver to the frequency at which the cable must be a half wavelength, and then tune downband until a pronounced null is heard.
5. Patiently and slowly trim the cable length, and then reshort the end after each cut until the null moves upband to the desired frequency.
6. To check the result, remove the short and terminate the cable with a carbon or other noninductive resistance in the 10-to-250-$\Omega$ range. Adjust $R1$ for best null at the desired frequency. The dial on $R1$ should indicate a resistance that is very nearly the value of the resistor used to terminate the cable.

**21–15**

## USING THE NOISE BRIDGE TO FIND THE VELOCITY FACTOR OF COAXIAL CABLE

We may rearrange Equation (21–13) to give us the velocity factor $V$ in terms of line length and frequency:

$$V = \frac{LF}{492N} \tag{21–14}$$

where   $V$ = the velocity factor and has a value between 0 and 1
   $L$ = the length in feet (ft)
   $F$ = the frequency in megahertz (MHz)
   $N$ = any integer (for convenience, set $N = 1$)

## Procedure

1. Cut a sample length $L$ of the coaxial cable being tested. It should have a length that is an electrical half wavelength at some frequency within the tuning range of the receiver. Measure the length in feet.
2. Calculate the free-space wavelength of the cable, using the formula $F = 492/L$. Tune the receiver to this frequency.
3. Connect one end of the cable to the noise bridge, and *short* the other end.
4. Set $R1$ on the noise bridge to a value that is slightly higher than zero.
5. Tune the receiver downband until a null is obtained.
6. Using the length $L$ found in Step 1 and the frequency found in Step 5, use Equation (21–14) to calculate $V$.

**21–16     SWEEP TECHNIQUES**

A frequency-modulated (FM) or sawtooth-modulated sweep generator can also be used to test transmission lines and antennas for resonant frequency. Figure 21–11(a) shows the equipment configuration to make this type of measurement. The signals from the sweep generator and a crystal controlled marker generator (used to identify points on the frequency display) are combined in a resistor-diode adder circuit. The output of the adder, a dc voltage, remains relatively constant at frequencies removed from resonance. But at frequencies near resonance, a null or dip is noted. A sample oscilloscope trace for a piece of coaxial cable shorted at the half-wavelength point is shown in Fig. 21–11(b). The marker "pip" in the null indicates that the cable is indeed cut to the correct frequency.

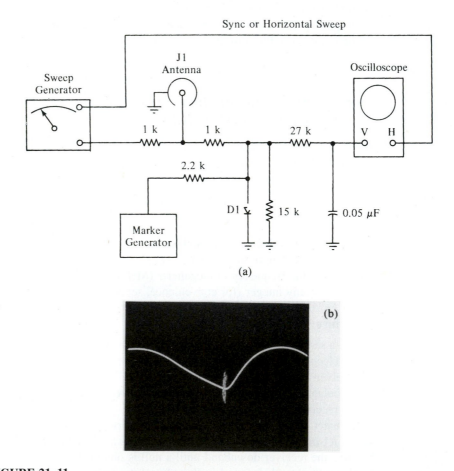

(a)

(b)

**FIGURE 21–11**
(a) Sweep testing resonant antennas, (b) resultant trace.

**21–17**      **TIME DOMAIN REFLECTOMETRY**

**Time domain reflectometry (TDR)** is a means for evaluating transmission lines from one end using a pulse generator and an oscilloscope. It works on the principle that a reflection from the far end of the transmission line, or from an anomaly (kink, short, or open) somewhere along the length of the transmission line, will create a pattern on the oscilloscope that is the algebraic sum of the forward and reflected pulses. Figure 21–12(a) shows the setup of the basic equipment for making TDR measurements on a transmission line. The signal source is a square wave generator or a pulse generator with an output impedance that matches that of the coaxial cable being tested. The signal source output is connected to the vertical input of the oscilloscope through one port of a coaxial T-connector; the remaining port of the T-connector is where the coaxial cable under test is connected. This oscilloscope display will be the combination of the forward pulse created by the signal generator and the reflected pulse (if any).

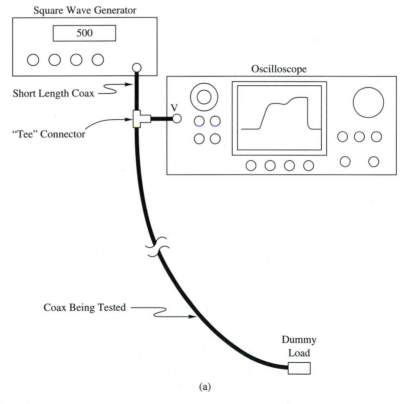

(a)

**FIGURE 21–12**
A method for doing time domain reflectometry.

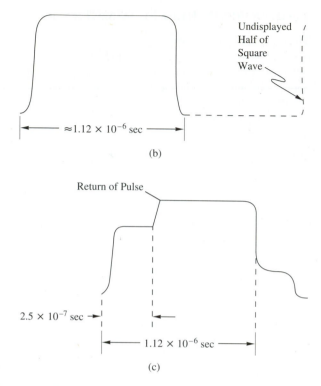

$\approx 1.12 \times 10^{-6}$ sec

Undisplayed Half of Square Wave

(b)

Return of Pulse

$2.5 \times 10^{-7}$ sec

$1.12 \times 10^{-6}$ sec

(c)

**FIGURE 21–12,** *continued*

Figure 21–12(b) shows the oscilloscope display when no coax is connected to the other T-port. This signal is the "raw" pulse from the signal generator. What is displayed is one-half of the square wave (note the dotted lines representing the missing half of the square wave). The time base of the oscilloscope and the frequency of the signal generator are adjusted so that only one-half of the square wave is shown given the approximate length of the cable.

In an experiment I performed when writing this material, I tested a 100-ft (30.5-m) length of RG-8/U 52-$\Omega$ coaxial cable. For purposes of calculation, the overall length was twice the physical length because the pulse must make a round trip; thus, our 30.5-m cable represented a round trip transit length of 61 m. For this length the appropriate frequency was between 450 and 550 kHz, and the time base of the oscilloscope was adjusted to 0.2 $\mu$s/cm. The width of the pulse on the screen was about 5.6 cm, so the duration of the pulse was 0.2 $\mu$s/cm $\times$ 5.6 cm = 1.12 $\mu$s = 1.12 $\times 10^{-6}$ s.

Figure 21–12(c) shows the pattern found when a 100-$\Omega$ load was connected to the end of the 61-m round trip length of this coaxial cable. The pulse shape changes because the reflected pulse is added to the forward pulse. Because the load impedance is higher than the correct impedance (100 $\Omega$ > 52 $\Omega$), the reflected pulse has the same polarity as the forward pulse, so it increases the amplitude of the displayed pulse. The reasons the displayed pulse is not twice the amplitude of the incident pulse are (a) attenuation in the

cable and (b) absorption of some of the forward energy by the mismatched load impedance.

In Fig. 21–12(c) the rise in the display pulse due to the return of the reflection occurs
at approximately 1.25 cm, as interpolated on the oscilloscope screen from the 0.2-cm
marking ticks. Thus, the return time of the pulse for the 61-m round trip was 1.25 cm ×
0.2 μs/cm = $2.5 \times 10^{-7}$ s.

In free space a radio signal travels at the speed of light ($c \approx 3 \times 10^8$ m/s). Thus, we
can expect the signal to travel 61 m in free space in $61/3 \times 10^8$ m/s = $2.03 \times 10^{-7}$ s. The
signal will travel more slowly in coaxial cable, and the ratio of the velocity in the cable
$v$ to the free-space velocity $c$ is called the **velocity factor** $V$ of the cable. For the type of
cable used for this experiment, the rated $V$ is 0.8 (according to the catalog). The ratio
$(2.03 \times 10^{-7}$ s$)/(2.5 \times 10^{-7}$ s$)$ is 0.81, which for the accuracy expected of this experiment
is quite good agreement.

The situation in Fig. 21–12(c) represents a VSWR of 2:1 because the load impedance
is twice the source impedance (100 Ω/50 Ω = 2:1). This situation was set up by design to
fulfill the purposes of the experiment and to make the oscilloscope waveform pho-
tographs in Figure 21–13. VSWR can be calculated from the trace, which is drawn in Fig.
21–12(c). The pedestal created by the reflected pulse was 4 cm above the baseline, while
the incident pulse was 3 cm. These values represent voltages, but because they are linear,
we can use just the relative heights to make VSWR calculations. Thus, the forward voltage pulse $E_F$ is 3 cm high, and the reflected voltage pulse $E_R$ is the difference between the
"back porch" (4 cm) and the incident pulse (3 cm), or 1 cm. VSWR can be calculated
from

$$\text{VSWR} = \frac{E_F + E_R}{E_F - E_R} \tag{21–15}$$

$$= \frac{3 + 1}{3 - 1} = \frac{4}{2} = 2:1$$

Figure 21–13 shows several different responses that represent different loads connected to the 30.5-m length of coaxial cable. The raw pulse is shown for reference in Fig.
21–13(a). This pulse, which has a frequency around 450 kHz, was from the 50-Ω output
of an Elenco model GF-8026 function generator.

The same pulse is shown in Fig. 21–13(b) with the coaxial cable being tested connected to the signal generator and oscilloscope and with the coax terminated in its characteristic impedance (50 Ω). The trace in Fig. 21–13(c) represents a 100-Ω load, which
makes for a 2:1 VSWR between the load and the cable. A VSWR of 6:1 is represented in
Fig. 21–13(d) and represents the situation when the 50-Ω cable is connected to a 300-Ω
load.

The traces in Figs. 21–13(a) through 21–13(d) represent a VSWR created when the
load impedance is higher than the cable impedance. In Figs. 21–13(e) and 21–13(f),
which represent loads of 22 and 6 Ω, respectively, notice that the "back porch" of the
pulse is a droop rather than a rise. This droop is caused by the fact that when the load
impedance is less than the source impedance, the reflected pulse is inverted.

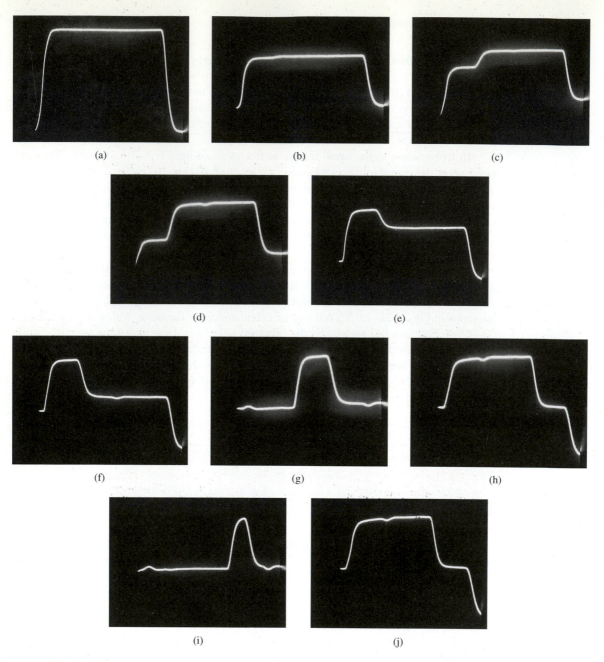

**FIGURE 21–13**

(a) The "raw" pulse before it is applied to the transmission line. (b) The pulse after the transmission line is connected. In this case the line is terminated in its characteristic impedance. (c) The pulse displayed when there is a 2:1 VSWR present and the load impedance is higher than the transmission line characteristic impedance. (d) The pulse with a 6:1 VSWR (300-$\Omega$ load, 52-$\Omega$ coaxial cable). (e) The pattern displayed when a 22-$\Omega$ load is connected to 52-$\Omega$ coaxial cable. (f) The pattern displayed with a 6-$\Omega$ load. (g) The pattern caused by shorting the load end of the coaxial cable. (h) The pattern caused by open-circuiting the load end of the coaxial cable. (i) The pattern when a short circuit exists halfway along the length of the coaxial cable. (j) The pattern when the time domain reflectometry setup is connected to a four-band ham radio vertical antenna.

Two opposite situations are shown in Figs. 21–13(g) and 21–13(h). In Fig. 21–13(g) the trace represents a load that is shorted (i.e., the load impedance is 0 Ω), while in Fig. 21–13(h) it is open (the load impedance is infinite).

In Fig. 21–13(i) the coaxial cable was shorted halfway, or at about 15.25 m (30.5 m round trip). Once you have enough experience, you can predict the location of the shorted or opened cable by noting the time for the pulse to reach it and return. A rule to follow is

$$L_{(m)} = \frac{V_c T_d}{2}$$
(21–16)

where $L_{(m)}$ = the length of the path in meters (m)
$V_c$ = the velocity factor of the coaxial cable
$T_d$ = the time for a round trip excursion of the pulse

For most common cables the velocity factors are as follows:

| Type of Cable | Value of $V$ |
| --- | --- |
| Polyethylene dielectric | 0.66 |
| Teflon dielectric | 0.70 |
| Polyfoam dielectric | 0.80 |

One final trace is shown in Fig. 21–13(j). I obtained this trace by connecting the transmitter end of the coaxial cable to my Cushcraft R-4 antenna. This antenna is a half-wave vertical with an impedance-matching network at the base of the antenna. Note that the antenna is fairly well matched. I suspect that the far right edge anomalies are due to the inductive and capacitive reactances in the impedance-matching network. Similar responses were noted when various LCR combinations were used on the 30.5-m test cable.

**21–18**     **POLAR PATTERNS**

An antenna may or may not be omnidirectional, depending upon its physical design. In fact, antenna **gain** is nothing more than directing the power in one or two directions instead of in all directions.

Polar patterns are often created in an anchroic test chamber in which the antenna, or a VHF model (i.e., scaled down in size), is surrounded by cones of RF absorptive material that reduces reflections (see Fig. 21–14). Alternatively, the antenna is in an outdoor antenna range where there are no objects for many wavelengths.

Figure 21–14 shows a model of an automobile with a pair of quarter-wavelength vertical monopoles mounted one-eighth wavelength apart on the trunk. This model is a 300-mHz scale-down of a 30-mHz (i.e., CB) cophased antenna system.

The method used to make the polar plot is diagrammed in Fig. 21–15(a). The antenna, or the model containing the antenna, is mounted on a motor-driven platform. The

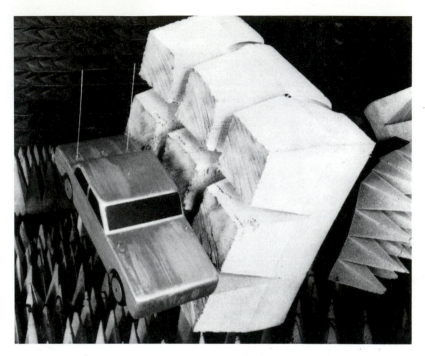

**FIGURE 21–14**
Antenna model in reflection-free test chamber (NASA photo—courtesy NASA).

platform position transducer commands a servomotor that causes a turntable chart recorder to rotate. Polar coordinate graph paper is mounted on the turntable, and a radio receiver tuned to the test frequency drives the pen position amplifier. A *distant* (i.e., many wavelengths) RF emitter (i.e., another antenna that is connected to a transmitter) sends out a signal whose strength is plotted as a function of radius on the graph paper. The servo system causes the paper to rotate as the model rotates, thereby creating the polar radiation plot of that antenna. Figure 21–15(b) and 21–15(c) show typical antenna patterns.

## 21–19      ANTENNA GAIN

The subject of antenna gain measurement usually provokes no small amount of controversy. The problem revolves around methodology and definitions, and these vary considerably from one manufacturer to another, so very different results are reported for the same general types of radio antennas. Some of this difference may be due to honest differences of opinion within the antenna engineering profession, but in many cases it appears to be the result of a little of the aforementioned "creative specification writing" by the company's advertising department. Higher gain for less money, after all, enhances customer appeal so much that some CB antenna companies now claim ridiculous forward-gain figures. We will endeavor to offer some sensible techniques that allow relative comparisons between different types of antennas.

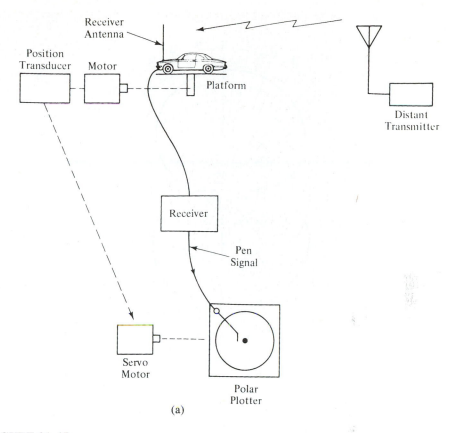

(a)

**FIGURE 21–15**
(a) Antenna range for plotting polar patterns.


Two different types of figures are often considered important in the comparison of gain antennas: **forward gain** and either **front-to-back** or **front-to-side** ratios (which depends upon antenna design).

The antenna gain is achieved by redirecting the energy from the transmitter in one direction, instead of in all directions (in the vertical monopole) or in two directions (in the horizontal dipole). The total energy in the field remains the same, but it is redistributed to increase the field intensity in the favored direction(s). The polar pattern shows us the favored directions, and the same pattern holds true for both transmit and receive conditions. In the receive condition the antenna pattern causes the receiver to be *more sensitive* in some directions than in others. In addition to increasing the signal arriving from the forward direction, it will also reject interfering signals arriving from the direction of the nulls in the polar pattern. This capability can be judged on a receiver. A single half-wavelength dipole is bidirectional (i.e., it has a figure-eight pattern), while a beam or yagi is unidirectional. Therefore, we would also measure the front-to-side ratio on a dipole and the front-to-back ratio on a beam.

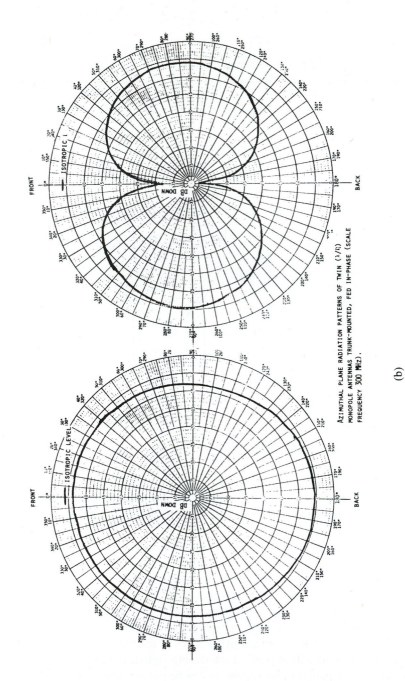

AZIMUTHAL PLANE RADIATION PATTERNS OF TWIN ($\lambda/l_1$) MONOPOLE ANTENNAS TRUNK-MOUNTED, FED IN-PHASE (SCALE FREQUENCY 300 MHz).

(b)

**FIGURE 21–15,** *continued*

(b) Polar patterns for several antennas (NASA photo—courtesy NASA).

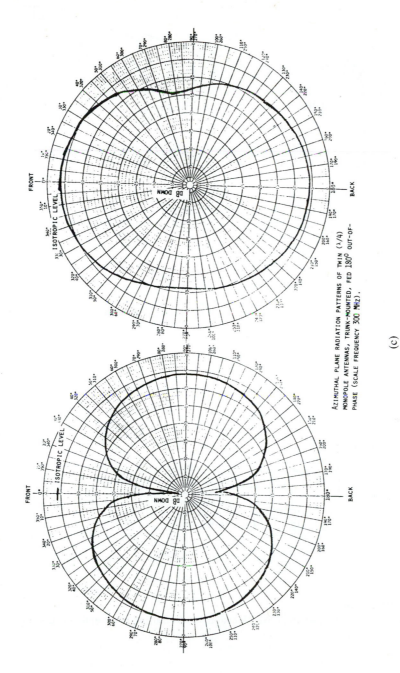

AZIMUTHAL PLANE RADIATION PATTERNS OF TWIN ($\lambda/4$) MONOPOLE ANTENNAS, TRUNK-MOUNTED, FED 180° OUT-OF-PHASE (SCALE FREQUENCY 300 MHz).

(c)

**FIGURE 21–15,** *continued*

(c) Polar patterns for several antennas (NASA photo—courtesy NASA).

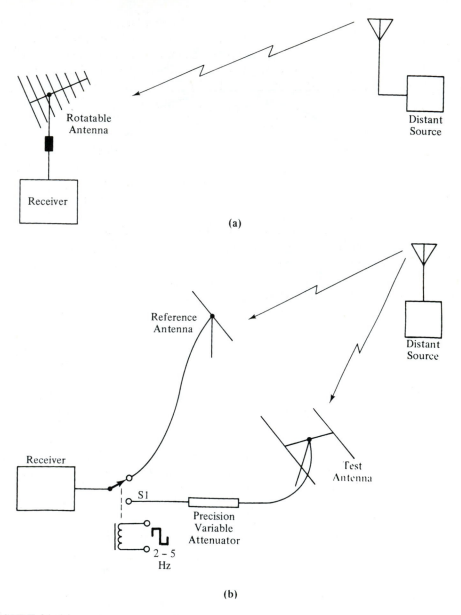

**FIGURE 21–16**
(a) Testing for polar pattern at fixed location, (b) measuring comparative antenna gain.

Figure 21–16(a) shows a test configuration for roughly measuring these ratios. The antenna being tested is installed in a way that allows it to be rotated. A constant output signal source is located downrange many wavelengths away. The signal strengths at the receiver are measured for two conditions: main lobe looking into the signal source and the null looking into the signal source. The levels obtained are compared, and the difference in decibels is noted.

The receiver's S-meter will give a rough indication in decibels, provided that its S-unit/dB calibration factor is known. A somewhat superior technique, however, is to use a precision RF attenuator that has calibrated, selectable steps. The attenuator is used in conjunction with a receiver S-meter, but the S-meter is used merely as a relative indicator. First note the precise S-meter reading when the antenna null is pointed toward the source. Next, insert the step attenuator into the line and point the antenna main lobe toward the signal source (i.e., note the point of maximum signal strength as the antenna is rotated). The signal level reported by the S-meter should have increased considerably. Now, adjust the step attenuator until the same S-meter is obtained as was found when the null was pointed at the antenna. The ratio is given by the amount of attenuation, in dB, that was required to bring the two signal strengths into equality.

Both of these methods, S-meter and attenuator reading, result in only an approximate result, so they should be repeated many times over the course of several days and then averaged. The principal problem is in keeping the intensity of the field from the RF emitter constant over the entire measurement period. This problem becomes especially acute when a distant broadcasting station is used as the source because of ionospheric problems and fading.

Antenna gain is measured by using a test setup similar to that of Fig. 21–16(b). Here we are referencing a test antenna to a standard antenna, such as a dipole, that is oriented in the same direction. Again, we use a distant RF emitter as the signal source, often a broadcasting station signal.

Switch $S1$ is an electrically driven coaxial relay that is pulsed to alternately connect both antennas to the receiver at a 2-to-5-Hz rate. A precision attenuator (as used above) is inserted in series with the transmission line of the higher-gain antenna.

Since the receiver input is alternating back and forth between two antennas with different gains, the S-meter will oscillate at the switching rate, attempting to measure the signal strength from first the reference antenna and then the test antenna. The degree of attenuation required to cause the S-meter to cease oscillating indicates the relative difference in dB between the two antennas. For example, if a three-element beam were compared with a dipole cut to the same frequency and oriented in the same direction, and if 4.6 dB of attenuation in the feed line from the beam were required to create equal signal levels, then the "gain" of the beam would be 4.6 dB *over a dipole* (which is sometimes written as 4.6 dBd).

# SUMMARY

**1.** Standing waves on a transmission line can be measured by using voltage, current, power, or impedances.

2. Antenna impedance can be measured by using a Wheatstone bridge or an RX noise bridge. Simple instruments, however, measure only the radiation resistance and ignore the reactive component of impedance.
3. The **velocity factor** of transmission lines causes the physical length of a half-wave line to be shorter than the calculated length (e.g., 492/*F*).
4. Antennas can be compared with each other to find **relative gain.**

## RECAPITULATION

Now go back and try to answer the questions at the beginning of the chapter. When you are finished, answer the questions and work the problems given below. Place a mark beside each problem or question that you cannot answer, and then go back to the text and reread appropriate sections.

## QUESTIONS

1. The impedance of an ideal antenna at resonance is equal to the _____ _____.
2. List two important properties of a transmission line: _____ _____ and _____.
3. The velocity of propagation in a transmission line is _____ than the velocity in free space.
4. Describe in your own words the meaning of **standing waves** as it applies to radio transmission lines.
5. Standing waves create _____ and _____ nodes and antinodes at half-wavelength intervals along the line.
6. Name four parameters that could be used to measure SWR.
7. SWR is dependent upon transmission line length. True or false?
8. Draw a circuit for an RX antenna impedance bridge.
9. Draw circuit for an RX noise bridge.
10. List two techniques for trimming a transmission line to half wavelength.

## PROBLEMS

1. An antenna shows a feed-point current of 1.6 A and a feed-point voltage of 26 V. Find its impedance at resonance.
2. Calculate the physical length (in feet) of a half-wavelength radiator in free space if the frequency is 18 mHz.
3. Calculate a practical length in feet of an 18-mHz half-wave dipole near the earth's surface.
4. Calculate the length (in feet) of a quarter-wavelength, 12-mHz, vertical radiator (at the earth's surface).

**5.** Calculate the physical length in feet required to make a half-wavelength stub at 22 mHz out of foam-filled coaxial cable.

**6.** The voltage along a transmission line is measured and is found to peak at 58 V and then drop to a minimum of 31 V at a point 9 ft away. Assume a perfect free-space transmission line to calculate (a) the SWR and (b) the operating frequency of the transmitter used to excite the transmission line.

**7.** The forward power at a point in a transmission line is 100 W, and the reflected power is 7 W. Find the VSWR.

**8.** The VSWR is measured as 3:1, and the forward power is 480 W. Calculate the reflected power.

## BIBLIOGRAPHY

Carr, Joseph J. *Practical Antenna Handbook,* 2nd ed. Blue Ridge Summit, PA: TAB/McGraw-Hill, 1994.

# 22

# Radio Receiver Measurement and Alignment

22–1

**OBJECTIVES**

1. To learn the fundamentals of radio receivers.
2. To learn the principal parameters of radio receivers.
3. To learn how to make the measurements of radio receiver performance.
4. To learn how to align a radio receiver.

22–2

**SELF-EVALUATION**

Before studying the material in this chapter, try to answer the questions given below. These questions test your knowledge of the subject. If you cannot answer a particular question, place a check mark beside it, and then look for the answer as you read the text.

1. Draw the block diagram for a simple superheterodyne receiver.
2. What is meant by receiver **sensitivity?**
3. Define receiver **selectivity** in your own words.
4. FM receivers use a ——————— sensitive demodulator.

22–3

**RECEIVER BASICS**

A radio or television receiver must do two jobs: (a) it must *intercept* the signal transmitted in the air and then *demodulate* it to recover the *information* impressed on the signal at the transmitter, and (b) it must *reject* unwanted signals. AM, CW, SSB, and FM receivers are functionally the same, except at the demodulator circuit and filtering bandwidth, so we will begin this discussion by considering an AM receiver.

The most elementary receiver is the crystal set, which in the simplest case consists of nothing more than a signal diode shunted across a pair of earphones. But for at least the past 70 years more complex varieties have been used.

An early attempt at receiver design preceded the diode used to demodulate AM signals, with a cascade chain of tuned RF amplifiers. This design, called the TRF receiver, faded because of technical difficulties in keeping several RF amplifiers tuned to the same frequency over a wide tuning range.

The superheterodyne design of Fig. 22–1 has been in standard use for over 60 years, and it proves to be the best design even today. In this type of receiver the RF signal captured by the antenna is heterodyned to another (usually lower) frequency called the **intermediate frequency (IF).** Most of the receiver's total gain is in the IF amplifier.

Figure 22–1 shows the usual stages found in a superheterodyne receiver. The RF amplifier receives the weak signal from the antenna and amplifies it before applying the signal to the mixer stage. The RF amplifier is a tuned stage, so it serves two functions other than amplification: (a) it improves the receiver's ability to reject unwanted signals, and (b) it prevents radiation of the local oscillator through the antenna.

The mixer stage and local oscillator (LO) are used to convert RF signals to IF. The LO signal is heterodyned against the RF to produce the IF. The mixer is the nonlinear element required for heterodyne action. The four frequencies found in the output of the mixer are the RF, LO, RF + LO, and RF − LO. Most receivers use the difference frequency between the LO and RF as the IF, so a tank circuit tuned to RF − LO will select only the mixer product designated as the IF and will reject the others. In some receivers a combination stage called a **converter** replaces the mixer/LO. Such a stage merely uses a single active element (i.e., a transistor) for both LO and mixer functions.

The IF amplifier is also tuned, but the tuning is not variable as it is in the RF and LO circuits; the IF remains tuned to one frequency (RF − LO). Since an amplifier that is tuned to but one frequency is easily neutralized, the IF amplifier can provide most of the gain in the receiver without oscillating. Low-cost AM band broadcast receivers may have only one IF amplifier stage, while certain communications receivers and FM broadcast receivers have four or five stages in cascade.

The detector or demodulator stage recovers the information modulated onto the RF signal at the transmitter. In AM receivers the detector is simply a diode rectifier and filter circuit. Since this detector is sensitive to amplitude variations, it is called an **envelope**

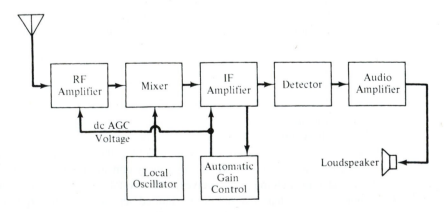

**FIGURE 22–1**
The superheterodyne receiver.

**detector.** The nonlinearity of the diode causes heterodyning between the AM signal's carrier and its sidebands, the difference frequency being the audio signal used to modulate the transmitter.

Single sideband (SSB) receivers must contend with the fact that the carrier was removed at the transmitter (Chapter 23), so it is not available for heterodyning against a sideband. The SSB receiver, then, must use a **product detector** that consists of a mixer and a fixed-frequency oscillator, followed by an audio frequency low-pass filter.

Frequency and phase modulation (FM and PM, respectively) must use a **phase-sensitive detector** to demodulate the carrier.

## 22–4 RECEIVER PARAMETERS

Certain specifications are common to all radio receivers, regardless of the frequency band that they cover. Some of these are **sensitivity, selectivity, image rejection,** and **noise figure.**

**Sensitivity** is a measure of a receiver's ability to pick up weak signals. Various definitions of sensitivity exist, leading to some confusion. One definition gives sensitivity as the level required across the antenna terminals to produce a certain amount of audio power at the output (usually 500 mW or 1 W).

Another definition holds that sensitivity is the signal level at the input required to produce a 10- or 20-dB reduction in output level (the noise level compared with the noise level at zero signal level).

**Selectivity** is the ability to reject unwanted signals. A receiver does not look at the entire radio spectrum at one time, but only at a narrow window centered on the frequency indicated on the dial. Selectivity is defined in terms of bandwidth, so ideally the receiver selectivity matches exactly the bandwidth of the transmitted signal.

An **image** is a spurious response due to the heterodyning process. An IF amplifier does not care whether the RF frequency is above or below the LO because it accepts the difference. If the LO is above the RF, then the IF is LO − RF. But a strong signal located at LO + IF also produces a difference frequency that is equal to the IF. The spurious signal is located at a frequency of twice the IF frequency from the RF, on the other side of the LO frequency, and the receiver must be able to reject that frequency.

Noise in a receiver (e.g., background hiss) is due largely to thermal agitation within the components. The **noise figure** in dB is the measure of how much noise is produced.

## 22–5 MEASURING SENSITIVITY

Recall that there are two definitions of sensitivity. In the following two sections we will discuss both methods of measuring sensitivity, describing circumstances under which each is appropriate.

Figure 22–2 shows the equipment required for both types of sensitivity measurement. A calibrated-output signal generator is connected to the antenna terminals, and a volt-

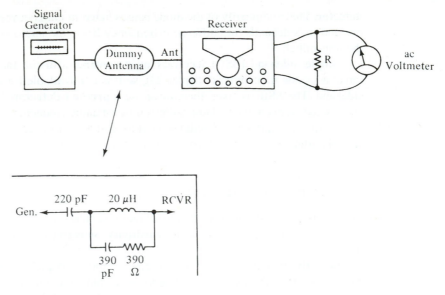

**FIGURE 22–2**
Measuring receiver sensitivity.

meter is connected across the output load. A load resistor $R$ is used in place of the speaker; its value should be equal to the nominal loudspeaker impedance.

A dummy antenna, in receiver alignment or testing, is a passive network that simulates a transmission line of infinite length.

## 22–6    AUDIO POWER METHOD

The **audio power method** for measuring sensitivity is used on AM receivers. Most manufacturers recommending this technique specify an AM signal with a certain percentage of modulation between 25% and 100% (usually 30% at 400 or 1000 Hz).

Audio power is measured by noting the output voltage across the speaker load and applying the relationship $E^2/R$. A typical manufacturer's specification will require 500 mW across an 8-$\Omega$ load, at 30% modulation of the carrier by a 400-Hz (or 1000-Hz) sine wave. This requirement means that we do the following:

### Procedure

**1.** Use a 400-Hz tone to modulate the generator 30%.
**2.** Find the RF input signal in *microvolts* ($\mu$V) that will cause the output load to dissipate 500 mW, as indicated by an rms output voltage of

$$E = (PR)^{1/2}$$
$$= [(0.5)(8)]^{1/2}$$
$$= (4)^{1/2} = \textbf{2 V}$$

3. Initially set the output control on the generator to zero, and then slowly advance the level higher until the output voltmeter reads 2 V rms.

Some receivers have an **automatic gain control (AGC)** circuit that senses the signal level and then creates a dc voltage proportional to the signal level. This voltage is fed back to the RF and IF stages to reduce their gain as signal strength increases. The AGC circuit must be disabled before any sensitivity measurement is possible. In some receivers an AGC-ON-OFF switch is provided. This switch should be set to OFF. If there is no switch, then refer to the manufacturer's service manual for instructions on disabling the AGC. Some will have you disconnect it from the AGC rectifier and clamp it to a given voltage (usually several volts negative).

22–7       ## THE QUIETING METHOD

The **quieting method** also uses the same equipment shown in Fig. 22–2, but the signal generator must be *unmodulated* as follows:

### Procedure

1. Set the signal generator output control to zero, and turn up the receiver volume and gain controls to maximum.
2. Measure the rms voltage across the load produced by the noise.
3. Increase the input signal level until the output voltage drops 10 dB (some use 20 dB) from its zero signal level. If a properly calibrated dB voltmeter is not available, then rearrange the standard formula for voltage dB to solve for $V_{final}$:

$$dB = 20 \log_{10} \left[ \frac{V_{initial}}{V_{final}} \right] \tag{22–1}$$

4. Read the sensitivity in microvolts from the output control of the signal generator.

### ■ Example 22–1

We want to measure the sensitivity of a receiver using the quieting method. We must determine the signal level required to produce 10 dB of quieting. Assuming that the zero signal output noise voltage is 5.6 V rms, calculate the voltage that should exist when the receiver is quieted −10 dB.

### *Solution*

$$10 = 20 \log_{10}(V_i/V_f) \tag{22–1}$$
$$10/20 = \log_{10}(V_i/V_f)$$
$$10^{(10/20)} = V_i/V_f$$
$$3.16 = V_i/V_f$$
$$V_f = V_i/3.16$$
$$V_f = (5.6)/(3.16) = \textbf{1.77 V rms}$$

An alternate quieting method for AM receivers is to measure the signal level required to produce an output signal 10 dB above the noise level from a 30% modulated AM signal.

**Procedure**

1. Connect the apparatus as in Fig. 22–2, set the signal generator output level to zero, and then adjust the receiver volume control to produce a readable output on the meter. If a calibrated audio VU or dB meter is available, then set the output level to 0 dB or 0 VU.
2. Set the signal generator modulation to 30% and adjust the output level until the receiver output level increases 10 dB. [If a dB meter is available, then use Equation (22–1).]
3. Read the sensitivity from the signal generator output level control or from the meter. (Again, it is necessary to disable the AGC of the receiver.)

**22–8**     SELECTIVITY MEASUREMENT

**Selectivity** is the bandwidth of a receiver and is a measure of the receiver's ability to reject unwanted signals. Most of a receiver's selectivity is due to the IF amplifier bandwidth; the RF and mixer bandwidth are somewhat broader. In most cases, therefore, selectivity is measured by injecting a signal at the IF into the system at the output of the mixer (or, alternatively, some manufacturers use the mixer RF input port to avoid detuning the first IF tank circuit). It is best to disable the LO regardless of which point is used for signal injection. Also, as before, disable the AGC circuit.

Figure 22–3 shows the equipment connection for selectivity measurement. The output indicator will be either a dc voltmeter across the AGC line or an RF voltmeter (i.e., a

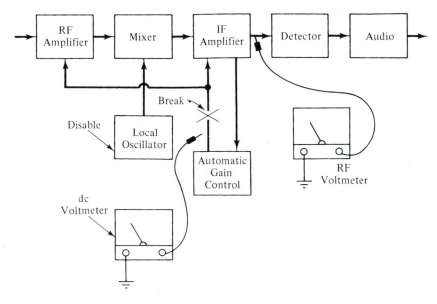

**FIGURE 22–3**
Measuring selectivity.

dc voltmeter with an RF detector probe) immediately preceding the receiver detector. If the receiver is an AM model, then an ac voltmeter across the speaker load may also be used. In the latter instance a modulated signal generator is needed.

### Procedure

1. Adjust the signal generator to a convenient output level, and then tune it to the IF, as indicated by a maximum output on the indicator.
2. Note the output voltage and the generator frequency. (*Note:* The output voltage obtained in Step 2 is used as the 0-dB reference level.)
3. Measure the output voltage at equally spaced frequencies above and below the IF, and record the voltages on a graph.
4. Plot the data *dB down-vs.-ΔF* in the form shown in Fig. 22–4(b). The selectivity is usually *defined* as the bandwidth in hertz or kilohertz between points where the response is −6 dB from the response at the IF.

The selectivity of receivers may be different in different reception modes, or there may be a means for varying the selectivity as needed. In any event, the technique above is terribly time-consuming. A superior technique is the sweep method described in the next section.

## 22–9    SWEEP METHOD

In the sweep method a sweep or FM signal generator is used to sweep back and forth across the IF. Figure 22–5(a) shows an appropriate equipment connection scheme. In some cases, incidentally, the sweep generator signal is at the RF and is applied to the antenna terminals of the receiver.

The audio output of the receiver, the marker, and the sweep signals are all applied to an **adder** before going to the vertical input of the oscilloscope. The marker oscillator(s) superimposes "pips" on the waveform [Fig. 22–5(b)] indicating precise frequency points.

In the sweep method we generally use the markers and the horizontal gain control of the oscilloscope to calibrate the oscilloscope screen in kHz/div. This calibration allows us to read bandwidth directly from the horizontal scale. The −6-dB points, incidentally, are the points where the *voltage* response drops to one-half the output level at the IF. These points are also known as the **half-power points.**

## 22–10    IMAGE RESPONSE

In Section 22.4 we defined the problem of image response in superheterodyne receivers. One of the most notable differences between poor and good radio receivers is this parameter. In fact, low-cost short-wave receivers are so poor in this respect that it is often difficult to discern which is the signal and which response is the image, especially in the over-15-mHz band.

Image response is reduced by using a sharply tuned, high-$Q$, RF amplifier "front end" and a high IF. A 9-mHz IF, for example, produces images $2 \times 9$ mHz $= 18$ mHz away from the RF, but a 455-kHz IF produces images only 0.91 mHz away from the RF signal. The measurement of the image response is as follows:

| −ΔF | E | dB | +ΔF | E | dB |
|---|---|---|---|---|---|
| 0 | 4 | 0* | +1 | 3.57 | −1 |
| −1 | 3.57 | −1 | +1.5 | 1.9 | −6.5 |
| −1.5 | 1.9 | −6.5 | +2 | 0.25 | −24 |
| −2 | 0.25 | −24 | +3 | 0.016 | −48 |
| −3 | 0.016 | −48 | +4 | 0.010 | −52 |
| −4 | 0.010 | −52 | +5 | 0.007 | −55 |
| −5 | 0.007 | −55 | | | |

*By definition

(a)

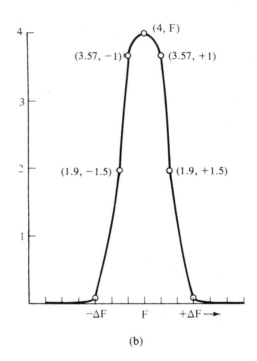

(b)

**FIGURE 22–4**
(a) Trial data, (b) data plotted.

**Procedure**

1. Tune the receiver and the signal generator to the same frequency. Adjust the generator output control for 50 μV or for some other setting that gives a readable output voltage across the load. Call this voltage V1.
2. Record the output voltage and signal generator attenuator setting (i.e., V1).
3. Tune the signal generator to a frequency equal to 2 × IF above or below the RF until the signal is heard in the output. Fine-tune the signal generator (*not* the receiver) for maximum output.

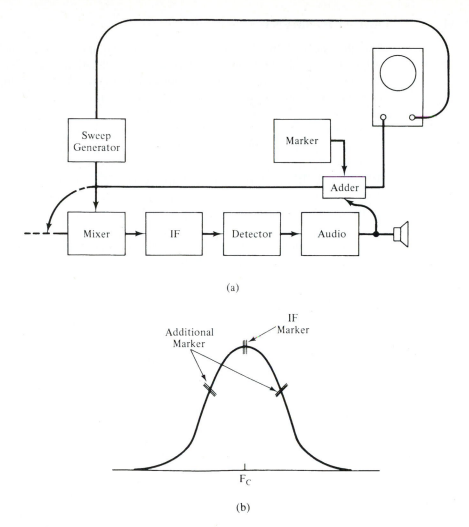

(a)

(b)

**FIGURE 22–5**
(a) Sweep alignment equipment setup, (b) sweep alignment IF response curve.

**4.** Adjust the generator output attenuator for the same receiver output as was obtained in Step 1. The generator output that creates this level, in microvolts, is $V2$. Apply the standard voltage decibel equation to find the image rejection in dB:

$$dB = 20 \log_{10}(V2/V1)$$

The result obtained is now the image response of the receiver in dB. Note that the image may occur above or below the initial RF signal, depending upon whether the LO is above or below the RF.

## 22–11     AM ALIGNMENT

The alignment of simple AM receivers uses an equipment setup such as in Fig. 22–6. Adjust the signal generator to produce a 30% modulation at a signal level that gives a readable output. Set the signal generator dial to a precisely known frequency within the receiver range. Set the receiver dial to the same point.

### Procedure

1. Adjust the local oscillator trimmer controls for maximum output signal.
2. Adjust the RF amplifier trimmer controls for maximum output signal.
3. Adjust the IF transformer tuning for maximum output signal.
4. Reduce the signal generator output level until noise is heard in the receiver output.
5. Repeat Steps 1 through 4 until no further improvement is possible.

The procedure given above is a generalization that will usually work on most receivers. Some manufacturers, however, specify their own procedure, especially on high-cost short-wave receivers.

## 22–12     FM RECEIVER ALIGNMENT

The FM receiver is similar to the AM receiver except in the demodulator. An FM receiver requires a phase-sensitive detector, whereas an AM receiver uses an envelope detector. Also, most broadcast and two-way communications FM receivers use three to five stages of IF amplification. Some models of more recent design use one or two integrated circuit gain blocks.

Many communications FM or PM receivers are dual conversion models such as that shown in Fig. 22–7. In a dual conversion receiver, the front end converts the RF signal to some high IF, usually 10.7 mHz. This high IF signal is then heterodyned against the output of a crystal oscillator to produce a low IF, usually 455 kHz.

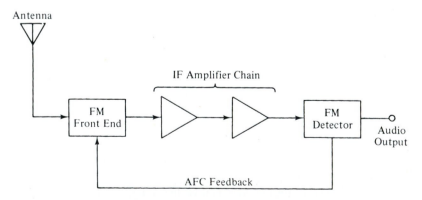

**FIGURE 22–6**
Block diagram of an FM receiver.

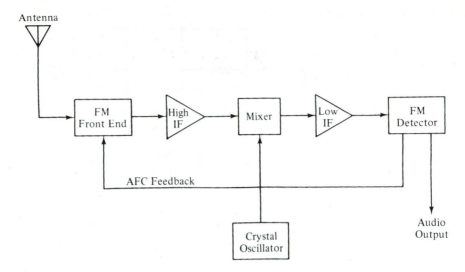

**FIGURE 22–7**
Block diagram of a dual conversion FM receiver.

One problem in alignment of FM receivers is to determine which of several popular FM detector circuits are used in the demodulator. Several types are in general use: **discriminator, ratio detector,** and **quadrature detector.**

Figure 22–8 shows the Foster-Seeley FM **discriminator.** In this circuit, signal voltages from $T1$ are added to signal voltages in the secondary. When the signal frequency is exactly equal to the frequency to which the secondary is tuned, the output voltage will be zero. FM deviation, whether caused by a shift in carrier frequency or by the process of frequency modulation, will produce a positive or negative output voltage, depending upon the direction of frequency shift.

The ratio detector is shown in Fig. 22–9. Two aspects of this circuit are different from the discriminator: (a) the diodes point in opposite directions, and (b) the circuit

**FIGURE 22–8**
FM discriminator.

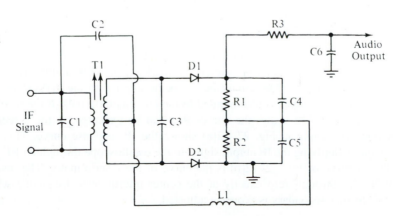

**FIGURE 22–9**
Ratio detector.

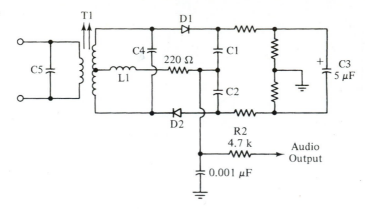

includes an electrolytic capacitor (*C*3). One of the functions of this capacitor is the suppression of amplitude modulation or noise signals. Because of *C*3, receivers using a ratio detector often do not have limiter amplifiers.

In a **ratio detector** the relative charges on capacitor *C*1 and *C*2 will have a ratio of 2:1 when the IF is precisely equal to the resonant frequency of the secondary of the transformer. Frequency deviation causes the ratio of the voltages across these capacitors to change, and this results in the audio output signal.

The last type of detector to be considered here is the **quadrature detector.** Although this type of detector was once popular in vacuum tube form, several IC models are available. The name of this detector is derived from the fact that two components of the IF signal, phased 90° apart, are combined in a synchronous detector circuit. A block diagram for a quadrature detector is shown in Fig. 22–10. The input stages are a wideband (i.e., untuned) high-gain limiter amplifier that is used to eliminate any AM or noise components on the IF signal. The output of this limiter consists of square waves, which are fed to a signal splitter. One output of the splitter is applied directly to one input of the synchronous detector, while the other is passed through a tank circuit that shifts its phase 90°. The output of the detector is integrated to recover the modulation audio.

**22–13**     **SWEEP ALIGNMENT**

The test equipment configuration for sweep alignment is shown in Fig. 22–11. The FM signal generator signal, a precise crystal marker generator, and the receiver audio output are summed in an adder circuit before being applied to the vertical input of an oscilloscope. The RF signal is also applied to the antenna terminals of the receiver.

Figure 22–12 shows the curves that can be expected on the screen of the oscilloscope. The curve in Fig. 22–12(a) shows the IF response curve. This curve would be found by applying the IF output signal to the oscilloscope through an RF voltmeter probe. The S-curve of Fig. 22–12(b) is the curve of the discriminator. The discriminator curve will pass through zero exactly at the center intermediate frequency when the detector transformer secondary is properly adjusted.

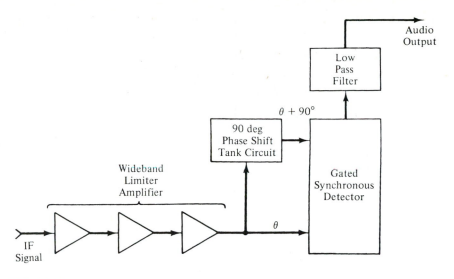

**FIGURE 22–10**
Block diagram of an IC quadrature detector.

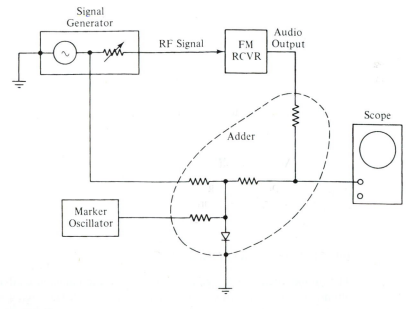

**FIGURE 22–11**
FM sweep alignment test equipment setup.

**531**

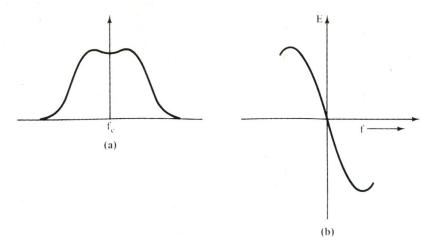

**FIGURE 22–12**
(a) FM IF response curve (dip should be no more than 10% of trace amplitude); (b) discriminator S-curve response.

In some alignment procedures a zero-center dc voltmeter is connected to the audio output line from the detector. The voltmeter will indicate zero output when the transformer is adjusted properly.

The quadrature detector is adjusted by using an ac voltmeter across the output of the receiver. The 90° phase shift network is then adjusted for maximum audio output.

Any type of FM receiver can be properly adjusted by using a total harmonic distortion analyzer (THD) at the receiver output. Adjust the FM detector transformer for minimum THD reading.

## 22–14        NONSWEPT ALIGNMENT

If only nonswept signal generators are available, it is still possible to perform the alignment. Connect the RF output of the signal generator to the antenna terminals of the receiver, and connect a zero-center dc voltmeter across the audio output of the detector, not the receiver. The proper connection point is prior to any ac coupling that is used. Set the signal generator to the precise center intermediate frequency, and then adjust the detector transformer for a zero volts indication. This technique is used primarily on discriminator detectors.

## 22–15        DUAL-SWEEP ALIGNMENT

The Sound Technology signal generators use the patented dual-sweep technique for FM alignment. In this method a low-frequency (i.e., 60-Hz) signal and a 10-kHz signal are used to modulate the FM oscillator. With proper oscilloscope connections, the output trace will appear as in Fig. 22–13. This trace contains points that are proportional to the

**FIGURE 22–13**
Dual-sweep trace (courtesy of Sound Technology, Inc.).

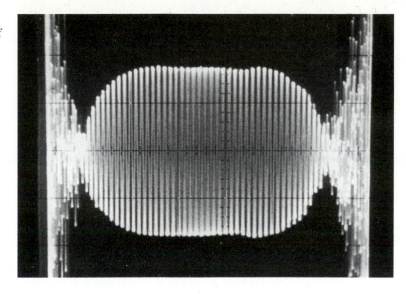

intermodulation distortion of the output signal, and this is directly related to the state of the IF alignment. The detector is adjusted for maximum flatness of the curve.

## SUMMARY

1. Four important receiver parameters are sensitivity, selectivity, image rejection ratio, and noise figure.
2. AM receivers use an envelope detector; SSB receivers use a product detector; and FM/PM receivers use a phase-sensitive detector.
3. Receiver sensitivity can be defined in terms of the signal required to produce a given audio output or the level required to produce a 10- or 20-dB quieting level.
4. Selectivity is usually measured as the bandwidth between −6-dB points on the IF response curve.

## RECAPITULATION

Now go back and try to answer the questions at the beginning of the chapter. When you are finished, answer the questions and work the problems given below. Place a check mark beside each problem or question that you cannot answer, and then go back to the text and reread appropriate sections.

## QUESTIONS

1. Define **selectivity** in your own words.
2. Define **sensitivity** in your own words.

3. Describe two methods for measuring sensitivity.
4. Describe a method for measuring selectivity.
5. There are two methods for specifying and measuring sensitivity. Do these two methods always yield the same numbers?
6. List three types of FM detector circuits.
7. In an FM detector the output voltage will be _____ when the input signal is unmodulated and precisely tuned to the center intermediate frequency.
8. In most receiver measurements or alignment procedures the _____ should be disabled.
9. The _____ stages of a receiver provide most of the gain and selectivity.
10. Spurious response to a signal removed from the RF signal by a factor of 2 × IF results from poor _____ rejection.

## PROBLEMS

1. A receiver being tested for sensitivity by the audio power method uses a 10-Ω speaker. Calculate the output voltage that indicates a 500-mW output power level.
2. We are testing receiver sensitivity by the quieting method using an unmodulated signal generator. If the zero signal noise level is 6.9 V rms, what level represents 10 dB of quieting?

# 23

# Spectrum Analyzers

## 23–1 OBJECTIVES

1. To learn the theory of operation of the spectrum analyzer.
2. To learn applications of spectrum analyzers.

## 23–2 SELF-EVALUATION

Before studying the material in this chapter, try to answer the questions given below. These questions test your knowledge of the subject. If you cannot answer a particular question, then look for the answer as you read the text.

1. A spectrum analyzer displays frequency and _____ information.
2. A spectrum analyzer is essentially a superheterodyne receiver with a _____
   _____ local oscillator frequency.
3. List three possible applications for the spectrum analyzer.
4. Draw the block diagram of a spectrum analyzer.

## 23–3 SPECTRUM ANALYZERS

The **spectrum analyzer** is an instrument that brings together a superheterodyne radio receiver with a swept-frequency local oscillator and an oscilloscope to present a display of amplitude vs. frequency. This instrument can be used to check the spectral purity of signal sources, to evaluate local electromagnetic interference (EMI) problems, to do site surveys prior to installing radio receiving or transmitting equipment, to test transmitters, and to perform a host of other applications. In this chapter you will learn the basics of spectrum analyzer operation, examine typical spectrum analyzer instruments, and learn a way to do laboratory experiments in spectrum analysis using low-cost, easily available components. You will also learn about computer software that will do spectrum analysis tasks on audio and near-audio waveforms.

## 23-4        BASICS OF SPECTRUM ANALYZERS

Figure 23–1 shows the relationship between the time and frequency domains. This example shows a waveform that is the summation of a sine wave fundamental and its 2nd harmonic. Figure 23–1(a) shows a three-dimensional coordinate system plotting the **amplitude vs. time** (*A-t*) data and **amplitude vs. frequency** (*A-f*) data.

Figure 23–1(b) shows the composite as viewed in the *A-t* plane. The trace produced is the classic waveform for 2nd harmonic interference on a fundamental. But the trace in Fig. 23–1(c) is a little different; it is an *A-f* bar graph showing the relative amplitudes of the two components. A **spectrum analyzer** is an oscilloscope that displays the *A-f* data.

The spectrum analyzer is actually a superheterodyne receiver (Fig. 23–2) in which the local oscillator is a sweep generator. A low-frequency sawtooth wave is applied to both the sweep oscillator and the horizontal deflection plates of the CRT, producing a horizontal deflection that is a function of frequency. The lowest frequency of the sweep oscillator is represented by the left side of the trace, while the highest frequency is represented by the right side. The sweep is from left to right.

The input signals are mixed with the local oscillator signal to produce the IF (i.e., difference) signal. The bandwidth of the IF amplifier is relatively narrow band, so the output signal at the detector will have a strength that is proportional to the frequency that the LO is converting to the IF at that instant. The display, then, will contain "poles" that represent the amplitudes of the various input frequency components.

Two examples of commercial spectrum analyzers are shown in Fig. 23–3. The example shown in Fig. 23–3(a) is the Hewlett-Packard model 8557A. It has a range of 10 kHz

**FIGURE 23–1**
The frequency-time domains.
(a) Three-dimensional coordinates showing time, frequency, and amplitude. The addition of a fundamental and its 2nd harmonic is shown as an example. (b) View seen in the *A-t* plane. On an oscilloscope only the composite $f_1 + 2f_1$ would be seen. (c) View seen in the *A-f* plane. Note how the components of the composite signal are clearly seen here (courtesy of Hewlett-Packard).

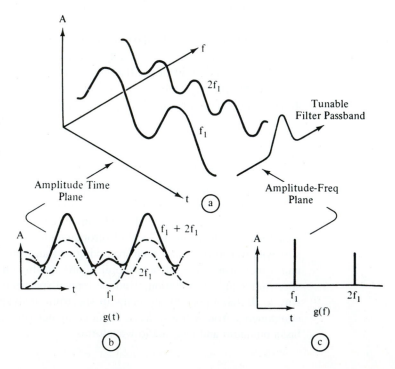

**FIGURE 23–2**

Block diagram of a spectrum analyzer (courtesy of Hewlett-Packard).

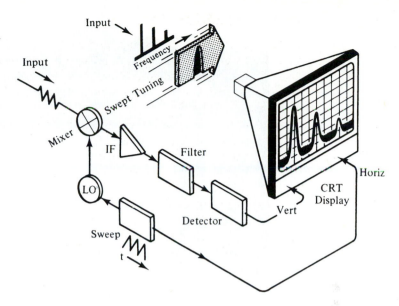

to 350 mHz and can accommodate signal levels between −117 and +20 dBm. This model has 12 calibrated sweep spans from 5 kHz/div to 20 mHz/div in a 1, 2, 5 sequence. The resolution varies from 1 kHz to 3 mHz, depending upon the range.

A somewhat wider-range instrument is shown in Fig. 23–3(b); it will cover 100 Hz to 1500 mHz, at input levels from −137 to +30 dBm. Notice that the parameters desired of the measurement are entered via a numeric keyboard.

Examples of spectrum analyzer traces are shown in Fig. 23–4. The trace in Fig. 23–4(a) shows the output spectrum of a 1500-mHz crystal controlled oscillator. The spectrum analyzer is set for 10-Hz resolution.

The results of a two-tone intermodulation distortion test on a 20-mHz transmitter are shown in Fig. 23–4(b). Note that the output is relatively pure, the 3rd-order IM products being greater than 85 dB down.

In Fig. 23–4(c) we see the spectrum of a 247-mHz transmitter, with the sidebands produced by a low-frequency modulating signal. The spectrum of a 1000-mHz AM transmitter is shown in Fig. 23–4(d). The sideband contains both the product due to the 2nd harmonic of the modulating signal as well as the products that are due to the signal itself.

The spectrum analyzer finds application in many areas of technology. Whether it is to check the output of a transmitter or to find the Fourier series of a signal from a vibration transducer, we may deduce much about the system being tested by using a spectrum analyzer.

## 23–5        LOW-COST SPECTRUM ANALYZERS

In recent years a new class of spectrum analyzers has become available. Although not in the same quality class as the instruments discussed earlier, they are satisfactory for many applications. For example, the lower-cost units will reveal whether or not a radio transmitter has an illegal harmonic of its fundamental operating frequency (which would inter-

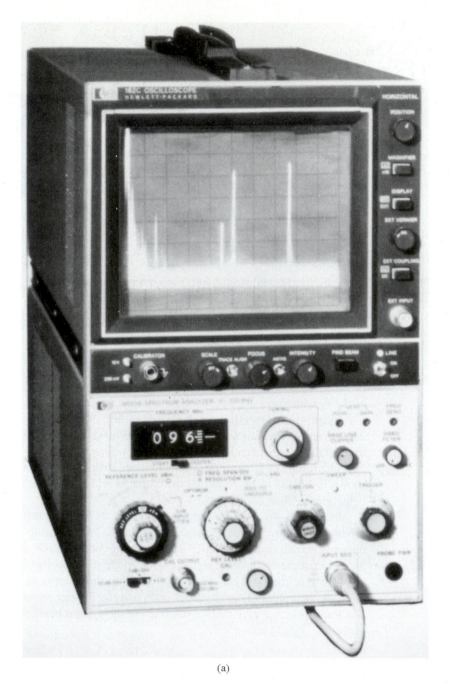

(a)

**FIGURE 23–3**

(a) 10-kHz-to-350-mHz spectrum analyzer (courtesy of Hewlett-Packard).

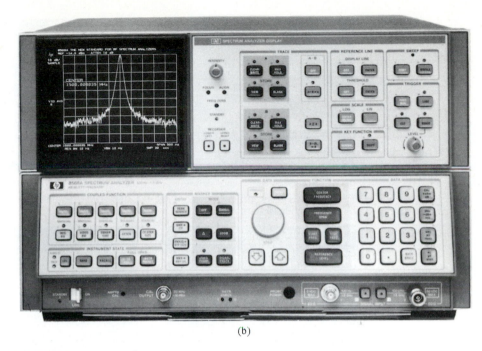

(b)

**FIGURE 23–3,** *continued*

(b) 100-Hz-to-1.5-GHz spectrum analyzer (courtesy of Hewlett-Packard).

fere with other stations operating on the harmonic frequency). Similarly, these instruments can be used for site surveys and EMI troubleshooting, especially if equipped with marker oscillators or other means for better calibrating the frequency display.

Figure 23–5 shows a simple, low-cost spectrum analyzer manufactured by Penntek Instruments. It is designed for use with an external X-Y oscilloscope and can be driven from an external sawtooth waveform generator, if desired. This instrument covers the VHF range from 30 MHz to 1300 MHz.

## 23–6 SPECTRUM ANALYZER EXPERIMENTS WITH LOW-COST COMPONENTS

It is possible to create a simple spectrum analyzer using simple, low-cost components. Figure 23–6 shows the block diagram for a workshop-built instrument. The swept-frequency front end of this spectrum analyzer is a voltage-tuned television or cable converter tuner. These instruments tune from 54 to 900 MHz with a voltage of 0 to 30 V. Television tuners are easily available on the market or can be salvaged from used TV receivers. The center output frequency of most of these tuners will be either 45 or 53 MHz, with the latter predominating. The bandwidth of the tuner will be 6 to 10 MHz.

A sawtooth generator that will create the 0-to-30-V waveform is needed. In most cases a standard laboratory or service shop sawtooth generator can be used if an external

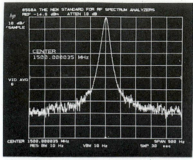

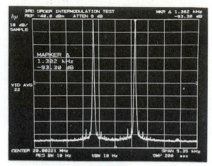

(a) 1.5-GHz crystal-based signal shown with 10-Hz resolution bandwidth. Note purity of signal and display.

(b) Two-tone test shows 3rd order IM products >85 dB down.

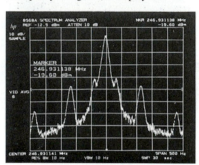

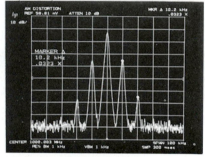

(c) Cavity-tuned VHF oscillator showing 30-Hz, 120-Hz, and 240-Hz sidebands.

(d) AM distortion: Marker Δ data indicates frequency and amplitude *difference* between modulation sideband and its harmonic.

**FIGURE 23–4**
Examples of spectrum analyzer traces (courtesy of Hewlett-Packard).

**FIGURE 23–5**
Low-cost spectrum analyzer (courtesy of Penntek Instruments).

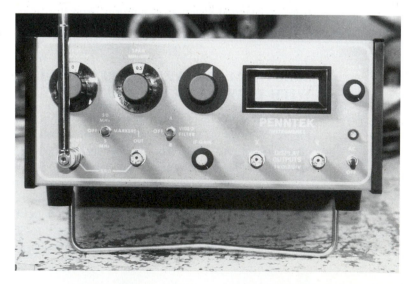

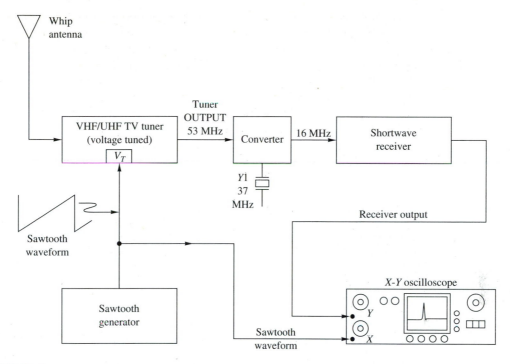

**FIGURE 23–6**
Block diagram of simple workshop-built spectrum analyzer.

dc amplifier with a +30 V dc output ceiling is provided. Use a sawtooth frequency of 20 to 50 Hz, and adjust the frequency to reduce flicker on the oscilloscope screen.

If a 53-MHz (or a 45-MHz) receiver is available, then you can use it to directly tune the output of the TV tuner. Otherwise, use a down-converter circuit, such as the NE-602AN IC converter, in which the 53-MHz frequency is down converted to the middle of the short-wave bands. In the case shown in Fig. 23–6, the desired converted frequency is 16 MHz, so a mixer with a local oscillator frequency of either 37 MHz or 69 MHz must be provided.

The short-wave receiver shown in Fig. 23–6 is the heart of the spectrum analyzer. The receiver must either have available or be modified to provide the automatic gain control (AGC) voltage to an output. Alternatively, the unfiltered output of the AM detector could be used. On most simple short-wave receivers this voltage should be easily available. The top of the volume control is sometimes a good point to tap, but consult the schematic for the receiver before making any modifications. The voltage desired is near dc, so there cannot be any coupling capacitors between the detector and the pick-off point.

A limitation on this form of spectrum analyzer is the resolution provided by the receiver intermediate frequency (IF) filtering. The principal difference between high-priced and low-priced spectrum analyzers is the frequency resolution. If the receiver has only AM filters, then the resolution will be 4 to 8 kHz, depending on the model receiver selected. For SSB receivers the filtering will be 2.2 to 2.9 kHz; for radioteletype (RTTY) receivers, 1.8 kHz; and for CW (Morse code) receivers, 250, 500, or 1000 Hz.

Figure 23–7 shows the results of using an experimental homemade spectrum analyzer kit offered by Science Workshop (Bethpage, NY). The instrument kit has a TV/cable tuner, a sawtooth waveform generator, and a fixed-tuned down converter and 10.7-MHz receiver. Figures 23–7(a) and 23–7(b) show the output of a crystal controlled test oscillator operating at 88 MHz. In Fig. 23–7(a) the sweep width of the spectrum analyzer is adjusted to show just this signal and its immediate sidebands. In Fig. 23–7(b) the sweep width of the spectrum analyzer is widened to reveal the fact that the signal gener-

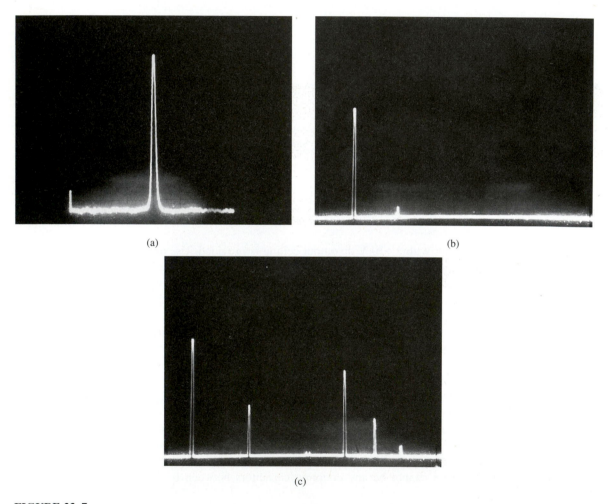

(a)                                                    (b)

(c)

**FIGURE 23–7**
(a) Spectrum analyzer display when the sensitivity is turned down. (b) Spectrum analyzer display when the sensitivity is higher. Note the emergence of a spurious response (harmonic).

ator is producing a 2nd-harmonic component (the small spike to right of the large spike). The scale is linear, so taking the ratio of the two spike lengths shows that the 2nd harmonic is only −19 dB (voltage) down from the main carrier signal.

Figure 23–7(c) shows the result of tuning the spectrum analyzer to the low end of the FM broadcast band. This pattern was taken near Washington, D.C. The spike on the far left was confirmed to be WAMU-FM (88.5 MHz), while the other spikes represent signals for about 3 MHz higher than 88 MHz.

**23–7**      ## SPECTRUM ANALYSIS SOFTWARE

It is possible to use computers to do audio spectrum analysis of signals, including signals received off the air if the computer is equipped with a sound board compatible with at least the 8-bit version of the SoundBlaster® board (or a similarly equipped Macintosh®. The sound board serves as an analog-to-digital converter to translate the analog audio signal into a series of digital data that can be digested by the computer.

The spectrum analysis software discussed here is Spectra Plus Ver. 2.0 by Pioneer Hill Software. Hardware requirements are rather modest by today's standards:

386 or later, IBM-compatible computer with 2 MB of memory minimum (or 4 MB for recording)

VGA monitor capable of at least 16 colors (which includes almost all monitors except monochrome)

hard disk space available of 1 MB (plus space for audio files, which can be large)

Windows® software

sound card (16-bit recommended, but 8-bit will work)

mouse or trackball

There are three modes and four functions. The modes are Real-Time, Recorder, and Post-Processing. The Real-Time mode accepts digitized audio directly from the sound card and then analyzes the waveform and displays the results. Although the program can be run indefinitely, the raw audio data cannot be saved in a disk file.

The Recorder mode digitizes the input audio signal and stores it on the hard drive in a .WAV file (the standard audio file format). Later these files can be analyzed in the Post-Processing mode or played back through an audio system.

The Post-Processing mode processes recorded audio data, whether from the Recorder mode or from other sources; it looks for a .WAV file. More of the functions of the software work with this mode than with Recorder or Real-Time.

The functions of Spectra Plus include Time Series, Spectrum, Spectrogram, and 3-D Surface Plot. Time Series is the ordinary amplitude-vs.-time display on an oscilloscope. In this mode the digitized audio is seen in the Volts/Time display on the computer screen, and it can be printed.

The Spectrum function analyzes the audio signal and then produces an amplitude-(volts)-vs.-frequency display (examples are given below).

The Spectrogram produces a display that has the frequency spectrum along the vertical axis and time along the horizontal axis. This type of display is useful for seeing the time history of the spectrum, that is, how the frequency content of a waveform changes over time.

The 3-D Surface Plot displays a three-dimensional perspective of the Spectrum over time. It contains amplitude information, frequency information, and time information.

**23–7–1**     ### Display Examples

Figures 23–8 through 23–10 show examples of software spectrum analyzer output printed on a standard office laser printer. The example in Fig. 23–8 is of a sound recording of an automobile engine. This pattern shows a large spike at a low frequency, which the soft-

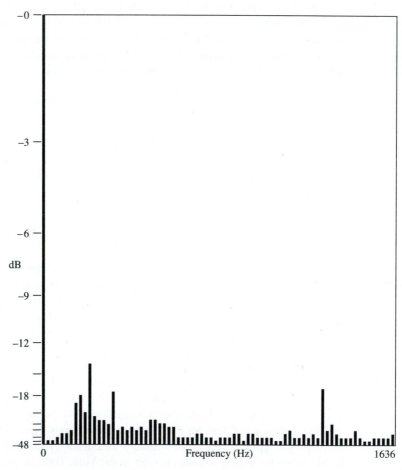

**FIGURE 23–8**
Sounding recording of an automobile engine.

ware takes as the 0-dB reference level. The other frequencies present are shown to the right of the large spike. Automotive engineers could use this type of display to find noise sources. It is likely that someday automobile mechanics will be trained to look for problems by examining the spectrum of an engine recording.

Figures 23–9 and 23–10 are plots of a radioteletype signal in which the MARK and SPACE (0 and 1 binary) levels are spaced 800 Hz apart. The frequency spectrum plot is shown in Fig. 23–9. Note that each spike has its own sidebands, which can be used to analyze the quality of the signal. A three-dimensional (3-D) plot of the same information is shown in Fig. 23–10. The vertical axis is the time plot, while the horizontal axis is the frequency plot. The amplitude data are shown in the third dimension (out of the page).

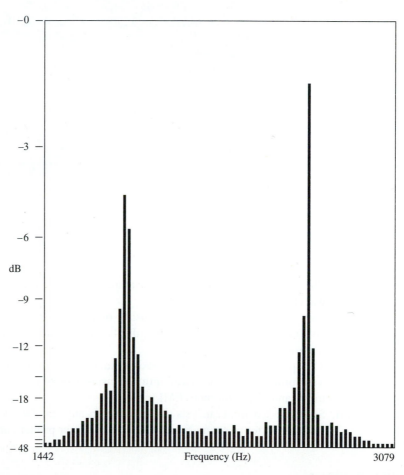

**FIGURE 23–9**
Frequency spectrum plot of radioteletype signal.

**FIGURE 23–10**
3-D plot of Fig. 23–9.

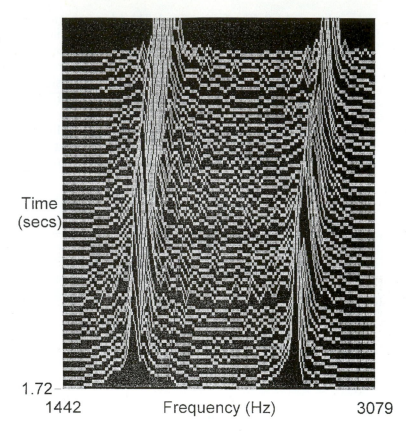

Time (secs)

1.72 —
1442          Frequency (Hz)          3079

## SUMMARY

1. The spectrum analyzer displays the amplitude-vs.-frequency spectrum of the measured waveform.
2. The basic spectrum analyzer consists of a swept-frequency, local oscillator, super-heterodyne receiver and an output display device such as an oscilloscope, a strip chart recorder, or a laser printer.
3. The spectrum analyzer can be used to examine the spectral purity of signal sources and transmitters, to do site surveys prior to installation of receiving and transmitting equipment, and to troubleshoot EMI problems.
4. Software spectrum analysis can be done on any standard computer equipped with an A/D converter or sound board. Output via a laser printer can be used for analysis.

## RECAPITULATION

Now go back and try to answer the questions at the beginning of this chapter. When you are finished, answer the questions and work the problems below. Place a mark beside each problem or question that you cannot answer, and then go back and reread the appropriate sections of the text.

## QUESTIONS

1. The spectrum analyzer displays _____ vs. _____ information. The spectrum analyzer is a _____ receiver with a swept-frequency local oscillator.

## PROBLEMS

1. Draw the block diagram of a spectrum analyzer based on a voltage-tuned TV tuner with a 45-MHz IF output. Assume that the TV/cable tuner used as the front end must send signal via a down converter to a 10.7-MHz fixed-tuned receiver. Specify the requirements of the receiver for 270 resolution and any modifications needed.
2. A spectrum analyzer has a linear amplitude scale. When the output of a 100-MHz VHF signal generator is observed, three spikes are observed: an 8.8-cm spike for the fundamental (100 MHz), a 2.3-cm spike for the 2nd harmonic (200 MHz), and a 0.2-cm spike for the 3rd harmonic (300 MHz). If the 100-MHz spike is the 0-dB reference, calculate the relationship in decibels of the 2nd and 3rd harmonics.

# 24

# Radio Transmitter Measurements

24–1        **OBJECTIVES**

1. To learn how to measure RF power.
2. To learn the different types of transmitter monitor instruments available.
3. To learn how AM and FM are measured.
4. To review frequency measurement techniques.

24–2        **SELF-EVALUATION**

Before studying the material in this chapter, try to answer the questions given below. These questions test your knowledge of the subject. If you cannot answer a particular question, place a check mark beside it, and then look for the answer as you read the text.

1. List two different types of RF wattmeters.
2. A _____ _____ meter consists of a discriminator and an output meter that is calibrated in units of frequency.
3. SSB transmitters are usually tested with a _____ _____ audio signal.
4. Thermocouple RF ammeters can be used to measure RF power up to approximately _____ MHz.
5. An _____ oscilloscope can be used as a linearity indicator for RF linear amplifiers.

24–3        **TRANSMITTER MEASUREMENTS**

Certain parameters must be measured on radio transmitters. For example, some measurements are required to assure that the transmitter meets FCC specifications and regulations. In other cases measurements are made to repair the equipment or to optimize the performance of newly installed transmitters. In still other cases manufacturers must make certain types of measurements to confirm engineering designs or to correct defects discovered in the testing.

In this chapter we will consider certain measurements that have been considered in other chapters. You are advised, however, to review Chapters 18, 20, and 22, and those sections of Chapter 9 that deal with frequency synthesizers.

**24–4**     ## RF POWER MEASUREMENTS

Power measurements at radio frequencies are sometimes considered difficult to achieve with any accuracy. But even crude, home-built instruments will yield ±25% accuracy, and certain low-cost professional equipment will yield better than ±5% accuracy.

Figure 24–1 shows a method often used in AM broadcasting stations, in which a thermocouple RF ammeter is connected in series with the load (i.e., the antenna or a dummy load). Such meters are inherently rms-reading devices, so they will measure the RF power from a wide variety of RF waveforms. Radio frequency power can be calculated from the conventional formula

$$P = I^2 R_L \qquad\qquad (24\text{–}1)$$

where     $P$ = the power delivered to the load
          $I$ = the line current to the load
          $R_L$ = the resistance of the load

■   **Example 24–1**

Calculate the power delivered to a 75-Ω load if the line current is measured as 1.8 A rms.

*Solution*

$$P = I^2 R_L \qquad\qquad (24\text{–}1)$$
$$= (1.8\ \text{A})^2 (75\ \Omega)$$
$$= (3.24\ \text{A}^2)(75\ \Omega) = \mathbf{243\ W}$$

---

**FIGURE 24–1**
Measuring RF power with a thermocouple RF ammeter.

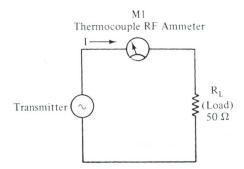

Radio frequency ammeters produce usable results at frequencies up to approximately 50 mHz or so. Most of these instruments are labeled "frequency compensated to 50 mHz" or some frequency in that region.

Radio frequency voltmeters are also used to make power measurements, although few of these instruments are actually rms-reading devices. Figure 24–2 shows an RF voltmeter that could be used to make RF power measurements. Resistor $R1$ is the load, while resistors $R2$ and $R3$ form a **voltage divider.** The voltage divider output is the voltage drop across $R3$, which is rectified by $D1$ and is used to charge $C2$ to the waveform's peak voltage. The rms voltage is equal to 0.707 times the peak voltage. The voltmeter, which is actually calibrated in watts instead of volts, measures a potential of

$$E_{\mathrm{m}} = \frac{ER3}{(R2 + R3)} \qquad\qquad (24\text{–}2)$$

where   $E_{\mathrm{m}}$ = the voltage across the meter
   $E$ = the peak voltage across the load
   $R2$ = the value of resistor 2
   $R3$ = the value of resistor 3

To find the RF power, we must first rewrite Equation (24–2) to solve for $E$:

$$E = \frac{E_{\mathrm{m}}(R2 + R3)}{R3} \qquad\qquad (24\text{–}3)$$

Once $E$ is determined, we may use the expression

$$P_{\mathrm{p}} = \frac{E^2}{R1} \qquad\qquad (24\text{–}4)$$

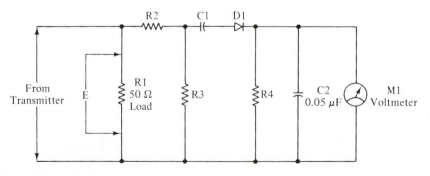

**FIGURE 24–2**
Measuring RF power by the voltage across the load.

to calculate the RF power dissipated in load resistor $R1$. But care must be used when measuring power with this type of circuit because the RF voltmeter is a peak-reading device. To find rms power, which is the important measurement, we use the expression

$$P = \frac{(0.707E)^2}{R1} \tag{24–5}$$

But this expression holds true only if the RF power waveform is sinusoidal. If any other waveform is used, then a correction factor other than 0.707 must be used, or the error must be tolerated.

An RF wattmeter using a toroidal core current transformer is shown in Fig. 24–3. This same type of instrument is also used in SWR measurements (Chapter 21). This type of meter also suffers from the necessity of ensuring a sinusoidal RF waveform.

Capacitors $C1$ and $C2$ are used to frequency-compensate the instrument so that it is useful over a wide frequency range, that is, 2 to 30 mHz.

There is no simple relationship between the RF power and the indication given by meter $M1$. Calibration, therefore, is best done by using an RF ammeter to correlate power and $M1$ readings.

The example of Fig. 24–2 showed a resistor voltage divider, but capacitor, inductor, and RC/RL dividers are also seen occasionally.

**FIGURE 24–3**
Toroidal core current transformer
RF wattmeter.

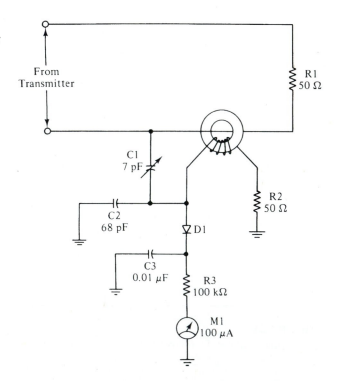

## 24–5     THERMAL METHOD

When any electrical energy is dissipated in a resistance, heat is created. In fact, textbooks define rms electrical values in terms of the amount of dc power required to generate the same amount of heat in a standard resistance. This same phenomenon gives us a technique for thermal RF power measurements.

Figure 24–4(a) shows the basic requirements for such an instrument. The load is not the antenna but a **dummy load,** that is, a standard-value, noninductive resistor. A thermal sensor is closely coupled to the load, and both are inside a thermal chamber that isolates them from outside heat sources.

An early version of this system placed the load inside an oil-filled, thermally insulated chamber, and a mercury thermometer measuring the oil temperature was visible from the outside of the chamber. The operator noted the temperature before and after the power was applied and then used a nomograph to calculate the power dissipated from the temperature data.

But this method is too cumbersome because it is necessary to allow the oil temperature to come to thermal equilibrium after RF power is applied. This procedure might take

**FIGURE 24–4**
(a) Thermal method for measuring
RF power, (b) test equipment setup
for measuring RF power using the
thermal method.

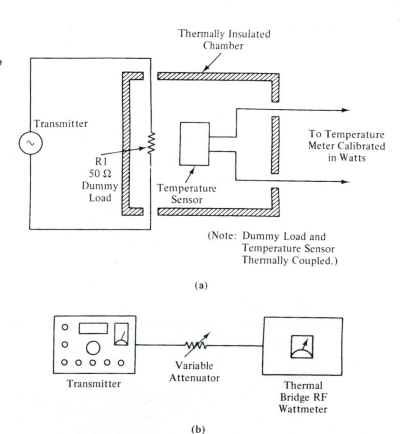

(a)

(b)

several minutes. Additionally, there is no direct power indication, so the user has to use a graph or an equation to interpret the results.

Modern thermal RF power meters use faster thermometry elements to make the measurement. All of the electronic thermal transducers of Chapter 13 have been used in this type of application. Being electronic, these thermometers can be calibrated directly in power units (i.e., watts).

Thermal sensors have two advantages: (a) they are inherently rms indicating, and (b) they will operate over a wide frequency range. One instrument, for example, boasts a frequency response from dc to the microwave range. The limitation on frequency response is a function of the properties of the dummy load.

Some thermal RF power meters have only a low-power-measuring range. One, for example, will measure RF power levels up to 100 mW. We can extend the power range by using a calibrated attenuator [Fig. 24–4(b)] between the transmitter and the thermal meter. The attenuator may be a precision step attenuator that can be adjusted, or it may be a fixed "barrel" attenuator. The latter type of attenuator uses a fixed-resistance network inside an extended RF connector. The attenuation factor is given in dB and is expressed by

$$dB = 10 \log_{10}\left(\frac{P1}{P2}\right) \tag{24–6}$$

where    dB is the notation for decibels
       $\log_{10}$ denotes the base-10 logarithms
       $P1$ = one power level (i.e., the input to the attenuator)
       $P2$ = the other power level (i.e., the attenuator output)

We may measure higher full-scale power levels by setting $P2$ equal to the wattmeter reading and then solving Equation (24–6) for $P1$ by taking 10 to a power equal to both sides of the equation:

$$dB = 10 \log_{10}(P1/P2) \tag{24–6}$$
$$dB/10 = \log_{10}(P1/P2)$$
$$10^{(dB/10)} = P1/P2$$

$$P1 = (P2)(10^{(dB/10)}) \tag{24–7}$$

■  **Example 24–2**

A thermal RF wattmeter connected as in Fig. 24–4(b) reads 84 mW when 15 dB of attenuation is used. Find the applied power.

*Solution*

$$P1 = (P2)(10^{(dB/10)}) \tag{24–7}$$
$$= (84 \text{ mW})(10^{(15/10)})$$
$$= (84 \text{ mW})(10^{1.5})$$
$$= (84 \text{ mW})(31.6) = 2654 \text{ mW} = \textbf{2.654 W}$$

The thermal RF wattmeter is used with a dummy load, so it will be able to measure transmitter output power. But to measure the RF power applied to an antenna system or other active load, we must use an RF wattmeter that may be inserted in the transmission line. Simple instruments based on circuits such as those in Figs. 24–2 and 24–3 will work only under the right conditions.

## 24–6    THRULINE®* WATTMETERS

When the Thruline® RF wattmeter is inserted into the transmission line of an antenna system, it presents so little loss that it may be left inserted during normal transmitter operations. Figure 24–5(a) and 24–5(b) show examples of the Thruline RF wattmeters that are popular with two-way radio servicers. The Thruline RF wattmeters use a directional coupler assembly, shown in Fig. 24–5(c), connected in series with the transmission line. The plug-in directional element can be rotated 180° to measure both forward and reflected power levels. A sampling loop and diode detector are contained within each plug-in element. The main RF barrel is actually a special coaxial line segment with a 50-$\Omega$ characteristic impedance. The Thruline sensor works due to the mutual inductance $M$ between the sample loop and the center conductor of the coaxial element. Figure 24–6 shows an equivalent circuit. The output voltage from the sampler ($e$) is the sum of two voltages, $e_r$ and $e_m$.

Voltage $e_r$ is created by the voltage divider action of $R$ and $C$ on transmission line voltage $E$. If $R$ is much less than $X_c$, then we may write the expression for $e_r$ as follows:

$$e_r = \frac{RE}{X_c} = RE(j\omega C) \qquad (24\text{–}8)$$

Voltage $e_m$, on the other hand, is due to mutual induction and is expressed by

$$e_m = I(j\omega) \pm M \qquad (24\text{–}9)$$

We now have the expression for both factors contributing to output voltage $e$.
We know that

$$e = e_r + e_m \qquad (24\text{–}10)$$

So, by substitution,

$$e = j\omega(CRE \pm M) \qquad (24\text{–}11)$$

If the sample element components are selected so that

$$CR = \frac{M}{Z_0} \qquad (24\text{–}12)$$

---

*"Thruline®" is a registered trademark of Bird Electronics.

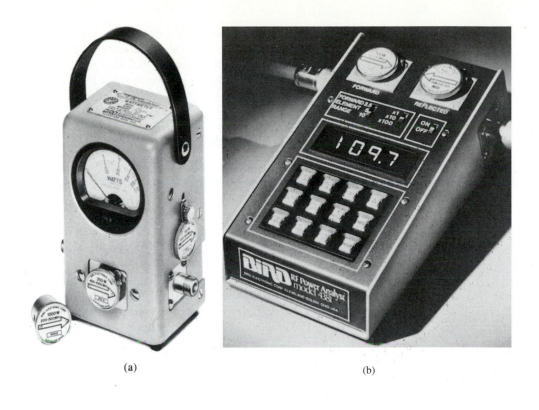

(a)

(b)

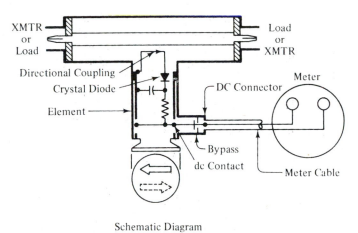

Schematic Diagram

(c)

**FIGURE 24–5**

(a) Bird model 43 Thruline® RF wattmeter, (b) digital readout Thruline® RF wattmeter, (c) construction of Thruline® RF wattmeter [(a)–(c) courtesy of Bird Electronics Corp.].

**FIGURE 24–6**
Equivalent circuit for Thruline®
RF wattmeter.

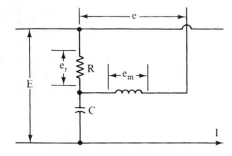

we may then write Equation (24–11) in the form

$$e = j\omega M \left[ \frac{E}{Z_0} \pm I \right] \tag{24–13}$$

by recognizing that at any given point in a transmission line, $E$ is the sum of the forward voltage $E_f$ and the reflected voltage $E_r$ and that the line current is equal to

$$I = \frac{E_f}{Z_0} - \frac{E_r}{Z_0} \tag{24–14}$$

We may specify $e$ in the forms

$$e = \frac{j\omega M(2E_f)}{Z_0} \tag{24–15}$$

and

$$e = \frac{j\omega M(2E_r)}{Z_0} \tag{24–16}$$

The output voltage $e$, then, is proportional to the mutual inductance and the frequency. But the manufacturer terminates $e$ in a capacitive reactance, so the frequency dependence is lessened (see Fig. 24–7). Each element is custom compensated, therefore, for a specific frequency and power range. Beyond the specified range for any given element, accuracy is not guaranteed.

The basic Thruline RF wattmeter is valid only in CW and unmodulated AM/FM systems. In many systems, however, peak power readings may be required. Figure 24–8 shows the envelope waveforms for several different types of transmitter emissions. For some of these waveforms, a peak-reading RF wattmeter is required. Bird Electronics makes the model 4311, based on its basic model 43 system, in which an IC amplifier-and-servo system finds the peak envelope voltage and uses it to derive the peak power value.

Consider an amplitude-modulated (AM) transmitter that is modulated 100% by a sinusoidal audio signal. If the transmitter is able to produce 100 W of unmodulated CW power, it produces a peak power in the modulation envelope of 400 W (i.e., $P_p = 4 \times P_a$).

**FIGURE 24–7**
Frequency response of various Bird model 43 elements (courtesy of Bird Electronics Corp.).

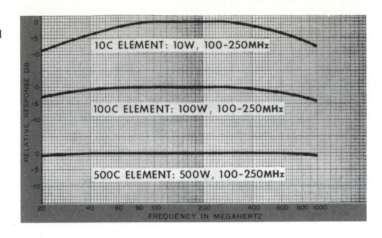

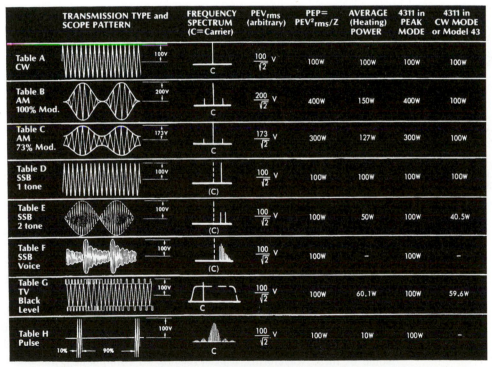

| TRANSMISSION TYPE and SCOPE PATTERN | | FREQUENCY SPECTRUM (C=Carrier) | PEV$_{rms}$ (arbitrary) | PEP= PEV$^2$rms/Z | AVERAGE (Heating) POWER | 4311 in PEAK MODE | 4311 in CW MODE or Model 43 |
|---|---|---|---|---|---|---|---|
| Table A CW | 100V | C | $\frac{100}{\sqrt{2}}$ V | 100W | 100W | 100W | 100W |
| Table B AM 100% Mod. | 200V | C | $\frac{200}{\sqrt{2}}$ V | 400W | 150W | 400W | 100W |
| Table C AM 73% Mod. | 173V | C | $\frac{173}{\sqrt{2}}$ V | 300W | 127W | 300W | 100W |
| Table D SSB 1 tone | 100V | (C) | $\frac{100}{\sqrt{2}}$ V | 100W | 100W | 100W | 100W |
| Table E SSB 2 tone | 100V | (C) | $\frac{100}{\sqrt{2}}$ V | 100W | 50W | 100W | 40.5W |
| Table F SSB Voice | 100V | (C) | $\frac{100}{\sqrt{2}}$ V | 100W | – | 100W | – |
| Table G TV Black Level | 100V | C | $\frac{100}{\sqrt{2}}$ V | 100W | 60.1W | 100W | 59.6W |
| Table H Pulse | 100V | C | $\frac{100}{\sqrt{2}}$ V | 100W | 10W | 100W | – |

In the table above, $Z_0$=50 ohms, PEP is Peak Envelope Power, and PEV is Peak Envelope Voltage. The PEV of the Carrier (or suppressed Carrier) C was arbitrarily chosen at 100 volts in all examples. PEV$_{rms}$=PEV/$\sqrt{2}$

**FIGURE 24–8**
Envelope waveforms (courtesy of Bird Electronics Corp.).

**24–7**

## DUMMY LOADS

It is illegal to perform some types of transmitter measurements with a live (i.e., radiating) antenna because of possible interference with other stations. In other cases it is simply not advisable to use an antenna, so **dummy loads** are used to absorb transmitter power.

A dummy load is a noninductive, pure resistance inside a shielded metal container (see Fig. 24–9). The value of the resistor is selected to match the impedance of the transmitter. Since most commercial antenna systems are designed to match 50 Ω, the majority of the dummy loads on the market have a value of 50 Ω.

An instrument called an **absorption wattmeter** is made by pairing a dummy load and an RF wattmeter. These instruments are often used on the test bench in making transmitter measurements or in certain applications in the broadcast industry.

**24–8**

## MODULATION MEASUREMENTS

**Modulation** is an information-carrying signal that modifies a radio carrier signal. The modulation may consist of audio speech signals or tones that represent analog or digital data.

In most communications applications one or more of three basic properties of the carrier signal are varied by the information signal: **amplitude, frequency,** or **phase.** We will consider amplitude modulation (AM) in one class, and frequency modulation (FM) and phase modulation (PM) in a class together.

**24–9**

## FM DEVIATION METERS

The relationship between the amplitude of an audio signal and the FM transmitter's output frequency is shown in Fig. 24–10(a). When the signal amplitude is zero, or when there is no signal present, the output frequency is $F_0$; this frequency is the assigned carrier frequency of the transmitter. When the audio signal amplitude begins increasing

**FIGURE 24–9**
Dummy load (courtesy of Bird
Electronics Corp.).

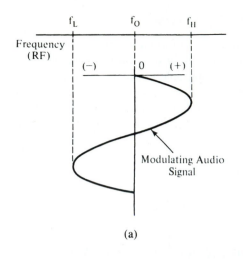

(a)

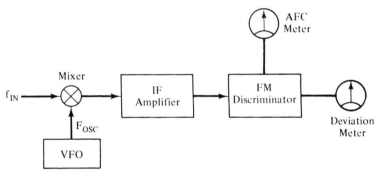

(b)

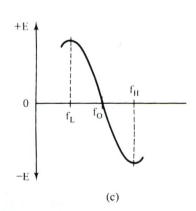

(c)

**FIGURE 24–10**
(a) Relative carrier frequency-vs.-modulating audio amplitude, (b) FM deviation meter, (c) FM discriminator curve.

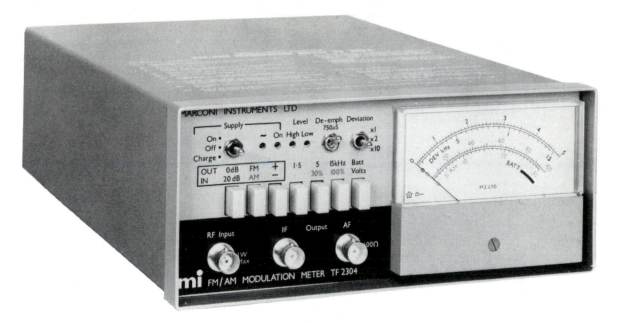

**FIGURE 24–10,** *continued*
(d) FM/AM modulation meter (courtesy of Marconi Instruments).

in the positive direction, the carrier frequency increases to $F_h$. Similarly, on negative excursions of the audio signal, the carrier frequency decreases to $F_L$. The measure of frequency modulation is **deviation,** which is defined as the frequency *change* $\Delta F$ from $F_0$ to either $F_h$ or $F_L$. Positive deviation, then, is the quantity $F_h - F_0$, and negative deviation is $F_0 - F_L$.

Figure 24–10(b) shows a typical deviation meter in block diagram form. It is basically an FM superheterodyne receiver with a dc voltmeter across the output of the discriminator detector. The **voltage-vs.-frequency** characteristic of a discriminator is the S-shaped curve of Fig. 24–10(c). The voltage on one side of the discriminator circuit will be positive an amount proportional to the positive deviation, while the voltage on the other half will be negative an amount proportional to the negative deviation. A commercial example is shown in Fig. 24–10(d).

An FM deviation meter with a variable-frequency oscillator will also include the automatic frequency control voltage from the discriminator to allow easier adjustment of the meter. The VFO is adjusted, with the transmitter turned on, for a zero output indication on the AFC meter. The transmitter is then modulated and the deviation measured.

The term "100% modulation" has physical meaning in AM systems, but in FM and PM systems it is defined by regulation or general agreement. In FM broadcasting, for example, a deviation of $\pm 75$ kHz is defined as 100% modulation, while in TV sound 25 kHz qualifies as 100%. Most two-way radio transmitters using FM/PM are considered 100% modulated if the deviation is $\pm 5$ kHz.

**24–10**     ## AM MEASUREMENTS

In an amplitude modulation system the audio signal adds to or subtracts from the RF carrier power. The relationship is shown in Fig. 24–11. Zone A represents the unmodulated carrier, and zones B and C show the maxima and minima created by a modulating audio sine wave. This modulation envelope can be viewed on an oscilloscope that is loosely coupled to the RF by a "gimmick," or a single-turn link. The oscilloscope time base is adjusted to show one or more cycles of the modulating frequency. The percentage of modulation is given by

$$\% \text{ modulation} = \frac{C - B}{C + B} \times 100\% \qquad (24\text{–}17)$$

■  **Example 24–3**

An oscilloscope with a $10 \times 10$ cm screen graticule is used to measure the percentage of modulation of an AM transmitter. It is found that dimension C [Fig. 24–11(a)] is 8.6 cm and dimension B is 2.4 cm. Calculate the percentage of modulation.

**FIGURE 24–11**
(a) AM waveform, (b) applying AM waveform to Y-time oscilloscope.

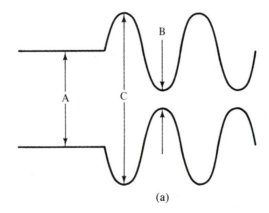

(a)

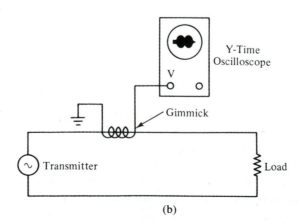

(b)

*Solution*

$$\% \text{ modulation} = (C - B)(100\%)/(C + B) \qquad \textbf{(24–17)}$$
$$= (8.6 - 2.4)(100\%)/(8.6 + 2.4)$$
$$= (6.2)(100\%)/(11) = \textbf{56\%}$$

Another method for measuring amplitude modulation is shown in Fig. 24–12(a). The RF envelope is applied to the vertical input of the oscilloscope, and a sample of the modulating audio is applied to the horizontal input. (An X-Y oscilloscope is needed.) The trapezoidal pattern of Fig. 24–12(b) results when a sinusoidal audio signal is applied to

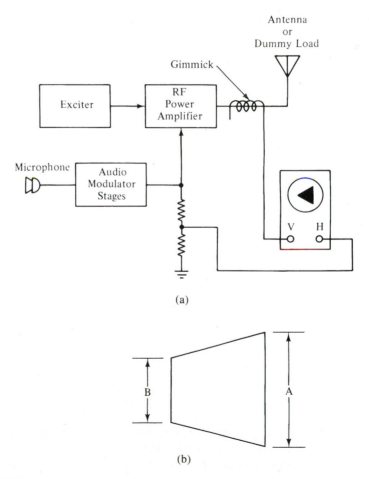

(a)

(b)

**FIGURE 24–12**
(a) Equipment connection for measuring AM modulation percentage by the trapezoid method, (b) AM trapezoidal trace.

**FIGURE 24–13**
Trapezoidal traces for (a) less than 100% modulation, (b) 100% modulation, (c) over 100% modulation.

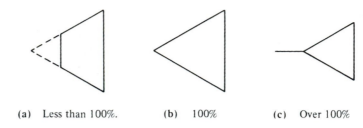

(a)  Less than 100%.        (b)    100%        (c)   Over 100%

the modulator. Using the notation of Fig. 24–12(b), we may compute the percentage of modulation from

$$\% \text{ modulation} = \frac{A - B}{A + B} \times 100\% \qquad \text{(24–18)}$$

■ **Example 24–4**

Calculate the percentage of modulation if dimension A occupies 4.6 divisions on the oscilloscope screen and dimension B occupies 1.4 divisions.

*Solution*

$$
\begin{aligned}
\% \text{ modulation} &= (A - B)(100\%)/(A + B) \qquad \text{(24–18)}\\
&= (4.6 - 1.4)(100\%)/(4.6 + 1.4)\\
&= (3.2)(100\%)/(6) = \mathbf{53\%}
\end{aligned}
$$

Figure 24–13 shows trapezoidal patterns for less than 100%, 100%, and greater than 100% modulation.

**24–11**        ## SSB MEASUREMENTS

A single sideband suppressed carrier (SSBSC, or simply SSB) transmitter is a special case of the AM transmitter. Figure 24–14(a) shows the frequency spectrum for a regular AM transmitter, in this case a 1000-kHz carrier modulated by a 1-kHz audio sine wave.

Amplitude modulation is a multiplication process in which the audio and the RF carrier signals are heterodyned together, resulting in the production of sum and difference frequencies. The sum and difference are RF frequencies and are called **sidebands.** In Fig. 24–14(a) we see the 1000-kHz carrier $F_0$, the sum frequency $F_2$ of 1001 kHz, and a difference frequency $F_1$ of 999 kHz.

In SSB transmitters, carrier frequency $F_0$ and *one* of the two sidebands are suppressed, so only one sideband is transmitted. The spectrum of an upper sideband (USB) transmitter is shown in Fig. 24–14(b); $F_0$ and $F_1$ have been suppressed, and only $F_2$ is transmitted.

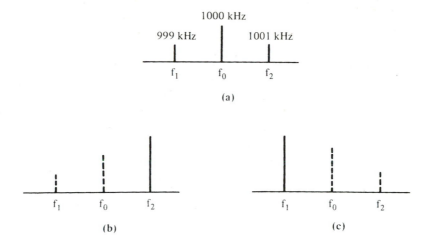

**FIGURE 24–14**
(a) AM signal, (b) USB signal, (c) LSB signal.

Similarly, the spectrum for a lower sideband (LSB) transmitter is shown in Fig. 24–14(c). In this case $F_0$ and $F_2$ are suppressed, and only $F_1$ is transmitted.

All stages of an SSB transmitter past the modulator must be linear amplifiers. As a result of this requirement, you must use a two-tone test signal when making SSB measurements. The two tones are used to modulate the SSB transmitter, and the output signal is inspected for the existence of intermodulation distortion products, or flattening of the waveform. Figure 24–15(a) shows a normal two-tone output wave, using a transmitter connected to an oscilloscope as in Fig. 24–11(b). Note that the SSB percentage of modulation has no real meaning, as it does in straight AM, and the transmitter audio control is

**FIGURE 24–15**
(a) Normal, (b) clipped.

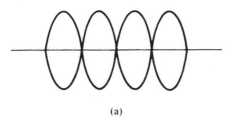

(a)

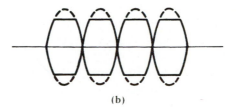

(b)

**FIGURE 24–16**
Two-tone "bow-tie" trapezoidal pattern.

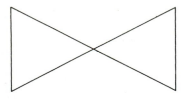

turned up until the output signal begins "flat-topping" [Fig. 24–15(b)]. The gain is then backed off until the flattening disappears.

A spectrum analyzer will show intermodulation distortion by the creation of tone products other than the two tones applied to the transmitter audio input.

Connection of the transmitter is similar to that shown in Fig. 24–12(a), in which the RF amplifier output is fed to the vertical input of the oscilloscope, and the audio signal is applied to the horizontal input of the oscilloscope. The pattern produced by a two-tone test signal will be the "bow-tie" of Fig. 24–16. Distortion will show up as bending of the ordinarily straight and sharp edges of the pattern. This technique is also used to test linear amplifiers.

Figure 24–17(a) shows another test configuration for checking the linearity of SSB power amplifiers. This technique uses Lissajous figures on an oscilloscope. The oscillo-

**FIGURE 24–17**
(a) Linearity testing of an RF power amplifier, (b) linear response, (c) Nonlinearity present.

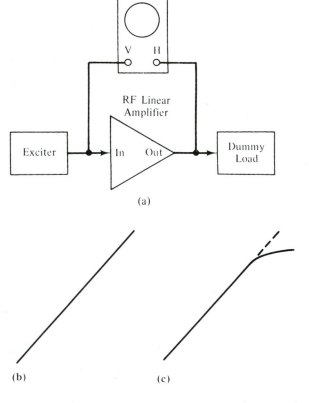

(a)

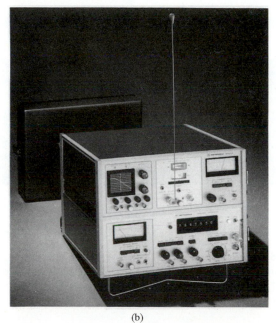

(b)

**FIGURE 24–18**
(a) and (b) Monitors using PLL frequency synthesizer [(a) courtesy of Cushman, (b) courtesy of Motorola Communications].

(c)

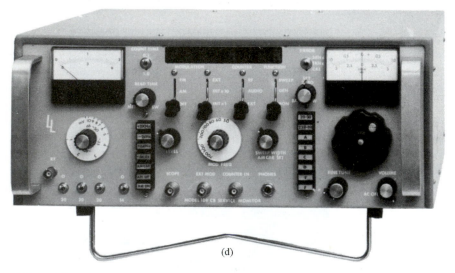

(d)

**FIGURE 24–18,** *continued*
(c) Monitor using a digital counter, (d) CB monitor [(c) courtesy of Motorola Communications, (d) courtesy of Lampkin Laboratories, Inc.].

scope gains are adjusted for approximately equal deflection amplitudes on both vertical and horizontal channels. A perfectly linear RF amplifier produces the pattern of Fig. 24–17(b), whereas nonlinearity shows up as a bending of the line, as in Fig. 24–17(c). Phase shift will show up as a thickening of the line or as an oval trace, depending upon the severity.

**24–12**  **COMMUNICATIONS MONITORS**

Communications monitors contain the principal instruments needed to service two-way radios (marine, landmobile, aeronautical, etc.) in one cabinet. These instruments typically contain a modulated signal generator, a frequency meter, modulation meters (AM percentage and FM deviation), and, sometimes, various standard IF alignment signals.

Two monitors that use the phase locked loop (PLL) frequency synthesizer are shown in Figs. 24–18(a) and 24–18(b). The Cushman model CE-5 is shown in Fig. 24–18(a), and a Motorola model is shown in Fig. 22–18(b). In both cases frequency is selected by switches on the front panel. The Cushman model is set to 125.000 mHz, while the Motorola is set to 164.5550 mHz. In both models a **kilohertz error** meter and an FM deviation meter are provided, along with an oscilloscope to view the modulation deviation characteristics of the transmitter. These instruments combine deviation and frequency measurement capability in one cabinet. Figure 24–18(c) shows a similar instrument, also by Motorola, in which the frequency meter section is a digital counter.

A Lampkin model 109 CB service monitor is shown in Fig. 24–18(d). This type of monitor has become popular in recent years because of the tremendous expansion in CB activity. The instrument contains a digital frequency counter and a signal generator covering commonly used frequencies (RF, IF, local oscillator, etc.).

## SUMMARY

1. Simple voltage bridge RF wattmeters produce good results if CW or FM signals are applied and if you have sinusoidal wave shape.
2. Thermal RF wattmeters and thermocouple RF ammeters produce true rms readings.
3. A dummy load is a shielded, nonreactive resistance with a value equal to either the expected load impedance of the transmitter or the surge impedance of the transmission line.
4. FM deviation meters are essentially FM receivers with a discriminator output voltmeter calibrated in units of frequency.
5. Amplitude modulation may be checked on an oscilloscope by viewing either the trapezoidal pattern or a modulation envelope display.

## RECAPITULATION

Now go back and try to answer the questions at the beginning of the chapter. When you are finished, answer the questions and work the problems given below. Place a mark beside each problem or question that you cannot answer, and then go back to the text and reread appropriate sections.

## QUESTIONS

1. How is RF power measured with an ammeter?
2. List two types of RF power measurements that yield an rms result.
3. Draw a schematic for two types of simple RF wattmeters. State any limitations of these instruments.
4. Describe in your own words what is meant by **dummy load.**
5. An instrument that mates a dummy load with an RF wattmeter is called an _____.
6. How is percentage of modulation defined for (a) AM and (b) FM?
7. In SSB transmitters a _____ _____ audio signal is used for modulation measurements.
8. A principal measurement of SSB transmitter linearity is the _____ distortion present in the output waveform.
9. A _____ Lissajous pattern gives an indication of SSB distortion.

## PROBLEMS

1. A transmitter causes 2.85 A of RF current to flow in a 36-$\Omega$ resistive load. Calculate the RF power level.
2. Calculate the rms power of a sine wave that has a peak potential of 52 V across a 50-$\Omega$ load.
3. A transmitter produces a reading of 98 mW in a thermal power meter when 22 dB of attenuation is used. Calculate the transmitter output power.
4. Calculate the percentage of modulation indicated on the RF envelope in Fig. 24–19(a).
5. Calculate the percentage of modulation indicated by the trapezoidal pattern of Fig. 24–19(b).

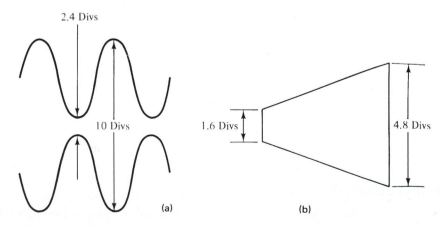

**FIGURE 24–19**
(a) RF envelope for Problem 4, (b) trapezoidal pattern for Problem 5.

# 25

# IEEE-488 General Purpose Interface Bus (GPIB) Instruments

**25–1**   **OBJECTIVES**

**1.** To learn the details of the IEEE-488 GPIB instruments.
**2.** To learn how to use IEEE-488 GPIB instruments.

**25–2**   **SELF-EVALUATION**

Before studying the material in this chapter, try to answer the questions given below. These questions test your knowledge of the subject. If you cannot answer a particular question, then look for the answer as you read the text.

**1.** The general purpose interface bus (GPIB) is used primarily in _____ _____ equipment applications.
**2.** Name the three major buses in the GPIB.
**3.** Name the two basic configurations for an IEEE-488 instrumentation system.
**4.** What is the function of the DIO1 through DIO8 lines?

**25–3**   **INTRODUCTION TO THE GPIB**

Automatic test equipment is now one of the leading methods for testing electronic equipment in factory production and troubleshooting situations. The basic method is to use a programmable digital computer to control a bank of test instruments. The program, often in a dialect of the BASIC language, turns the various instruments on and off and then evaluates the results as measured by other instruments.

The bank of equipment (see Fig. 25–1 for an example) can be configured for a special purpose or for general use. For example, we could select a particular line-up of equipment needed to test a broadcast audio console and provide a computer program to make the various measurements: gain, frequency response, total harmonic distortion, and so forth. Alternatively, we could make a generalized test set. This method is selected by a number of organizations who have many different electronic devices to test. There will be

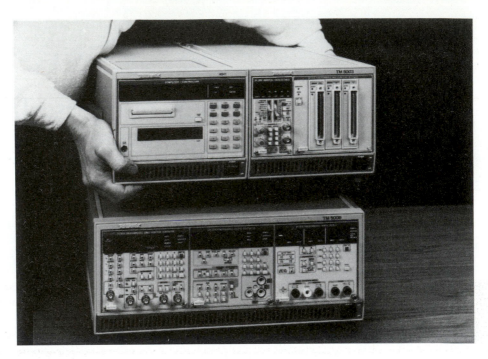

**FIGURE 25–1**
Automatic test equipment system using GPIB (courtesy of Tektronix, Inc.).

a main bank of electronic test equipment, adapters to make the devices under test interconnect with the system, and a special program for each type of equipment. Such an
approach makes for a cost-effective system of test equipment.

Previously, the main problem in attempting to make automatic test equipment (ATE)
was that it was impossible to use unmodified off-the-shelf commercial test instruments.
Thus, many of the best and most useful pieces of test equipment could not be used at all
or had to be extensively modified by the ATE maker before they could be used. The problem was programmability—how to make a signal generator respond to the computer commands. In 1978, the Institute of Electrical and Electronic Engineers (IEEE) released their
specification titled *IEEE Standard Digital Interface for Programmable Instrumentation,*
or IEEE-488 as it is called in the trade. This specification provides details for a standard
computer interface between a computer and instruments. It also calls out ASCII codes and
mnemonics for program instructions. The IEEE-488 bus is also called a **general purpose
interface bus (GPIB).** The Hewlett-Packard interface bus (HPIB) is a proprietary version
of the IEEE-488 bus. The main purpose for the IEEE-488/GPIB is automatic test equipment, both generalized and specific.

Test instrumentation that is intended for GPIB service will have a 24-pin "blueline"
connector on the rear panel. This connector is one of the Amphenol blueline series not
unlike the 36-pin connector used for parallel printer interface on microcomputers. There
will also be a GPIB ADDRESS DIP switch on the rear panel, usually near the connector.

The purpose of the switch is to set the 5-bit binary address where the instrument is located in the system; it determines whether or not the device is a listener only or a talker only and certain other details.

## 25–4      GPIB BASICS

The IEEE-488/GPIB specification provides technical details of the standard bus. The logic levels on the bus are generally similar to TTL: a LOW is less than or equal to 0.800 V, while a HIGH is greater than 2.0 V. The logic signals can be connected to the instruments through a multiconductor cable up to 20 m (66 ft) in length, provided that an instrument load is placed every 2 m. This specification works out to a cable length in meters twice the number of instruments in the system. Most IEEE-488/GPIB systems operate unrestricted to 250 kilobytes per second or faster with certain specified restrictions.

There are two basic configurations for the IEEE/GPIB system (Fig. 25–2): linear and star. These configurations are created with the cable connections between the instruments and the computer. The linear configuration is basically a daisy chain method in which the tap-off to the next instrument is taken from the previous one in the series. In the star configuration the instruments are connected from a central point.

There are three major buses in the IEEE-488/GPIB system. Each line in each bus has a circuit similar to that shown in Fig. 25–3. Besides the shunt protection diode and stray

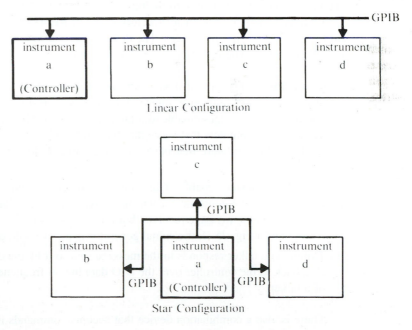

**FIGURE 25–2**

GPIB system configurations (courtesy of Tektronix, Inc.).

**FIGURE 25–3**
IEEE-488/GPIB input/output circuits.

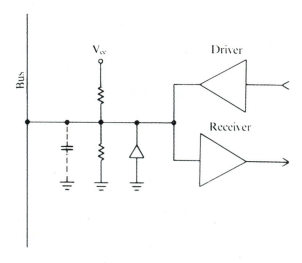

capacitance, there are also pull-up and pull-down resistors that effectively determine the standardized input impedance. Connected to the bus line are receiver and driver circuits. These similar-to-TTL logic elements provide input or output to the instrument. The driver is an output and will be a tristate device; that is, it is inert until commanded to turn on. A tristate output will float at high impedance until turned on. The receiver is basically a noninverting buffer with a high-impedance input. This arrangement of drivers and receivers provides low loading to the bus.

Figure 25–4 shows the basic structure of the IEEE-488/GPIB. There are three buses and four different types of devices. The devices are controllers, talker only, listener only, and talker/listener. These devices are defined as follows:

**Controllers.** This type of device acts as the brain of the system and communicates device addresses and other interface messages to instruments in the system. Most controllers are programmable digital computers. Both Hewlett-Packard and Tektronix offer computers that serve this function, and certain other companies produce hardware and software that permit other computers to act as IEEE-488/GPIB controllers.

**Listener.** A device capable of listening will receive commands from another instrument, usually the controller, when the correct address is placed on the bus. The listener acts on the message received but does not send back any data to the controller. As shown in Fig. 25–4, the signal generator is an example of a listener.

**Talker.** The talker responds to the message sent to it by the controller and then sends data back to the controller over the DIO data bus. A frequency counter is an example of a talker.

There is also a combination device that accepts commands from the controller to set up ranges, tasks, and so forth, and then returns data back over the DIO bus to the controller. An example is a digital multimeter (DMM). The controller will send the DMM

**FIGURE 25–4**
IEEE-488/GPIB structure.

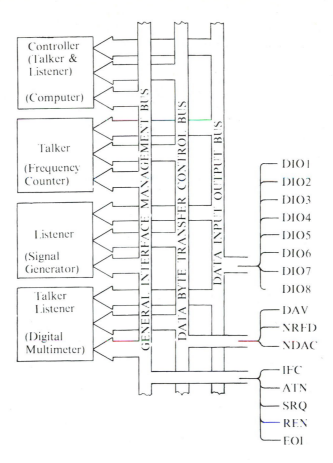

commands that determine whether the data are ac or dc; whether the data are volts, milliamperes, or ohms; and what specific range the data have—the device is thus acting as a listener. When the measurement is made, the DMM becomes a talker and transmits the data measured back over the DIO bus to the controller.

The IEEE-488/GPIB system has three major buses: general interface management (GIM), data byte transfer (DBT), and data input/output (DIO). These buses operate as follows:

**DIO bus.** The data input/output bus is a bidirectional 8-bit data bus that carries data, interface messages, and device-dependent messages between the controller, talkers, and listeners. This bus sends data asynchronously in byte-serial format.

**DBT bus.** The data byte transfer bus controls the sending of data along the DIO bus. There are three lines in the DBT bus: data valid (DAV), not ready for data (NRFD), and not data accepted (NDAC). These signal lines are defined as follows:

*DAV* The data valid signal indicates the availability and validity of the data on the line. If the measurement is not finished, for example, the DAV signal will be false.

*NRFD.* The not ready for data signal lets the controller know whether or not the specific device addressed is in a condition to receive data.

*NDAC.* The not data accepted signal line is used to indicate to the controller whether or not the device accepted the data sent to it over DIO bus.

**GIM bus.** The general interface management bus coordinates the system and ensures an orderly flow of data over the DIO bus. It has the following signals: interface clear (IFC), attention (ATN), service request (SRQ), remote enable (REN), and end or identify (EOI). These signals are defined as follows:

*IFC.* The interface clear signal is used by the controller to place all devices in a predefined quiescent or standby condition.

*ATN.* The attention signal is used by the controller/computer to let the system know how data on the DIO bus lines are to be interpreted and which device is to respond to the data.

*SRQ.* The service request signal is used by a device on the system to ask the controller for attention. This signal is essentially an interrupt request.

*REN.* The remote enable signal is used by the controller to select between two alternate sources of device programming data.

*EOI.* The end or identify signal is used by talkers for two purposes. It will follow the end of a multiple byte sequence of data in order to indicate that the data are now finished. It is also used in conjunction with the ATN signal for polling the system.

The 7-bit binary signals used in the IEEE-488/GPIB system for ASCII and GPIB message codes are shown in Fig. 25–5.

The signals defined above are implemented as conductors in a system interface cable. Each IEEE-488/GPIB-compatible instrument will have a female 36-pin Amphenol-style connector on the rear panel. The pinout definitions are as follows:

| Pin No. | Signal Line | Pin No. | Signal Line |
|---------|-------------|---------|-------------|
| 1 | DIO1 | 13 | DIO5 |
| 2 | DIO2 | 14 | DIO6 |
| 3 | DIO3 | 15 | DIO7 |
| 4 | DIO4 | 16 | DIO8 |
| 5 | EOI | 17 | REN |
| 6 | DAV | 18 | ground (6) |
| 7 | NRFD | 19 | ground (7) |
| 8 | NDAC | 20 | ground (8) |
| 9 | IFC | 21 | ground (9) |
| 10 | SRQ | 22 | ground (10) |
| 11 | ATN | 23 | ground (11) |
| 12 | shield | 24 | logic ground |

# ASCII & GPIB CODE CHART

| BITS B4 B3 B2 B1 | B7 B6 B5: 0 0 0 CONTROL | 0 0 1 CONTROL | 0 1 0 NUMBERS SYMBOLS | 0 1 1 NUMBERS SYMBOLS | 1 0 0 UPPER CASE | 1 0 1 UPPER CASE | 1 1 0 LOWER CASE | 1 1 1 LOWER CASE |
|---|---|---|---|---|---|---|---|---|
| 0 0 0 0 | NUL (0/0/0) | DLE (20/10/16) | SP (40/20/32) | 0 (60/30/48) | @ (100/40/64) | P (120/50/80) | ` (140/60/96) | p (160/70/112) |
| 0 0 0 1 | SOH (1/1/1) GTL | DC1 (21/11/17) LLO | ! (41/21/33) 1 | 1 (61/31/49) 17 | A (101/41/65) 1 | Q (121/51/81) 17 | a (141/61/97) 1 | q (161/71/113) 17 |
| 0 0 1 0 | STX (2/2/2) | DC2 (22/12/18) | " (42/22/34) 2 | 2 (62/32/50) 18 | B (102/42/66) 2 | R (122/52/82) 18 | b (142/62/98) 2 | r (162/72/114) 18 |
| 0 0 1 1 | ETX (3/3/3) | DC3 (23/13/19) | # (43/23/35) 3 | 3 (63/33/51) 19 | C (103/43/67) 3 | S (123/53/83) 19 | c (143/63/99) 3 | s (163/73/115) 19 |
| 0 1 0 0 | EOT (4/4/4) SDC | DC4 (24/14/20) DCL | $ (44/24/36) 4 | 4 (64/34/52) 20 | D (104/44/68) 4 | T (124/54/84) 20 | d (144/64/100) 4 | t (164/74/116) 20 |
| 0 1 0 1 | ENQ (5/5/5) PPC | NAK (25/15/21) PPU | % (45/25/37) 5 | 5 (65/35/53) 21 | E (105/45/69) 5 | U (125/55/85) 21 | e (145/65/101) 5 | u (165/75/117) 21 |
| 0 1 1 0 | ACK (6/6/6) | SYN (26/16/22) | & (46/26/38) 6 | 6 (66/36/54) 22 | F (106/46/70) 6 | V (126/56/86) 22 | f (146/66/102) 6 | v (166/76/118) 22 |
| 0 1 1 1 | BEL (7/7/7) | ETB (27/17/23) | ' (47/27/39) 7 | 7 (67/37/55) 23 | G (107/47/71) 7 | W (127/57/87) 23 | g (147/67/103) 7 | w (167/77/119) 23 |
| 1 0 0 0 | BS (10/8/8) GET | CAN (30/18/24) SPE | ( (50/28/40) 8 | 8 (70/38/56) 24 | H (110/48/72) 8 | X (130/58/88) 24 | h (150/68/104) 8 | x (170/78/120) 24 |
| 1 0 0 1 | HT (11/9/9) TCT | EM (31/19/25) SPD | ) (51/29/41) 9 | 9 (71/39/57) 25 | I (111/49/73) 9 | Y (131/59/89) 25 | i (151/69/105) 9 | y (171/79/121) 25 |
| 1 0 1 0 | LF (12/1A/10) | SUB (32/1A/26) | * (52/2A/42) 10 | : (72/3A/58) 26 | J (112/4A/74) 10 | Z (132/5A/90) 26 | j (152/6A/106) 10 | z (172/7A/122) 26 |
| 1 0 1 1 | VT (13/1B/11) | ESC (33/1B/27) | + (53/2B/43) 11 | ; (73/3B/59) 27 | K (113/4B/75) 11 | [ (133/5B/91) 27 | k (153/6B/107) 11 | { (173/7B/123) 27 |
| 1 1 0 0 | FF (14/1C/12) | FS (34/1C/28) | , (54/2C/44) 12 | < (74/3C/60) 28 | L (114/4C/76) 12 | \ (134/5C/92) 28 | l (154/6C/108) 12 | \| (174/7C/124) 28 |
| 1 1 0 1 | CR (15/1D/13) | GS (35/1D/29) | - (55/2D/45) 13 | = (75/3D/61) 29 | M (115/4D/77) 13 | ] (135/5D/93) 29 | m (155/6D/109) 13 | } (175/7D/125) 29 |
| 1 1 1 0 | SO (16/1E/14) | RS (36/1E/30) | . (56/2E/46) 14 | > (76/3E/62) 30 | N (116/4E/78) 14 | ^ (136/5E/94) 30 | n (156/6E/110) 14 | ~ (176/7E/126) 30 |
| 1 1 1 1 | SI (17/1F/15) | US (37/1F/31) | / (57/2F/47) 15 | ? (77/3F/63) UNL | O (117/4F/79) 15 | _ (137/5F/95) UNT | o (157/6F/111) 15 | DEL (RUBOUT) (177/7F/127) |
| | ADDRESSED COMMANDS | UNIVERSAL COMMANDS | LISTEN ADDRESSES | | TALK ADDRESSES | | SECONDARY ADDRESSES OR COMMANDS | |

## KEY

| octal | 25 | PPU | GPIB code |
|---|---|---|---|
| | **NAK** | | ASCII character |
| hex | 15 | 21 | decimal |

**Tektronix** ®
COMMITTED TO EXCELLENCE

REF: ANSI STD X3. 4-1977
IEEE STD 488-1978
ISO STD 646-1973

TEKTRONIX STD 062-5435-00    4 SEP 80
COPYRIGHT © 1979, 1980 TEKTRONIX, INC. ALL RIGHTS RESERVED

**FIGURE 25–5**
ASCII and GPIB codes (courtesy of Tektronix, Inc.).

## SUMMARY

The IEEE-488 general purpose interface bus (GPIB) can be used to marry various pieces of standard commercial test equipment, computers, and customized equipment into a system that is capable of providing specified tests and measurements under the control of a computer program, often in BASIC. The GPIB makes either special purpose or generalized automatic test equipment (ATE) possible with a minimum of effort.

# APPENDIX

# Integration and Differentiation

The mathematical operations of **integration** and **differentiation** are used extensively in both electronic instruments and the processes they support. These operations are part of the mathematics known as **calculus.** Although you will not need a calculus course to understand the material in this book, the concepts of calculus are used occasionally, so it is prudent to become familiar with the basic ideas and notation.

**Integration** is the operation for finding the area under a curve, and the result is called the **integral.** If the curve is a straight horizontal line, such as line $AC$ shown in Fig. A–1(a) running from $X_1$ to $X_2$, then the task is easy. The area under line is a rectangle or square, and its area is the product

$$A = (Y_1 - 0) (X_2 - X_1) \qquad \text{(A–1)}$$

[*Note:* If the base is at a location other than zero, then the "0" in Equation (A–1) is replaced with the new value of $Y$.]

Similarly, if the curve defines some regular geometric shape, we can either construct or look up the formula for the area of the figure and replace the parameters with the correct values of $X$ and $Y$. Books of tables of solved integrals have been published, so it is not necessary to solve each problem if a published version can be shown to have the same form. For line $AB$ shown dotted in Fig. A–1(a), it is possible to break the area into a rectangle ($X_1$-$A$-$C$-$X_2$) and a triangle ($ABC$). We know from geometry that the area of a triangle is equal to one-half the base-height product ($\frac{1}{2}BH$), so the area comprising the region $X_1$-$A$-$B$-$C$-$X_2$ is

$$A = (Y_1 - 0) (X_2 - X_1) + \frac{(Y_2 - Y_1) (X_2 - X_1)}{2} \qquad \text{(A–2)}$$

But what if the curve is not so regular? Most of the curves derived from electronic instrument data are not so well behaved as to be straight lines or regular plane geometric figures. For example, consider the curve in Fig. A–1(b). We can still use rectangles to find the area under the curve, but we construct a large number of small rectangles, calculate their individual areas, and then sum them together.

**FIGURE A–1**

(a) The integral is easy to calculate because the area is made up of regular geometric shapes. (b) When the area is irregularly shaped, it is divided into small rectangles whose areas are summed.

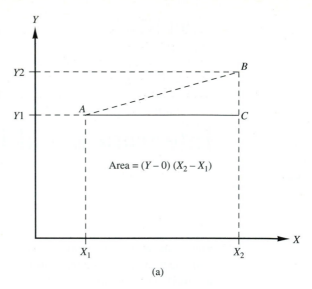

$$\text{Area} = (Y - 0)(X_2 - X_1)$$

(a)

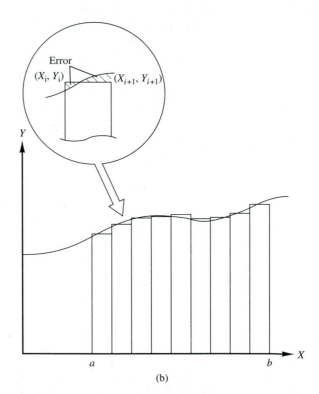

(b)

Few of the rectangles will exactly contain the line $Y = F(X)$. Some rectangles are oversized, while others are undersized. If there are enough rectangles, and if they are each small enough, then the rectangles that tend to underestimate the area under that region of the curve are usually balanced by those that tend to overestimate the area. What happens as the width of the rectangles goes to zero ($X_2–X_1 \rightarrow 0$), that is, is in the **limit,** is one of the central tenets of calculus and will be discussed early in any competent course on the subject.

If we want to find the area under the curve $Y = F(X)$ in Fig. A–1(b), we need to know that the area of any one curve is

$$A = (Y_{i+1} - Y_i)\ (X_{i+1} - X_i) \qquad (A\text{--}3)$$

Thus, the total area under the curve from $a$ to $b$ is

$$A = \sum_{i = a}^{b} (Y_{i+1} - Y_i)\ (X_{i+1} - X_i) \qquad (A\text{--}4)$$

In the notation of calculus, the equation might be written

$$A = \int_{a}^{b} F(X)\ dx \qquad (A\text{--}5)$$

The $\int$ symbol tells us that the integration is performed on curve $F(X)$, that is, a function of $X$, over the interval $a$ to $b$, and $dx$ tells us that the integral is calculated with respect to $X$.

In some cases you will want to find the average value of the curve. You will find the integral using an equation such as (A–5) and then divide by the interval:

$$\overline{Y} = \frac{1}{b - a} \int_{a}^{b} F(X)\ dx \qquad (A\text{--}6)$$

In many cases, especially where the equation for $F(X)$ is known, you would solve the problem analytically rather than graphically. But when the equation is not known, graphical methods come to the rescue. In electronic circuits you can find the time-average integral by using either a resistor-capacitor (RC) low-pass filter or an operational amplifier circuit called a **Miller integrator** (Fig. A–2). The Miller integrator consists of an inverting operational amplifier circuit with a resistor $R$ in series with the inverting input −IN and a capacitor $C$ in the feedback loop between output and the −IN terminal of the operational amplifier $A1$. In both the RC passive and Miller integrators, the time constant of the RC network should be long compared with the period or duration of the signal being integrated.

**Differentiation** is the process for finding the **rate of change** of a curve, and the result is called the **derivative** of the curve. For a straight-line curve [Fig. A–3(a)] of the

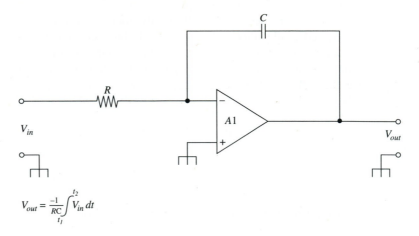

$$V_{out} = \frac{-1}{RC} \int_{t_1}^{t_2} V_{in}\, dt$$

**FIGURE A–2**
Miller integrator circuit.

form $Y = bx \pm a$, where $b$ is the slope and $a$ is the $Y$-axis intercept, the derivative is merely the slope $b$, which is the product of rise over run:

$$b = \frac{Y_2 - Y_1}{X_2 - X_1} = \frac{\Delta Y}{\Delta X} \tag{A–7}$$

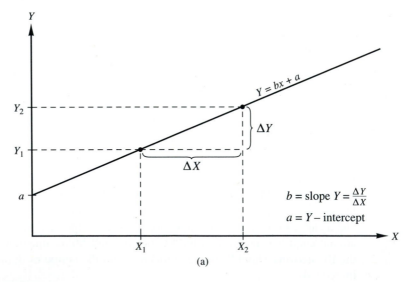

(a)

**FIGURE A–3**
(a) Straight-line curve of form $Y = bx \pm a$.

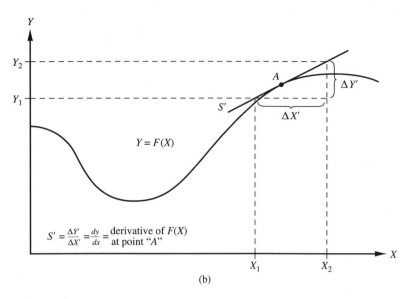

$$S' = \frac{\Delta Y'}{\Delta X'} = \frac{dy}{dx} = \begin{array}{l} \text{derivative of } F(X) \\ \text{at point "}A\text{"} \end{array}$$

(b)

**FIGURE A–3,** *continued*
(b) When the curve is not a straight line.

When the curve is not a straight line, we can find the instantaneous derivative at any given point [*A* in Fig. A–3(b)] by taking the slope of a line $S'$ tangent to the point. In this case the term $\Delta Y/\Delta X$ is usually replaced by the term $dY/dX,$ which is the proper calculus notation.

# Index